195 Current Topics in Microbiology and Immunology

Springer-Verlag Berlin Heidelberg GmbH

Clostridial Neurotoxins

The Molecular Pathogenesis of Tetanus and Botulism

Edited by C. Montecucco

With 28 Figures

Springer

Cesare Montecucco
Università degli Studi di Padova
Dipartimento di Scienze Biomediche Sperimentali
Via Trieste, 75
35121 Padova
Italia

Cover illustration: Tetanus and botulinum neurotoxins intoxication of nerve cells. These toxins bind to the presynaptic membrane (step 1) and are internalized by endocytosis (step 2). The lumen of the synaptic vesicle becomes acidic following the operation of a vacuolar-type ATPase proton pump. This induces a structural transition of the toxin which becomes able to insert in the vesicle membrane and to translocate the L chain (red) in the cytosol (step 3). Inside the neuron cytosol, L blocks neuroexocytosis via selective zincdependent proteolysis of a protein component of the neuroexocytosis apparatus (step 4). Tetanus neurotoxin and botulinum neurotoxins B, D, F and G cleave specifically VAMP (violet), while types A and E cleave SNAP-25 (yellow) and botulinum neurotoxin type C cleaves syntaxin (green).

Cover design: Künkel+Lopka, Ilvesheim

ISBN 978-3-642-85175-9 ISBN 978-3-642-85173-5 (eBook)
DOI 10.1007/ 978-3-642-85173-5

Library of Congress Catalog Card Number 15-12910

Softcover reprint of the hardcover 1st edition 1995

Typesetting: Thomson Press (India) Ltd, Madras
SPIN: 10128478 27/3020/SPS – 5 4 3 2 1 0 – Printed on acid-free paper

Preface

Tetanus has been known from the very beginning of medical literature since it was first described by Hyppocrates of Cos in the fifth century B.C. For 24 centuries it was considered a neurological disease until the breakthrough of CARLE and RATTONE (1884) who demonstrated its infectious etiology. Following the establishment of purified cultures of *Clostridium tetani* (KITASATO 1889), FABER (1890), and TIZZONI and CATTANI (1890) demonstrated that the disease is actually an intoxication caused by a proteic neurotoxin. This toxin was shown by BRUSHCHETTINI (1892) to move retroaxonally and to act at the spinal cord level. Soon thereafter VAN ERMENGEN (1897) demonstrated that botulism is also due to intoxication with a protein toxin produced by bacteria of the genus *Clostridium*. These bacteria and their spores and ubiquitous, and the majority of them do not produce neurotoxins. The selective advantage of producing such potent toxin is still a matter of speculation (see Popoff, this volume).

The next major advance was the discovery that tetanus neurotoxin[1] can be converted by formaldehyde treatment to a nonpathogenic but still fully immunogenic form, and that this can be used successfully as a vaccine to prevent tetanus (RAMON and DESCOMBEY 1925). Similar vaccines (toxoids) can be prepared with botulism neurotoxins (see MIDDLEBROOK and BROWN, this volume). The prevention of tetanus by vaccination (see Galatzka and Gasse, this volume) is one of the great successes of basic research coupled with an efficient public medicine service. Tetanus toxoid is now a standard immunogen in studies of molecular and cellular immunology (see CORRADIN and WATTS, this volume).

Botulism is of minor health concern with respect to tetanus, although it is much more frequent than that gathered from the number of patients admitted to hospitals, and can be caused by seven different serotypes of botulinum neurotoxin (see HATHEWAY). The great discovery that botulism neurotoxins block neurotransmitter release at the neuromuscular junction (BURGEN et al. 1949) paved the way for the introduction of botulism neurotoxins as therapeutic agents (SCOTT 1973) in the treatment of a variety of diseases characterized by an unbalanced contraction of opposing muscles (see CARDOSO and JANKOVIC, this volume). Here we have an astonishing example of how basic research can convert a terrible poison, one even considered for possible military use, into an agent beneficial to human health. Botulinum neurotoxin serotype A is currently used almost invariably, but other serotypes have recently been introduced into clinical use, and their preparation is discussed here (see SHONE and TRANTER, this volume).

The development of neuromuscular junction preparations and of sophisticated electrophysiological and electron microscopic techniques has allowed the study of the mechanism of action of these neurotoxins at the cellular level. Thus, it was demonstrated that these toxins bind the neuromuscular junction and penetrate cells via a multistep process involving intracellular vesicles (see HALPERN and NEALE, this volume). Moreover, it was found that tetanus toxin and botulinum neurotoxins B, D, and F act differently from types A and E, and that the light chain is the one responsible for the blockade of neurotransmitter release (see POULAIN et al., this volume). At the same time studies on the mechanism of neuroexocytosis have been progressing rapidly (see BENFENATI and VALTORTA, this volume). Together with determination of the sequence of genes encoding for the clostridial neurotoxins this has opened the way to recent major advances in the understanding of the molecular basis of botulism and tetanus. Indeed, it was these recent discoveries that led to the compilation of this volume. At present the sequences of all these neurotoxins are known, and they are discussed here together with the data available on the proteins associated with botulinum neurotoxins (see MINTON, this volume).

Approximately a century after their identification, the molecular mechanism of action of these neurotoxins was discovered: they are intracellular metalloproteinases. Tetanus and botulinum neurotoxin type B were demonstrated to be zinc endopeptidases specific for VAMP/synaptobrevin, a synaptic vesicle membrane protein (SCHIAVO et al. 1992a, b). Following this breakthrough and the molecular characterization of a 20S multiprotein complex involved in vesicle docking and fusion (SÖLLNER et al. 1993), the targets of all the neurotoxin types were rapidly identified: A and E act on SNAP-25, C on syntaxin, and D, F, and G also on VAMP (BLASI et al. 1993; SCHIAVO et al. 1993). The metalloproteinase activity of clostridial neurotoxins is discussed here by SCHIAVO et al., this volume.

These recent findings define the molecular pathogenesis of tetanus and botulism and they provide the basis for future investigations aimed at the therapeutic management of these diseases. One line of research is particularly promising, and this is based on previous work of the pharmaceutical industry aimed at identifying new inhibitors of metalloproteinases, enzymes involved in a variety of physiological and pathological functions. Another line of research is related to the use of clostridial neurotoxins as tools in the study of vesicle trafficking within the cell. Development of engineered neurotoxins able to penetrate non-neuronal cells appears particularly promising (HALPERN and LEPPLA, manuscript in preparation). Another major goal in this field is the identification of the receptor of these neurotoxins. Based on the potentiating effect of muscular activity on the intoxication with botulinum neurotoxins, their receptors are expected to be proteins of the lumen of small synaptic vesicles. Identification of the receptor of tetanus neurotoxin responsible for its retroaxonal transport is particularly awaited since this will throw light on the membrane trafficking of motorneurons. At the same time these studies will provide the molecular basis for the development of these neurotoxins as neurospecific carriers of biological agents.

Padova CESARE MONTECUCCO

---

[1] Throughout this volume tetanus neurotoxin has been abbreviated as TeNT, rather than TeTx as is frequently found in the literature. This abbreviation has been chosen to avoid the very frequent confusion with retrodotoxin, the sodium channel blocker. Botulinum neurotoxin is abbreviated BoNT. The presence in the abbreviation of the letter N (which refers to neuro-) also helps in avoiding the confusion with tetanolysin and botulinolysins, two cholesterol-binding cytolysins, also produced by clostridia, which are not neurospecific toxins

References

Blasi J, Chapman ER, Link E, Binz T, Yamasaki S, Decamilli P, Südhof T, Niemann H, Jahn R (1993) Botulinum neurotoxin A selectively cleaves the synaptic protein SNAP-25. Nature 365: 160–163

Bruschettini A (1892) Sulla diffusione del veleno del tetano nell'organismo. Riforma Medica 8: 256–259

Burgen ASV, Dickens F, Zatman LJ (1949) The action of botulinum toxin on the neuromuscular junction. J Physiol [Lond] 109: 10–24

Carle A, Rattone G (1884) Studio esperimentale sull'etiologia del tetano. G Accad Med Torino 32: 174–179

Faber K (1890) Die Pathogenie des Tetanus. Berl Klin Wochenschr 27: 717–720

Kitasato S (1889) Über den Tetanusbacillus. Z Hyg 7: 224–234

Ramon G, Descombey PA (1925) Sur l'immunization antitétanique et sur la production de l'antitoxine tétanique. C R Soc Biol 93: 508–598

Schiavo G, Poulain B, Rossetto O, Benfenati F, Tauc L, Montecucco C (1992a) Tetanus toxin is a zinc protein and its inhibition of neurotransmiter release and protease activity depend on zinc. EMBO J 11: 3577–3583

Schiavo G, Benfenati F, Poulain B, Rossetto O, Polverino de Laureto P, DasGupta BR, Montecucco C (1992b) Tetanus and botulinum-B neurotoxins block neurotransmitter release by a proteolytic cleavage of synaptobrevin. Nature 359: 832–835

Schiavo G, Rossetto O, Catsicas S, Polverino de Laureto P, DasGupta BR, Benfenati F, Montecucco C (1993) Identification of the nerve-terminal targets of botulium neurotoxins serotypes A, D and E. J Biol Chem 268: 23784–23787

Söllner T, Whiteheart SW, Brunner M, Erdjument-Bromage H, Geromanos S, Tempst P, Rothman JE (1993a) SNAP receptors implicated in vesicle targeting and fusion. Nature 362: 318–324

Tizzoni G, Cattani G (1890) Untersuchungen über das Tetanusgift. Arch Exp Pathol Pharmakol 27: 432–450

Van Ermengem E (1987) Über einem neuen anaeroben Bacillus und seine Beziehungen zum Botulisms. Z Hyg Infekt Krankh 26: 1–56

List of Contents

List of Contributors

(Their addresses can be found at the beginning of their respective chapters.)

Ecology of Neurotoxigenic Strains of Clostridia

M.R. Popoff

Unité des Toxines Microbiennes, Institut Pasteur, 28 rue du Dr. Roux, 75724 Paris Cedex 15, France

1 Introduction

Botulism and tetanus are diseases of humans and animals which are characterized by specific neurological disorders. These diseases have been known since antiquity. The first clinical descriptions are from Greek and Roman authors several centuries B.C. The bacterial origin, however, was established only in last century.

The Italians CARLE and RATTONE (1884) showed that tetanus can be transmitted to animals. Rabbits injected with pustule suspension from a patient suffering of tetanus developed the disease. An emulsion of sciatic nerve from these tetanic rabbits was injected into other rabbits which also died of tetanus. KITASATO (1889) obtained a pure culture of the toxigenic micro-organism and FABER (1989), TIZZONI and CATTANI (1989) showed that the culture filtrate contained a factor responsible of the tetanus symptoms. Active immunization with formaldehyde-treated tetanus toxin was undertaken by RAMON and DESCOMBEY (1925).

Botulism (Latin: *botulus*, sausage) refers to an acute food poisoning, accompanied by flaccid paralysis and caused by the ingestion of spoiled sausages. The bacterial origin of this food intoxication was determined by VAN ERMENGEM (1897), who isolated an anaerobic, spore-forming bacillus from the remains of raw ham and the livers of three dead members of a music club who had been affected. Culture filtrates of the isolates were administered to various experimental animals and produced paralytic symptoms resembling those of human botulism and caused the death. The bacteriological and toxiocological investigations of various outbreaks of botulism in men and animals led to the differentiation of seven toxinotypes of the botulinum neurotoxin according to their antigenic properties (HATHEWAY 1989).

Botulinum neurotoxin (BoNT) and tetanus neurotoxin (TeNT) cause all the symptoms of botulism and tetanus, respectively. The pathology of the neurotoxigenic *Clostridium* is the simplest model of bacterial pathogenicity. One bacteria produces one toxin which induces all the specific disorders of the disease and death. Since the first evidence of TeNT by the two Italian investigators, much research has been devoted to the purification and biochemical and genetic characterization of these neurotoxins; however, its mechanism of action has remained obscure. A century later another Italian research group solved the riddle of neurotoxin activity at the molecular level (SCHIAVO et al. 1992) These discoveries provide more precise information on the pathogenicity of these diseases and on important processes in cell function, including the exocytosis of neurotransmitter vesicles and exocytosis in other cells (see SCHIAVO et al., this volume).

This review describes the neurotoxin-producing organisms, their habitat, and their transmission to human and animals.

2 Characterization of Neurotoxigenic Clostridia

2.1 *Clostridium tetani*

The TeNT-producing *Clostridium* displays homogeneous bacteriological characteristics and belongs to a uniform group forming the *C. tetani* species.

2.1.1 Morphological and Cultural Characteristics

The cells of *C. tetani* are usually 0.3–0.6 µm in width and 3–12 µm in length. The length of the rods may vary considerably. They are gram positive in young cultures but lose the gram coloration after 24 h of incubation. *C. tetani* is usually highly motile by peritrichous flagella, which are responsible for the swarming growth on agar medium. Some strains, however, are nonmotile and nonflagellated.

The enlarged, terminal, and spherical spores lead to the typical appearance of drum stick. The sporulation rate is variable according to the strains. At pH 7 or above and at temperature near 37°C, sporulation starts within 24h and continues for 4–12 days or more. It does not occur at high temperature (above 41°C) and is also slow or poor when pH is less than 6. Composition of the culture medium also affects sporulation (Bytchenko 1981). Spores generally survive moderate heating (75–80°C for 10 min) but are usually destroyed after exposure to 100°C for 1 h.

C. tetani spores generate under anaerobic or aerobic conditions. Outgrowth of *C. tetani*, however, which follows spore germination, depends upon a low oxidation reduction potential (Smith 1975).

C. tetani is a strictly anaerobic bacterium and forms colonies on the surface of agar medium only in an anaerobic environment. Motile strains swarm over the entire surface of the agar, leading to a transparent film. Discrete colonies (2–5 mm) can be obtained with media containing 3%–4% agar. On blood agar, colonies are slightly raised, semitranslucent, gray, with an irregular margin, and surrounded by a narrow zone of hemolysis. *C. tetani* grows fairly well on the usual media containing peptones or tissue extracts.

Most biochemical tests used for *Clostridium* identification are negative: no carbohydrate acidification, no proteolysis, no production of lipase and lecithinase (Table 1). Gelatin is liquified slowly (2–7 days), the peptone used in the basal medium is of considerable importance in evaluating the ability of *C. tetani* to liquify the gelatin. H_2S and indole are usually produced (Smith 1975).

2.1.2 Genetic Characteristics

The G+C content of *C. tetani* is 25%–26% (Cato et al. 1986). This species has been classified in the *Clostridium* group II by 23S rRNA homology (Johnson and Francis 1975). The *Clostridium* genus encompasses more than 100 species, which display a wide range of phenotypes and genotypes (Hippe et al. 1992). Recent phylogenetic analysis using 16S rRNA comparison has indicated that the *Clostridium* genus should be restricted to the homology group I defined by

Table 1. Characteristics of neurotoxin-producing clostridia

	C. tetani	*C.botulinum* Group I	*C.botulinum* Group II	*C.botulinum* Group III	*C. argentinense*	*C. butyricum*	*C. baratii*
Toxin type	TeNT	BoNT/A,B, F	BoNT/B,E, F	BoNT/C,D	BoNT/G	BuNT/E	BaNT/F
Spore position[a]	T	ST	ST	ST	ST	ST	ST
Motility	+	+	+	+	+	+	–
Proteolysis	–	+	–	–	+	–	–
Gelatin hydrolysis	+	+	+	+	+	–	–
Lipase	–	+	+	+	–	–	–
Lecithinase	–	–	–	–	–	–	+
Esculin hydrolysis	–	+	–	–	–	+	+
Indole production	+	–	–	–/+	–	–	–
H_2S production	+	+	–	–/+	+		
Acidification of							
Glucose	+	+	+	+	–	+	+
Fructose	–	–	+	–	–	+	+
Sucrose	–	–	+	–	–	+	+
Fermentation end products[b]	A B p l s	A B iV ib ic vp	B A l	B P A v l s	A B iv ib	B A F l s	B A L f p s
Growth temperature							
Optimum	35°–38°C	30°–40°C	25°–30°C	30°–37°C	30°–37°C	30°–37°C	30°–37°C
Interval	14°–43°C	10°–48°C	3.3°–45°C	15°–50°C	20°–45°C	10°–50°C	20°–50°C
Spore resistance to heat[c]		104°–121°C D 6–0.2	77°–82°C D 4–0.3	104°C D 0.02–0.9	82°–120°C D 5.9–5.9		
Related non-neurotoxigenic species	*C. tetanomorphum* *C. cochlearium*	*C. sporogenes* *C. putrificum`*		*C.oedematiens*	*C. subterminale C. hastiforme*	typical *C. butyricum* strains	typical *C. baratii* strains

[a] Spore position: T, terminal; ST, subterminal.
[b] Fermentation endproducts: A, acetic; B, butyric; C, caproic; L, lactic; F, formic; P, propionic, S, succinic; V, valeric; IB, isobutyric; IC, isocaproic; IV, isovaleric. Uppercase letters indicate major products; lowercase letters indicate minor products (less than 1 mEq/100ml; CATO et al. 1986).
[c] Spore resistance to heat: D, time in minutes required to inactivate 90% of the population at a specified temperature (KIM and FOEGEDING 1993).

JOHNSON and FRANCIS (1975; LAWSON et al. 1993). According to these data, *C. tetani* should be classified in a different genus. However, a study based on restriction maps of 16S rRNA showed that *C. tetani* belongs to the same cluster as *C. perfringens*, *C. sporogenes*, and *C. botulinum* C and G, which are members of the homology group I of Johnson and Francis (GURTLER et al. 1991).

Ten toxigenic and three nontoxigenic *C. tetani* strains studied by NAKAMURA et al. (1979) were homogeneous in DNA/DNA hybridization (85%–93% similarity). *C. tetani* is culturally and biochemically similar to *C. cochlearium* and *C. tetanomorphum* but can be distinguished from them by DNA comparison (NAKAMURA et al. 1979; WILDE et al. 1989). *C. cochlearium* and *C. tetanomorphum* are nontoxic and are difficult to differentiate from the nontoxic *C. tetani* strains on the basis of bacteriological characteristics; they do not liquify gelatin or only weakly, and *C. tetanomorphum* acidifies glucose and maltose, in contrast to *C. tetani* (CATO et al. 1986).

2.2 Botulinum Neurotoxin Producing Clostridia

Clostridia able to produce a BoNT display heterogeneous bacteriological characters and are divided into several species and groups. The taxonomic position of the *C. botulinum* species was originally based on only one phenotype, the production of BoNT, and nontoxic variant strains, although genetically related to *C. botulinum*, were assigned to different species such as *C. sporogenes* and *C. subterminale*. It appeared soon that BoNT falls into seven different toxin types (A, B, C1, D, E, F, and G) according to their antigenic properties. However, all BoNT types induce typical symptoms of flaccid paralysis in experimental animals similar to those observed in human disease. Physiological differences among *C. botulinum* strains later became apparent, but the production of the various BoNT types are not necessarily correlated with the *C. botulinum* strain phenotypes. The species were divided into four physiological groups (SMITH 1975). Group IV includes *C. botulinum* G strains and nontoxic strains previously identified as *C. subterminale* and *C. hastiforme*, is metabolically distinct from the other groups, and has been assigned to a different species termed *C. argentinense* (SUEN et al. 1988a).

The taxonomic position of *C. botulinum* became more ambiguous since it was found that BoNT can be produced by *Clostridium* strains clearly distinct from the already defined *C. botulinum* and biochemically and genetically related to different species such as *C. butyricum* and *C. baratii*, organisms implicated in human botulism. The toxinogenic property of the *Clostridium* strains remains an important characteristic, mainly in clinical aspects, but this single property is insufficient to define a bacterial species. Moreover, the botulinogenic parameter is unstable, and atoxic variants have been described and seem to be transferable in a wide variety of *Clostridium* species. Based on taxonomic considerations, distinct species names should be assigned for each physiological and genetic group, and the BoNT production should be clearly indicated for each strain (HATHEWAY 1990).

2.2.1 Morphological Characteristics

BoNT-producing *Clostridium* are usually straight or slightly curved rods 0.6–2 x 2–22 μm. They are usually motile and peritrichous, except for *C. baratii* which is non-motile. Spores are oval and subterminal and swell the cell. Type G strains sporulate poorly, and *C. butyricum* spores are central to subterminal and usually do not swell the cell (CATO et al. 1986).

Cultures in the usual anaerobic liquid media are abundant with the production of gas. Growth temperatures are indicated in Table 1. Surface colonies on blood agar plates and incubated in anaerobic conditions are circular (1–6 mm in diameter) to irregular, with a scalloped or lobated margin, translucent to semi-opaque, and gray-white. They are surrounded by a narrow zone of partial hemolysis: those of *C. butyricum* are nonhemolytic.

2.2.2 Physiological Characteristics

Group I strains (*C. botulinum* A; proteolytic stains of types B and F) are characterized by proteolytic activity and by nonacidification of carbohydrates (Table 1). Glucose is acidified weakly. The cell wall contains glucose as sugar. Optimal temperature of growth is 37°C. Thermoresistance of spores is usually high (above 120°C), but can vary according to the strain (SMITH 1975).

Group II strains (*C. botulinum* E; glucidolytic strains of types B and F) acidify various carbohydrate substrates (amygdalin, dextrin, fructose, galactose, glucose, glycogen, maltose, ribose, sorbitol, sucrose, and trehalose but not lactose, mannitol, melibiose, or salicin) and produce mainly butyric and acetic acids from trypticase yeast extract glucose broth. They hydrolyze gelatin but are nonproteolytic (CATO et al. 1986). The strains of this group contain glucose and galactose as the cell wall sugars. They have a lower optimal temperature or growth (around 25°–30°C), and can grow and produce toxins at very low temperature (Table 1). Spores are only moderately resistant to heat despite 10 min at 90°C (SMITH 1992). Neurotoxins produced by group II strains, mainly *C. botulinum* E, are not fully activated by endogeneous protease(s), and trypsinisation enhances the toxicity.

Group III strains (*C. botulinum* C and D) are not or only very slightly proteolytic. They ferment glucose, glycerol, inositol, ribose, and xylose. Their cell wall contains no sugar or only traces of glucose (SMITH 1975). This group is heterogeneous in terms of different biochemical properties (fermentation of several sugars, production of indol and hydrogen sulfide). A subdivision into four subgroups have been proposed (OGUMA et al. 1986).They grow at higher temperatures than the other *C. botulinum* strains (optimal temperature for growth, 30°–37°C), and most strains grow well at 45°C (CATO et al. 1986). Spore resistance to heat is intermediate.

C. argentinense strains (also referred as group IV) are proteolytic, hydrolyze gelatin, and do not ferment any of the usual carbohydrates. In contrast to the strains of the *C. botulinum* groups, *C. argentinense* strains do not produce a lipase. Optimal temperature or growth is 30°–37°C. Phenotypic differences have been observed between toxic and nontoxic *C. argentinense* strains using cellular fatty analysis and multilocus enzyme electrophoresis (ALTWEGG and HATHEWAY 1988; GHANEM et al. 1991). Toxigenic *C. butyricum* strains are phenotypically related to the typical strains of this species. They are nonproteolytic, do not hydrolyze gelatin, and acidify strongly various carbohydrates including glucose, cellobiose, fructose, galactose, glycogen, lactose, maltose, mannose, melibiose, raffinose, ribose, salicin, starch, sucrose, trehalose, xylose, and pectin. They produce neither lipase nor lecithinase (SMITH 1992). Toxigenic and nontoxigenic *C. baratii* strains display similar morphological and biochemical characteristics. *C. baratii* is phenotypically closely related to *C. perfringens*. Both species are nonmotile, with thick, straight rods (0.5–1.9 x 1.6–15 μm). Strains sporulate poorly in the usual culture medium. They produce a lecithinase but not a lipase.

C. baratii is readily differentiated from *C. perfringens* by not hydrolyzing gelatin (CATO et al. 1986)

2.2.3 Genetic Relatedness

All the BoNT-producing clostridia have a G+C content between 26% and 28% (CATO et al. 1986) and belong to group I of 23S rRNA homology (JOHNSON and FRANCIS 1975) which contains the type species *C. butyricum*. A 16S rRNA comparison has corroborated that group I strains are homogeneous and form a distinct phylogenetic branch. It has been proposed that the genus *Clostridium* should be retained only for this taxon (HIPPE et al. 1992; LAWSON et al. 1993).

Genetic studies using DNA/DNA hybridization have shown that the physiological *C. botulinum* groups constitute four different DNA homology clusters (Table 2; LEE and RIEMANN 1970a,b; NAKAMURA et al. 1977). In each group all the strains regardless of toxin type are closely related, but strains of the same or different toxin types and from different groups show low relatedness. Recently, analysis of 16S rRNA sequences confirm the previously identified genetic groups of *C. botulinum* (LAWSON et al. 1993).

Nontoxigenic strains phenotypically related to *C. botulinum* group I are included in the *C. sporogenes* species. Some strains display high DNA homology with *C. botulinum* group I strains, and others do not (LEE and RIEMANN 1970a; NAKAMURA et al. 1977). Comparison of 16S rRNA sequences shows that *C. sporogenes* is phylogenetically related to *C. botulinum* group I and is genealogically almost identical with the type strain of *C. putrificum* (99.7% 16S rRNA sequence similarity; LAWSON et al. 1993).

Strains of saccharolytic *C. botulinum* types B, E, and F (group II) are homogeneous in DNA/DNA hybridization (LEE and RIEMANN 1970a) and form a single phylogenetic group (99.6%–99.7% sequence similarity of 16S rRNA) which is remote from proteolytic *C. botulinum* group I strains (HUTSON et al. 1993). A probe constituted of a 0.5-kbp cloned DNA fragment was found specific or *C. botulinum* group I strains and did not hybridize with *C. sporogenes* or with *C. botulinum* group II, III, or IV strains (MCKINNEY et al. 1993).

Some genetic diversity has been found among strains of group III by DNA/DNA hybridization and by comparison of the 16S rRNA sequences (LAWSON et al. 1993; LEE and RIEMANN 1970a; NAKAMURA et al. 1983). *C. botulinum* C can be divided into at least two genetic subgroups (Table 2). One represented by strain Stockholm, is related to *C. oedematiens* B and to *C. haemolyticum* (also refered as *C. oedematiens* D) but not to *C. oedematiens* A; the other studied *C. botulinum* C strains show low relatedness with *C. botulinum* C Stockholm and the various types of C. *oedematiens* (NAKAMURA et al. 1983). Non-toxigenic variants display similar biochemical characteristics as those of the toxigenic parental strains (OGUMA et al. 1986).

C. botulinum G (group IV) and three nontoxic strains, two previously identified as *C. subterminale* and one as *C. hastiforme*, are phenotypically and genetically

Table 2. Summarized data on DNA/DNA homology (%) between botulinum neurotoxin producing clostridia[a] (from LEE and RIEMANN 1970a, b; NAKAMURA etal. 1977, 1983; SUEN et al. 1988a,b; TAKESHI et al. 1991)

	Group I	*C. Sporogenes*	Group II	Group III	*C. oedematiens A*	*C. oedematiens B*	*C. haemolyticum*	*C. argentinense*	*C. subterminale*	*C. hastiforme*	*C. butyricum*	*C. baratii*
Group I	69–100	48–100	11–19	5–13				8	6	2	6	3–6
C. sporogenes	81–91 66–73	66–74 89–100	12									
Group II	8–19	12	68–100									
Nontoxic strains of group II	4–10		58–78									
Group III	5–13		5–12	39–100								
Stockholm C strain				100[a] 66–74[b]	42	86–96	84–89					
Other C strains				91–100[b]	37–38	68–70	59–67					
C. oedematiens A				37–42	80–100	28–49	26–38					
C. argentinense	18–19		18–19					75–100	6–53	8–24		
Toxic *C. butyricum*	6	5	8–12					9	8	3	78–97	15
Toxic *C. baratii*	10	7	8–9					7	8	4	12	86

[a] With *C. botulinum* Stockholm C strain.
[b] With other *C. botulinum* C strains.

closely related and have been classified in a new species called *C. argentinense* (SUEN et al. 1988a). This species is genetically distinct from the other BoNT-producing *Clostridium*. *C. argentinense* strains are phenotypically related to *C. subterminale* and *C. hastiforme*, but they constitute distinct DNA hybridization groups (SUEN et al. 1988a).

The atypical toxigenic *C. butyricum* and *C. baratii* strains are genetically related to the type strains of these species and not with the other BoNT-producing *Clostridium* (Table 2).

The two toxigenic *C. butyricum* strain isolated from infant botulism in Italy (AURELI et al. 1986; MCCROSKEY et al. 1986) display 78% and 97% DNA homology with the nontoxigenic *C. butyricum* strains and are poorly related to the other *C. botulinum* strains (SUEN et al. 1988b; TAKESHI et al. 1991). The 16S rRNA sequence of toxigenic *C. butyricum* strains is identical to that of nontoxigenic *C. butyricum*; these strains form a phylogenetic branch distinct from that of *C. botulinum* E (HUTSON et al. 1993).

In the same way the toxigenic *C. baratii* strain BL2990 is genetically related to the *C. baratii* type of strain (86% DNA homology) and not with the other BoNT-producing *Clostridium* (Table 2; SUEN et al. 1988b). Sequence analysis of 16S rRNA confirms the high phylogenetic relatedness of toxigenic *C. baratii* strains with the type strain of this species, and their remote relationship with the *C. botulinum* group I and II (HUTSON et al. 1993).

3 Other Toxins

Beside the neurotoxins, *C. tetani* and BoNT-producing clostridia also produce various toxins and enzymes (e.g., lipase, lecithinase, protease, gelatinase) as the other toxigenic clostridia. The most important are the C2 toxin, C3 exoenzyme, and hemolysins.

3.1 C2 Toxin

The C2 toxin is produced by *C. botulinum* C and D strains. This toxin consists of two independent peptide chains termed C2-I (55 kDa) and C2-II (105 kDa). Both polypeptide chains are required for cell toxicity. The component C2-II is synthesized as a propetide which is proteolytically activated in its mature form (88 kDa; Ohishi 1987). This component displays a hemagglutinating activity. C2-II is involved in recognition of a receptor on cell surface, which has not been yet identified, and allows the internalization of C2-I into the cytosol (Considine and Simpson 1991). The enzymatic component C2-I is an ADP-ribosyltransferase, whose the substrate is the cellular nonmuscular G actin. ADP-ribosylation takes place at Arg177 of actin (Aktories et al. 1986; Vandekerckhove et al. 1987). This modification of actin results in depolymerization of the F actin filaments of the cytoskeletal stress fibers and in rounding up of the cell (Aktories and Wegner 1989). C2 toxin is not a neurotoxin; it interferes with noradrenaline release in PC12 cells by ADP-ribosylating actin and not by acting on specific exocytosis components as botulinum neurotoxins do (Matter et al. 1989).

C2 toxin increases vascular permeability, induces hemorrhage and edema in the lungs, and leads to hypotonic effects and finally death (Considine and Simpson 1991). This binary toxin also has an enterotoxin activity that causes marked distension of ligated intestinal loop in mice (Ohishi 1983) and degenerative and necrotic changes of the intestinal mucosa (Ohishi and Odagiri 1984).

3.2 C3 Exoenzyme

The exoenzyme C3 is an ADP-ribosyltransferase also produced by *C. botulinum* C and D strains and is distinct from the C2 toxin (Aktories et al. 1987). It consists of a single peptide chain (26 kDa). Exoenzyme C3 catalyzes the ADP-ribosylation of a small GTP binding protein (21–24 kDa) called Rho and, to a lesser extent, the Rac protein. Rho and Rac are parts of the Ras superfamily of proteins which are involved in the regulation of cell actin polymerization (Chardin et al. 1989). No binding component able to internalize the exoenzyme C3 into the cells, such as the C2-II of the C2 toxin, has been found. Thus, the cytotoxic activity of C3 exoenzyme is low; the minimum concentration causing a cytotoxic effect is 2–4 µg/ml. The incidence of the C3 exoenzyme in natural disease is unknown.

3.3 Hemolysins

The tetanolysin produced by *C. tetani* is an oxygen-sensitive hemolysin which is related functionally and antigenically to streptolysin O, pneumolysin, listeriolysin and hemolysins produced by eight clostridial species including theta toxin of *C. perfringens*, tetanolysin, and botulinolysin. These hemolysins belong to the thiol-activated cytolysin family (ALOUF and GEOFFROY 1991). Tetanolysin consists of a single polypeptide chain (48 kDa). It lyzes a wide variety of cells: erythrocytes, leukocytes, macrophages, fibroblasts, HeLa cells. Cytolytic activity is inhibited by oxidizing conditions, cholesterol, and related sterols.

Intravenous injection of tetanolysin to mice induces pulmonary edema and death. In rabbits and monkeys intravascular hemolysis effects have been observed, as has cardiotoxicity in monkeys (electrocardiographic changes, heart rate decrease). The participation of tetanolysin in tetanus disease has not been recognized (HATHEWAY 1990).

C. botulinum strains produce botulinolysin, which belongs to the thiol-activated cytolysin family (ALOUF and GEOFFROY 1991). Botulinolysin from a *C. botulinum* type C strain has been characterized. It is a protein of 58 kDa, which exhibits cytotoxic and lethal effects (HAQUE et al. 1992). The relevance of botulinolysin to *C. botulinum* infection is unknown.

4 Neurotoxin Genes

4.1 Localization

Plasmids of various sizes and bacteriophages have been found in *C. botulinum* in group I and II strains, but toxigenicity has not been associated with the presence of these genetic elements (EKLUND et al. 1989; WEICKERT et al. 1986). Many strains of *C. botulinum* A and of *C. sporogenes* contain plasmids. *C. botulinum* A strains naturally free of plasmids and strains cured of plasmids by growth on medium containing sodium deoxycholate were still toxic. No correlation between toxigenicity and the presence of any single plasmid in *C. botulinum* A could be found (WEICKERT et al. 1986). Recently, in *C. botulinum* A, B, E, and F the neurotoxin genes have been cloned from chromosomal DNA, indicating that these genes are localized on the bacterial chromosomal DNA (Table 3; BINZ et al. 1990a; EAST et al. 1992; POULET et al. 1992; WHELAN et al. 1992). In neurotoxigenic *C. butyricum* strains the BoNT/E gene has been found in a plasmid, using polymerase chain reaction amplification (HAUSER et al. 1992), but by DNA/DNA hybridization this gene seems to be localized on the chromosomal DNA (ZHOU et al. 1993).

In *C. botulinum* group III strains it has been shown clearly that BoNT is encoded by bacteriophages. *C. botulinum* C and D strains cured of their phages no longer produce BoNT/C1 or BoNT/D, respectively, but they do continue to produce the C2 toxin. Such organisms could be converted into neurotoxigenic strains C or D by reinfection with phages obtained from *C. botulinum* C or D (EKLUND et al. 1971, 1989; EKLUND and Poysky 1974). Recently, the BoNT/C1 and BoNT/D genes have been cloned and sequenced from purified phage DNA of *C. botulinum* C-468 and *C. botulinum* D-1873 respectively (BINZ et al. 1990b; HAUSER et al. 1990). The C2 toxin gene has been localized on the bacterial chromosome by DNA hybridization (unpublished data). In *C. oedematiens*, which is related to *C. botulinum* group III strains, the alpha toxin is also bacteriophage encoded. *C. botulinum* C and D strains which have lost their phages could be also converted in *C. oedematiens* A by infection with phage NA1 isolated from *C. oedematiens* (EKLUND et al. 1974). The presence of these bacteriophages in recipient organism of group III determine the toxinotype.

In *C. tetani* and *C. argentinense* the neurotoxin gene has been localized on a large plasmid (51 and 73 MDa, respectively; Table 3; EKLUND et al. 1988; FINN et al. 1984).

4.2 Transfer

The high homology between TeNT and BoNT (reviewed in HAUSER et al. 1994; NIEMANN 1991; see Middlebrook and Brown, this volume) and the fact that different *Clostridium* species can produce BoNT and some strains display combinations of BoNT (A and B, A and F, B and F; HATHEWAY 1993b) suggest strongly that TeNT and BoNT genes derive from a common ancestor and can be transferred between *Clostridium* strains.

As discussed above, bacteriophages carrying BoNT/C1 and BoNT/D genes can be transferred in various *Clostridium* host strains and determine the toxinotype of these strains. A pseudolysogenic relationship exists between these phages and hosts. During culture a proportion of the bacteria, depending on the strain and on the growth conditions (temperature, salinity) are lyzed and release free bacteriophages; other bacteria lose their bacteriophages and can be

Table 3. TeNT- and BoNT-producing clostridia, neurotoxin gene localization, and neurotoxin amino acid length

	Toxin Localization	Gene	Amino Acid Length
C. tetani	TeNT	Plasmid	1315
C. botulinum A	BoNT/A	Chromosome	1296
C. botulinum B	BoNT/B	Chromosome	1291
C. botulinum C	BoNT/C1	Phage	1291
C. botulinum D	BoNT/D	phage	1276
C. botulinum E	BoNT/E	Chromosome	1251
C. butyricum E	BuNT/E	Chromosome	1251
C. botulinum F	BoNT/F	Chromosome	1274
C. baratii F	BaNT/F	?	1268
C. botulinum G	BoNT/G	Plasmid	1297

reinfected with free bacteriophages. Such lysogeny and reinfection cycles probably occur in the environment (soil, intestinal tract of birds and animals) and account for isolation of nontoxigenic or low toxin producer variants (EKLUND et al. 1987).

The exoenzyme C3 gene is also harbored by the bacteriophages carrying BoNT/C1 and BoNT/D genes. In *C. botulinum* C-468, C-St, and D-1873 strains, the exoenzyme C3 gene has been localized in a 21.5-kbp fragment which is probably a mobile DNA element related to the site-specific transposon family of Tn*554*. This transposable element accounts for the presence of identical exoenzyme C3 gene in different bacteriophages in *C. botulinum* C and D (HAUSER et al. 1993).

Analogous transposable elements to that encompassing the exoenzyme C3 gene could also be involved in the BoNT gene transfer between *C. botulinum* strains. Therefore the 3′ nucleotide regions of the BoNT/A gene in *C. botulinum* A strains 62A and NCTC2916 show that the 97 nucleotides downstream from the BoNT/A gene are identical in both strains, and that the following 3′ nucleotides are totally unrelated (BINZ et al. 1990b; THOMPSON et al. 1990). This 97-nucleotide stretch could be part of a mobile DNA element encompassing the BoNT/A gene. This observation also indicates that the chromosomal localization of the BoNT/A gene is not identical in these two *C. botulinum* A strains and reinforces the idea that the BoNT/A gene is localized on a mobile DNA element.

Neurotoxigenic *C. butyricum* strains probably derive from nontoxigenic *C. butyricum* strains by acquisition of BoNT/E gene from *C. botulinum* E (POULET et al. 1992). The BoNT/E gene and its flanking regions are absent in nontoxigenic *C. butyricum* strains, suggesting a possible gene mobilization by mobile DNA element (HAUSER et al. 1992). This gene has been transferred from a neurotoxigenic *C. butyricum* strain to a nontoxigenic *C. botulinum* E strain by a protocol resembling transduction with a defective phage (ZHOU et al. 1993).

In *C. tetani* and *C. botulinum* G gene transfer was achieved by mobilizing the large plasmids harboring TeNT and BoNT/G genes respectively. However, up to now, only nontoxigenic variants free of plasmids have been obtained (EKLUND et al. 1988; FINN et al. 1984).

5 Ecology of Neurotoxin-Producing Clostridia

C. tetani and BoNT-producing Clostridium are, as the other clostridium, bacteria present in the environment. They are sporulating organisms, and spores present a resistance that enables them to survive for very long periods under extreme conditions (e.g., heat, dryness, radiation, chemicals, oxygen). Clostridium are therefore ubiquitous and are widely distributed in the environment. However, spore germination and cell division take place in anaerobic conditions and when nutritional requirements are available. This restricts the habitat of the clostridium to anaerobic or low oxygen tension areas and those containing sufficient amounts of organic material. Due to their physiological properties (tolerance or extreme sensitivity to oxygen, requirement of particular pH, temperature and substrate for growth or spore germination), the distribution of the various *Clostridium* species in the nature is not uniform (Hippe et al. 1992). Saccharolytic *Clostridium* species such as *C. butyricum* are able to grow on carbohydrates and are found mainly in vegetable and fruit decomposition. Proteolytic and gelatinolytic *Clostridium* species such as *C. botulinum* and *C. tetani* are preferentially associated with animal cadaver decomposition and soil or sediment rich in organic material.

Their interaction with mammalian organisms is accidental and occurs by oral route or a breaking of the integrity of the tegument. They can be found in the digestive tract of healthy human or animals, and after death they participate in cadaver decomposition.

5.1 In the Environment

5.1.1 *C. tetani*

C. tetani is ubiquitous organism commonly found in soil samples throughout the world. The frequency of its isolation varies according to the individual investigation. Surveys in Japan, Canada, Brazil, and the United States have yielded 30%–42% positive samples (Smith and Williams 1984).

Several factors influence the frequency of *C. tetani* isolation from soil samples, such as pH, temperature, moisture, and the presence of large or poor organic material amounts. Germination and multiplication of *C. tetani* have been observed preferentially in neutral or alkaline soil, when the temperature was above 20°C and humidity reached 15% (Smith 1975).

Geographical distribution of *C. tetani* shows that it is more widespread in the south than in the north. The incidence of tetanus is higher in warm countries (West and Central Africa, Southeast Asia, India, Pacific Islands, and southern United States) than in more northerly areas (Canada, Norway, United Kingdom, Finland, Sweden; Smith 1975). This bacterium can be found in intestinal content of animals, but it does not represent a significant part of the normal digestive flora.

C. tetani has also been frequently isolated from human feces, but recent investigations have failed to detect its presence (George and Finegold 1985). Various surfaces and objects contaminated with soil particles, dust, or feces may contain *C. tetani*, and the presence of toxigenic bacteria has been established in the hospital environment, in catgut, cotton wool, dust, and air samples of surgical wards, skin surfaces of men, and wounds (Bytchenko 1981).

5.1.2 BoNT-Producing Clostridia

C. botulinum is widespread in soils and in sediments in lake and seas in most parts of the world. However, the various toxinotypes of *C. botulinum* are not equally distributed. Some toxinotypes are restricted to particular ecological areas. The factors responsible for the geographical distribution of toxinotypes are still poorly understood. In general, toxinotypes A, B, E, F, and G seem to have their principal habitat in soil and in sea and fresh water sediments. Toxinotypes A and B occur more frequently in soil samples, and the regional distributions of these differ. Toxinotype E is more prevalent in sea or lake sediments and fish than in soil. Toxinotypes C and D seem to be obligate parasites in animals and birds. Cadavers of animals dying of botulism are the main source of these organisms. They are seldom encountered in soil samples, except in the areas with a high incidence of animal botulism (SMITH 1975).

C. botulinum is not found usually in the digestive tract of healthy humans. It can be found, however, in intestinal content of animals, particularly *C. botulinum* C and D. Healthy animal carriers are observed mainly in areas where botulism is frequent.

5.1.2.1 *C. botulinum* A and B

The most documented ecological distribution of *C. botulinum* is in the United States. Numerous investigations have surveyed the prevalence of *C. botulinum* (reviewed in DODDS 1993b; HAUSCHILD 1989). Types A and B are generally found in soil samples which are neutral to alkaline and low in organic matter, and not or rarely from aquatic sediment (Table 1). *C. botulinum* type A is predominant in the western United States (west of the Missouri and Mississipi Rivers), in soil that is neutral to alkaline (average pH 7.5) with a lower than average organic content. In contrast, type B prevails largely in the eastern United States (Table 4). This toxinotype was recovered in slightly more acidic soil samples (average pH 6.25) with a higher level of organic matter and mainly from cultivated soils (pasture, fields; SMITH 1978). Other investigations have reported the prevalence of type B in cultivated samples, but it should be noted that this type is rare in soils contaminated with manure or animal feces. Soil fertilization is not considered as a significant factor in the *C. botulinum* type B frequence in cultivated soils (DODDS 1993b).

The incidence of *C. botulinum* A and B is very low from aquatic (sea or fresh water) sediments and soil samples in northern North America (Alaska, Canada; Table 4).

Table 4. *C. botulinum* in the environment in the United States (from DODDS 1993b; HAUSCHILD 1989)

	Positive samples(%)	Type (%)				
		A	B	C/D	E	F
Eastern US soil	19	12	64	12	12	0
Western US soil	29	62	16	14	8	0
Great Lakes and east coast (North of 36°) sediment, shore soil	26	0	1	6	93	0
West coast (north of 40°) sediment, shore soil	43	6	1	0	92	1
East coast (south of 40°) sediment	5	7	7	49	37	0
West coast (south of 36°) sediment	10	44	50	0	0	6

Fewer investigations have been performed in Central and South America. Surveys from soil samples in Argentina, Brazil, and Paraguay have demonstrated the presence of *C. botulinum* A and B, with a prevalence of type A (Table 5).

C. botulinum type B is the most common type from soil and sediment samples in Central and South Europe (United Kingdom, Ireland, Netherlands, France, Switzerland, Italy), and from soil samples in Denmark and Ireland (Table 6). In the United Kingdom and Ireland *C. botulinum* is much more frequent in lake and loch sediments than in soil samples; type B is predominant.

C. botulinum types A and B are also widespread in Asia including the former Soviet Union, China, and Taiwan (Table 7). Soil from southern regions with warm climate or subtropical and tropical areas show a higher incidence of *C. botulinum* A and B. Populated regions are much more contaminated than desert areas. The distribution of types A and B varies locally and does not correspond to large geographical area as in North America.

Table 5. *C. botulinum* in the environment in Europe (from DODDS 1993b; HAUSCHILD 1989)

	Positive samples(%)	Type (%)					
		A	B	C/D	E	F	G
South America (Brazil, Paraguay, Argentina, Falkland Islands)	28	33	33	13	0	20	1
Central Africa (Kenya)	25	89	0	11	0	0	0
Southern Africa	12	0	50	50	0	0	0
Indonesia, Java	12	5	17	73	5	0	0
Australia, New Zealand	29	50	0	50	0	0	0

Table 6. *C. botulinum* in the environment (soil, sediments) in South America, Africa, Indonesia, Australia, and New Zealand (from DODDS 1993b; HAUSCHILD 1989)

	Positive samples(%)	Type (%)					
		A	B	C/D	E	G	F
Sweden, Norway, Denmark, Greenland, Iceland, Baltic coast (shore soil, sediment)	43	3	2	0	95	0	0
United Kingdom, Ireland (sediment, soil) Denmark, Iceland, Netherlands (soil)	35	0	67	19	14	0	0
France, Italy (soil)	4	29	71	0	0	0	0
Switzerland	44	19	56	3	0	19	3

Table 7. *C. botulinum* in the environment in Asia (from DODDS 1993b; HAUSCHILD 1989)

	Positive samples(%)	Type (%)				
		A	B	C/D	E	F
Former Soviet Union, Caspian Sea (sediment, soil)	15	8	18	2	72	0
China, Thailand, Bangladesh (sediment, soil)	69	16	14	61	8	1
Japan (soil, sediment)						
North	12	0	0	0	99	1
South	28	0	0	92	8	0

Few reports concern the incidence of *C. botulinum* in Africa. In Kenya the soil samples that have been examined seem heavily contaminated, with a predominance of type A (Table 5). In contrast, the contamination in southern Africa seems to be low; *C. botulinum* B has been identified in three soil samples out of 102 (Knock 1952).

In Australia and New Zealand, *C. botulinum* types A and B have been found, but their presence in the environment in rare, reflecting a low incidence of human botulism in these areas.

5.1.2.2 *C. botulinum* E

The distribution of type E is more regional than that of other types. *C. botulinum* E is found mainly in northern areas of the northern hemisphere: northern North America (Alaska, Canada, northern United States), northern Europe (Iceland, Greenland, Denmark, Norway, Sweden, coast of Baltic sea), and northern Asia (northern part of former Soviet Union, Caspian sea, northern Japan islands of Hokkaido and Northern part of Honshu) from soil that is in contact with water (Tables 4, 6; Smith 1992). The ability of *C. botulinum* E to grow at very low temperature (Table 1) reflects its prevalence in areas with cold temperature. Its frequency decreases considerably in southern areas with warm weather. This type is found mainly in aquatic environment (sediment, soil from the shores of lakes, seas, and rivers). Organic matter content, salinity, and other unindentified factors influence the distribution of *C. botulinum* E, which varies greatly from one region to other. It is commonly found in the Great Lakes in North America, especially in Lake Michigan and particularly in the Green Bay area. *C. botulinum* E is also very common in the Pacific northwest. Sediment samples of the Lake Washington near Seattle contained 18–25 *C. botulinum* E organisms per gram. These particular areas seem to represent the principal habitat of *C. botulinum* E, where it can grow and multiply (Smith 1975).

5.1.2.3 *C. botulinum* C and D

C. botulinum types C and D are widespread through the world. *C. botulinum* C is found mainly in muds, sediments of marshes, ponds, and seashores, where botulism in waterfowl is endemic. Outbreaks of botulism in birds and the presence of *C. botulinum* C in their environment have been reported in the United States, United Kingdom, Denmark, Netherlands, France and Japan. Intestinal contents and cadavers of susceptible birds seem to be the principal habitat of *C. botulinum* C. This type has also be detected in soil from warm areas such as in Indonesia, Bangladesh, and Thailand (Dodds 1993b; Hauschild 1989; Smith and Sugiyama 1988).

C. botulinum D is more frequently associated with botulism in animals (e.g., ruminants, horses). Carcasses of these animals and also those of small animals (e.g., rodents) constitute the most common source of this organism. *C. botulinum*

D can also be identified in soil samples, where animal botulism is common (southern and central Africa, Australia, the Americas, Europe; SMITH and SUGIYAMA 1988).

5.1.2.4 *C. botulinum* F and G

Types F and G are much less frequently encountered than the other types. Since the first identification of *C. botulinum* type F from a homemade liver paste, responsible for a human botulism outbreak on the Danish island of Langeland (MOLLER and SCHEIBEL 1960), this type has been demonstrated in marine sediments on the western coast of the United States (EKLUND and POYSKY 1965) and in marine or fresh water sediments in Brazil, Venezuela, and Indonesia (reviewed in HAUSCHILD 1989). *C. botulinum* G was first isolated from a soil sample in Argentina (GIMENEZ and CICCARELLI 1970) and has been identified from necropsy specimens in cases of unexplained death in adults and infants (SONNABEND et al. 1981) and from five soil samples of 41 in close association with cultivated land in Switzerland (SONNABEND et al. 1987b).

5.1.2.5 *C. butyricum*

C. butyricum is a very widespread bacteria in the environment, including soil, fresh water and marine sediments, cheese, rumen of healthy calves, animal and human feces, and more rarely from clinical specimens (blood, urine, respiratory tract, pleural cavity, abdomen, wound and abcesses; CATO et al. 1986). *C. butyricum* has been involved in certain cases of necrotizing enterocolitis in newborns. Artificial feeding and intestinal stasis support abundant proliferation of *C. butyricum* in the intestinal content. The bacterial overgrowth is accompanied by large production of hydrogen and organic acids, mainly butyric acid, which have been recognized as the main virulence factors of *C. butyricum* in the neonatal necrotizing enterocolitis (POPOFF 1990; POPOFF et al. 1985).

The two *C. butyricum* strains which produce BoNT/E were isolated from two cases of infant botulism in Rome (McCROSKEY et al. 1986). Extensive studies have been carried out in the vicinity of Rome, and no strains producing BoNT/E were isolated; only *C. botulinum* A and B strains were found in 9.6% of 52 soil samples (CRETI et al. 1990). The source of the toxigenic *C. butyricum* strains remains a mystery (CRETI et al. 1990).

5.1.2.6 *C. baratii*

C. baratii is isolated from soil, sediments, normal and rat feces, and occasionally from war wounds, peritoneal fluid, infection of the eyes, ear, and prostate (CATO et al. 1986). The source of the toxigenic *C. baratii* producing BoNT/F responsible for one case of infant botulism and two cases of adult botulism in the United States has not been elucidated (HATHEWAY 1993b).

5.2 In Food

This section deals only with *C. botulinum*.

5.2.1 Fish and Sea Food

In general, the contamination of fish and other aquatic animals by *C. botulinum* reflects that of the sediments of the respective areas (Table 8). *C. botulinum* E is predominant in fish and aquatic invertebrates from North America, Europe, and Asia. The incidence of *C. botulinum* depends on the area and on type of fish. In North America the level of contamination is high in fish from the Pacific northwest (5% of ocean salmon, 17% of other fish, and 23% of salmon from rivers; DODDS 1993a), and for the Great Lakes and adjacent bays (1 to 60% of fish) (HAUSCHILD 1989). In the Pacific northwest 8% of coho salmon, 17% of soleand cod, 23% of steelhead trout, 24% of sockeye salmon, 31% of oysters, 23% of clams, and 18% of crabs are contaminated by *C. botulinum* (CRAIG et al. 1968). Gills and intestines of fish are the most heavily contaminated parts. The presence of *C. botulinum* in gills indicates an environmental contamination. In general, *C. botulinum* E does not multiply in the intestines of living fish. However, when

Table 8. Prevalence of human botulism and food origin (from HAUSCHILD 1989, 1993)

	Average out-breaks per year	Type (%)				Food (%)			
		A	B	E	Others	Meat	Fish	Fruit/ vege-tables	Others
Canada and Alaska	4	7	10	83	0	59	36	4	1
United states (Alaska not included)	14	69	28	3	0.4 (F)	8	7	75	10
Argentina	2	77	8	0	15 (AF, A+F)	29	21	36	14
Northern Europe (Denmark, Norway)	2	0	24	74	2(F)	58	42	0	0
Western, central, and southern Europe (Belgium, France, Germany, Poland, former Czecho-slovakia, Hungary, Italy, Spain, Portugal)	37	5	86	2	7 (A+B, AB, B+C)	70	8	19	3
Former Soviet Union, Iran	14	16	21	63	0	10	82	8	0
China	38	93	5	1	<1 (A+B)	10	0	86	4
Japan	3	2	2	96	0	0	99	1	0

the fish die of any cause, *C. botulinum* can grow and produce BoNT. Dead fish in the sediments contribute to the persistence and multiplication of *C. botulinum* in the aquatic environment (DODDS 1993b). Botulism outbreaks with mortality have been attributed to the ingestion of dead fish containing high titers of BoNT (EKLUND et al. 1982).

The incidence of *C. botulinum* in fish and aquatic invertebrates is also high in the Baltic Sea, Swedish and Danish coastal waters, Caspian sea, and Japanese lakes. Type E is predominant. In the United Kingdom *C. botulinum* C is preferentially detected in most fish farms, followed by types B, E, and F (CANN et al. 1975). Type C is also predominant in fish and aquatic animals from Indonesian waters (DODDS 1993a).

The contamination in prepared fish depends on the level of contamination of freshly caught fish, the type of preparation, and the processing stage. The highest contamination is found in salted fish from the Caspian Sea (HAUSCHILD 1989) and dressed rockfish from California (DODDS 1993a). The presence of *C. botulinum* has been found in brined, frozen, and vacuum-packed and in frozen and packaged fishes. Smoking reduces the incidence of *C. botulinum*. In general, *C. botulinum* E is the predominant type in prepared fish, type A is most commonly found in fish from California, and type C followed by types D, A, B, and F in fish from Indonesia and tropical and subtropical waters (DODDS 1993a; HAUSCHILD 1989).

5.2.2 Meats

The incidence of *C. botulinum* is lower in meat and meat products than in fish, and it is lower in North America than in Europe (Table 8). Animals may carry *C. botulinum* in their intestinal tract, which can lead to meat contamination during meat processing. The identified toxinotypes are A and B, followed by C and more rarely E. Meat mainly from cattle, pig, and chicken are contaminated by *C. botulinum*. The incidence varies according to the investigation and to the sampling (DODDS 1993a). ROBERT and SMART reported that 2%–6% of vacuum-packed bacon samples of 140 in two investigations were contaminated by *C. botulinum*, and in one occasion 20 of 138 samples were positive (65% contained type C, 15% type A, and the rest were not typed; ROBERTS and SMART 1976).

5.2.3 Fruit and Vegetables

Raw fruits and vegetables may be contaminated by *C. botulinum* as a result of soil contamination. *C. botulinum* has been identified in many fruits and vegetables, but mainly in those harvested from the soil. The use of manure as fertilizer may affect the level of contamination. Cultivated mushrooms in Canada have been found to be highly contaminated by *C. botulinum* B, and in other areas this product was free or contained low level of *C. botulinum* (DODDS 1993a).

Contamination of fruits and vegetables reflects that of the soil. It is due mainly to *C. botulinum* A and B. Type A is predominant in fruits and vegetables from California, type B in vegetables from Italia, Germany, and other European countries, and both types A and B are detected in the former Soviet Union (Table 8; DODDS 1993a).

5.2.4 Other Foods

Honey from various areas (United States, Argentina, Canada, China, Hungary, Japan, Mexico, Spain) may contain *C. botulinum*. The level of contamination in usually low (1–10 spores/kg), but the honey samples associated with infant botulism contained 10^3–10^4 spores per kilogram. Type A was detected more frequently than type B, C, or D (DODDS 1993a). The incidence of *C. botulinum* in other foods such as dairy products or prepared foods (boil-in-bag foods, vacuum-packed foods, pressurized foods, dehydrated and freeze-dried foods) is very low (DODDS 1993a).

6 Epidemiology of Tetanus and Botulism

6.1 Tetanus

Tetanus has always been associated with wound infection. It has not been reported that tetanus results from ingestion of preformed toxin in food or by absorption of tetanus toxin produced by *C. tetani* in the intestine. This organism can grow and produce toxin in the intestinal tract of germ-free rats (WELLS and BALISH 1983), and it has been found in human and animal feces (SMITH and WILLIAMS 1984). The absence of digestive tetanus such as botulism is not well understood. This may be due to poor absorption of tetanus toxin by the intestinal mucosa or to its unstability in the digestive tract (e.g., low pH, protease degradation; HATHEWAY 1993a). Unlike BoNT, TeNT is not associated with nontoxic proteins to form a complex which is probably involved in the protection of BoNT.

C. tetani is an anaerobic bacteria, and it cannot grow in normal tissue. It has been shown that at normal pH of the tissue, an oxidation-reduction potential of 10 mV or less is needed for spore germination, while at the most acidified environment of traumatized tissue germination can take place at much higher oxidation-reduction potential (DEZFULIAN 1989). Often this is provided by a small amount of necrosis induced by soil, clothing, or other foreign material. Concomitant infection by other bacteria may also provide conditions suitable for the growth of *C. tetani* (SMITH 1975). Tetanus is frequent in subtropical and tropical countries, mainly those with a low level of hygiene.

Domestic animals can contract tetanus. This disease is frequent and severe in horses. Susceptibility to TeNT depends on the animal species. Birds are relatively resistant and cold blood animals quite resistant. The most susceptible animal species, in decreasing order, are: horse, sheep, goat, mouse, rat, rabbit, monkey, guinea pig, dog, cat, pigeon, and hen (DEZFULIAN 1989; SMITH 1975). The difference in susceptibility to tetanus is not well understood; it may correspond to differences in receptor amounts at the nerve ends. The susceptibility to tetanus also depends on environmental factors. Frogs are much more resistant if toxin treatment is performed at 15°C than at 23°C (SMITH 1975). Temperature may influence the fixation and/or internalization of TeTx at the nerve ends.

6.2 Human Botulism

Unlike tetanus, botulism results from ingestion of preformed BoNT in foods or from oral contamination, and rarely from wound infections. Four forms of botulism are recognized according to the mode of acquisition: foodborne botulism, infant botulism, botulinal toxi-infection, and wound botulism.

6.2.1 Foodborne Botulism

Foodborne botulism is due to ingestion of food in which *C. botulinum* has developed and produced sufficient amounts of BoNT. This form is the most common in adults. Usually several persons are intoxicated at the same time by consumption of a common contaminated food. Types A, B, and E are almost always involved. Only six botulism incidents with type F, two with type C, and one with type D have been reported (HATHEWAY 1993b). It is not known why types C and D botulism occur so rarely in humans. *C. botulinum* types C and D seem unable to grow and produce toxin in foods for humans (HATHEWAY 1993b). Type G has been isolated from autopsy specimens, but its implication in human botulism has not been confirmed (Sonnabend et al. 1981, 1987b).

The incidence and type of botulism depend on the occurrence of *C. botulinum.* in the environment and subsequently in foods, and on cooking practices. The geographical distribution of botulism types corresponds globally with that of the different toxinotypes of *C. botulinum* in the environment. As shown in Table 8, type E botulism is found mainly in the colder regions of the northern hemisphere (Alaska, Canada, Greenland, Scandinavia, and the northern parts of the former Soviet Union, Iran, and Japan). The foods involved in these outbreaks are mostly fish and aquatic animals (HAUSCHILD 1989, 1993). Types A and B botulism generally occur in temperate countries. Type A is predominant in the western United States, Argentina, and China, and the most frequently implicated foods are fruits and vegetables. In Europe (excluding northern countries), type B is prevalent, and meat has been recognized as the main source of these outbreaks, except in Italy and Spain where vegetables are more frequently incriminated.

In the majority of outbreaks, botulism is due to home-canned, home-fermented products, or home-processed low-acid vegetables. Preservation of fish and marine mammal meat by fermentation without adequate fermentable carbohydrates is the most common source of *C. botulinum* E. Home-canned or home-preserved fruit and vegetables are responsible for the majority of outbreaks in the United States, Spain, Italy, and China. Most of the implicated foods are vegetables home-preserved in oil in Italy and fermented vegetables in China. Home-cured smoked ham is the major cause of botulism in several European countries in particular in France, Belgium, Germany, and Portugal (HAUSCHILD 1989, 1993).

6.2.2 Infant Botulism

In contrast to foodborne botulism, infant botulism is not due to ingestion of preformed toxin in food but to ingestion of *C. botulinum* spores that germinate, multiply, and produce BoNT in the infant's gastrointestinal tract. The minimum infective dose for human infant has been estimated at 10–100 *C. botulinum.* spores (ARNON 1989). Moderate to high toxin level (10–10^5 mouse lethal dose per gram) and viable *C. botulinum* are recovered from stools of affected infants for a long period after the onset of the symptoms, as long as 158 days. Toxin is seldom found in the serum of these patients. In foodborne botulism toxemia is common, and toxin usually disappears from the stools within several days and is rarely detectable in serum after 2 weeks.

Infant botulism has been identified in many developed countries including Argentina, Australia, Canada, Chile, the former Czechoslovakia, United Kingdom, France, Italy, Japan, Spain, and Sweden, but most cases have been recorded in the United States (up to 90%), particularly in California. It is presumed that contamination occurs with *C. botulinum* spores from the environment. The only food implicated so far is honey (ARNON 1989).

Most of the *C. botulinum* strains involved in infant botulism belong to the group I (*C. botulinum* A and proteolytic *C. botulinum* B strains). One of the *C. botulinum* B strains also produces minor amounts of BoNT/F. The other implicated *C. botulinum* strains include one *C. botulinum* type C, two BoNT/E producing *C. butyricum*, and one BoNT/F producing *C. baratii* strains (HATHEWAY 1993b). This could indicate that *C. botulinum* strains from group I are more able to colonize the infant's digestive tract. Although belonging to the same group, the *C. botulinum* strains from infant botulism and from food-borne botulism may yet be different. It has been found that *C. botulinum* A strains isolated from infant botulism in Japan produce a BoNT/A complex which does not contain a hemagglutinin component, as classical foodborne botulism strains do, and a slightly different BoNT/A (90% amino acid sequence identity with BoNT/A from classical foodborne botulism strains; WILLEMS et al. 1993). Host susceptibility factors are also involved in colonization of the digestive tract by *C. botulinum.* In animal experimental models normal intestinal microflora of adult mice prevents the growth of *C. botulinum* in the digestive tract. Conventional infant mice are susceptible to colonization with *C. botulinum* in a limited age range (7–13 days; SUGIYAMA 1979). In human infants most botulism cases have been recorded in newborns between 1 and 6 months of age at onset (ARNON 1989). Moreover, nutritional factors are also involved, such as breast feeding or formula feeding and subsequent intestinal content pH and composition of the digestive microflora [formula-fed infants have less acidic feces (pH 5.9–8) and lower numbers of *Bifidobacterium* than infants fed human milk (pH 5.1–5.4)]. Infant botulism with more chronic symptoms seems to be correlated to breast feeding, and infant botulism associated with sudden infant death to formula feeding (ARNON 1989).

6.2.3 Botulinal Toxi-infection in Adults

A new category of botulism called "of undetermined etiology" has been proposed to include patients older than 1 year of age in which no food or wound is recognized as the source of toxin. *C. botulinum* and its toxin have been found in these patient's stools for a long period (3 weeks and in some cases up to 6 months; HATHEWAY 1993a). It has been concluded that this form of botulism in adults results from digestive tract colonization by *C. botulinum* and of BoNT production in the intestine, as in infant botulism (CHIA et al. 1986; MCCROSKEY and HATHEWAY 1988).

C. botulinum types A and B from group I are generally involved. One case with *C. botulinum* Type F and two cases with BoNT/F producing *C. baratii* have been documented (HATHEWAY 1993a; SONNABEND et al. 1987a). Surgical modification of the bowel, achlorhydria, and antimicrobial use have been proposed as predisposing factors. Moreover, chronic inflammation and necrotic lesions of the intestinal mucosa due to underlying disease, could support the growth of *C. botulinum* (TACKET and ROGAWSKI 1989).

6.2.4 Wound Botulism

Wound botulism, as tetanus, results from colonization of a wound with *C. botulinum* and subsequent local production of BoNT. However, the occurrence of wound botulism is much lower than that of tetanus, despite the fact that the population is unimmunized. A total of 47 laboratory-confirmed cases of wound botulism have been recorded by the Centers for Disease Control (WEBER et al. 1993). The *C. botulinum* strains involved in wound botulism belonged to the toxinotypes A or B from group I. This indicates that *C. botulinum* strains have a lower ability to grow in wounds than *C. tetani*. Among the *C. botulinum* strains those from group I seem the most potent to multiply in organism. The *C. botulinum* strains could exhibit different potentiality to grow in animal or human organisms. Thus *C. botulinum* E has never been found to colonize digestive tract or wound; it could be unable to multiply and/or produce BoNT at body temperature (HATHEWAY 1993a).

6.3 Animal Botulism

Botulism is common in animals; usually types C and D are involved. In domestic animals botulism leads to economic losses and also constitutes a possible risk factor of transmission to human.

Bovines are susceptible to BoNT/C1 and BoNT/D, but more frequently type D is encountered. The disease is contracted by eating carcass or chewing on bones contaminated with BoNT ("lamsiekte" in South Africa, "bulbar paralysis" in Australia, and "loin disease" in the southwestern United States), by drinking water from ponds (especially in Africa) or ingestion of forage poisoned by diffusion of BoNT from carcass decomposition of small animals or birds. Where botulism is common, animals (cattle and other) often carry *C. botulinum* in their digestive tract, and when they die for any reason, the carcass is colonized by *C. botulinum* and becomes highly toxic. More recently, poultry litter, often containing chicken carcass, has been found to transmit botulism to cattle (SMITH and SUGIYAMA 1988).

In the Netherlands type B botulism outbreaks have been observed in cattle. The animals were contaminated by eating brewer's cereal containing *C. botulinum* B and BoNT/B (BREULINK et al. 1978). *C. botulinum* B has been recovered in the rumen and feces of the affected animals and also in the pasture on which the cattle were feeding (NOTERMANS et al. 1981).

In France type D botulism has been found responsible for important losses in cattle since 1982. The animals were contaminated by grazing on pasture containing poultry litter. Battery-reared hens received commercial feedstuff containing meat meal prepared in local stock yards from died animals, even those suspected of botulism. The heat treatment was insufficient in some stock yards. Hens are relatively resistant to BoNT/D (SMITH and SUGIYAMA 1977), and no outbreaks of botulism have been observed in hen farms. *C. botulinum* type D was identified in meat meal, feedstuff for hen, litter, carcass of hen dying from unknown reasons, and cattle grazing on pasture manured with hen litter and dying with paralytic symptoms (POPOFF 1989; POPOFF et al. 1986).

Botulism is also frequent in domestic and wild birds all over the world. Almost all species of birds are susceptible to botulism, usually type C. An excellent review covers this topic (EKLUND and DOWELL 1987). Type C botulism outbreaks have been observed in battery-reared turkeys and chickens in the United States, United Kingdom (SMART 1983; SMART and ROBERTS 1977), France (unpublished), in water birds in the United Kingdom (DORLAND et al. 1977; SMITH and OLIPHANT 1983), in Netherlands, the Americas, Australia, and Japan (EKLUND and DOWELL 1987), and in wild ducks in France (GOURREAU et al. 1986).

Equine botulism occurs sporadically throughout the world and is commonly due to types C and D, but some type B outbreaks have been reported (HAAGSMA et al. 1990; SMITH and SUGIYAMA 1988).

Botulism has also been identified in various other animals, including minks, sheep, monkeys, dogs, see lions, and lions (DODDS 1993b; SMITH and SUGIYAMA 1988).

6.4 Concluding Remarks

TeNT- and BoNT-producing clostridia are widespread in the environment (soil, sediment, etc.), which constitutes their main habitat. *C. tetani* strains form an homogeneous group based on phenotypic and genotypic characters and produce only one type of TeNT. Until now no variant of TeNT has been observed.

However, BoNT-producing clostridia belong to different species and groups, and produce a diversity of BoNT. Seven toxinotypes are recognized according to their antigenic properties, and sequence analysis has revealed additional variants. At least 11 different BoNT sequences have been described. The various physiological groups of neurotoxigenic clostridia have distinct geographical distributions. *C. tetani* is widespread but is found mainly in warm countries. *C. botulinum* E is predominant in aquatic environments in northern areas; *C. botulinum* A in soil in the western United States, South America, and China; *C. botulinum* B in the eastern United States, United Kingdom, and continental Europe; and *C. botulinum* C and D in soil and cadavers in warm regions. These physiological groups also display differing abilities to colonize organisms. *C. tetani* grows easily in wounds whereas *C. botulinum* group I strains are the most able to colonize the digestive tract.

TeNT and BoNT do not represent a selective advantage for the growth or survival of clostridia since nontoxic derivatives also grow, in addition to toxigenic strains. The selective pressures involved in the diversity of BoNT genes and their transfer to different *Clostridium* species remain mysterious.

References

Aktories K, Wegner A (1989) ADP-ribosylation of actin by clostridial toxins. J Cell Biol 109: 1385–1387

Aktories K, Bärmann M, Ohishi I, Tsuyama S, Jakobs KH, Habermann E (1986) Botulinum C2 toxin ADP-ribosylates actin. Nature 322: 390–392

Aktories K, Weller U, Chhatwal GS (1987) Clostridium botulinum type C produces a novel ADP-ribosyltransferase distinct from botulinum C2 toxin. FEBS Lett 212: 109–113

Alouf JE, Geoffroy C (1991) The family of the antigenically-related cholesterol-binding ('sulphydryl-activated') cytolytic toxins. In Alouf JE, Freer JH (eds) Sourcebook of bacterial protein toxins. Academic, London, pp 147–186

Altwegg M, Hatheway CL (1988) Multilocus enzyme electrophoresis of clostridium argentinense (Clostridium botulinum toxin type G) and phenotypically similar asaccharolytic Clostridia. J Clin Microbiol 26: 2447–2449

Arnon SS (1989) Infant botulism. In: Finegold SM, George WL (eds) Anaerobic infections in humans. Academic, San Diego, pp 601–609

Aureli P, Fenicia L, Pasolini B, Gianfranceschi M, Mc Croskey LM, Hatheway CL (1986) Two cases of type E infant botulism caused by neurotoxigenic Clostridium butyricum in Italy. J Infect Dis 154: 207–211

Binz T, Kurazono H, Popoff MR, Eklund MW, Sakaguchi G, Kozaki S, Krieglstein K, Henschen A, Gill DM, Niemann H (1990a) Nucleotide sequence of the gene encoding Clostridium botulinum neurotoxin type D. Nucleic Acid. Res. 18: 5556

Binz T, Kurazono H, Wille M, Frevert J, Wernars K, Niemann H (1990b) The complete sequence of botulinum neurotoxin type A and comparison with other clostridial neurotoxins. J Biol Chem 265: 9153–9158

Breulink HJ, Wagenaar G, Wensing TH, Notermans S, Poulos PW (1978) Food poisoning in cattle caused by ingestion of brewer's grains contaminated with Clostridum botulinum B. Neth J Vet Sci 103: 303–311

Bytchenko B (1981) Microbiology of tetanus. In: Veronesi R (ed) Tetanus: important new concepts. Excerpta Medica, Amsterdam, pp 28–39

Cann DC, Taylor LY, Hobbs G (1975) The incidence of Clostridium botulinum in farmed trout in Great Britain. J Appl Bacteriol 39: 331–336

Carle A, Rattone G (1884) Studio esperimentall sull' etiologia del tetano. G Accad Med Torino 32: 174–179
Cato EP, George WL, Finegold SM (1986) Genus Clostridium. In: Sneath PHA, Mair NS, Sharpe ME, Holt JG (eds) Bergey's manual of systematic bacteriology. Williams and Wilkins, Baltimore, pp 1141–1200
Chardin P, Boquet P, Madaule P, Popoff MR, Rubin EJ, Gill DM (1989) The mammalian G protein rhoC is ADP-ribosylated by Clostridium botulinum exoenzyme C3 and affects actin microfilaments in Vero cells. EMBO J 8: 1087–1092
Chia JK, Clark JB, Ryan CA, Pollack M (1986) Botulism in an adult associated with food-borne intestinal infection with Clostridium botulinum. N Engl J Med 315: 239–241
Considine RV, Simpson LL (1991) Cellular and molecular actions of binary toxins possessing ADP-ribosyltransferase activity. Toxicon 29: 913–936
Craig JM, Hayes S, Pilcher KS (1968) Incidence of Clostridium botulinum type E in salmon and other marine fish in the Pacific northwest. Appl Microbiol 16: 553–557
Creti R, Fenicia L, Aureli P (1990) Occurrence of Clostridium botulinum in the soil of the vicinity of Rome. Curr Microbiol 20: 317–321
Dezfulian M (1989) Animal models of botulism and tetanus. In: Simpson LL (ed) Botulinum neurotoxin and tetanus toxin. Academic, San Diego, pp 335–350
Dodds KL (1993a) Clostridium botulinum in foods. In: Hauschild AHW, Dodds KL (ed) Clostridium botulinum: ecology and control in foods. Dekker, New York, pp 53–68
Dodds KL (1993b) Clostridium botulinum in the environment. In: Hauschild AHW, Dodds KL (eds) Clostridium botulinum: ecology and control in foods. Dekker, New York, pp 21–51
Dorland ED, Moryson CJ, Smith GR (1977) Avian botulism and the high prevalence of Clostridium botulinum in the Norfolk Broads. Vet Rec 100: 106–109
East AK, Richardson PT, Allaway D, Collins MD, Roberts TA, Thompson DE (1992) Sequence of the gene encoding type F neurotoxin of Clostridium botulinum. FEMS Microbiol Lett 96: 225–230
Eklund MW, Dowell J (1987) Avian botulism. Thomas, Springfield
Eklund MW, Poysky F (1965) Clostridium botulinum type F from marine sediments. Science 149: 306
Eklund MW, Poysky FT (1974) Interconversion of type C and D strains of Clostridium botulinum by specific bacteriophages. Appl Microbiol 27: 251–258
Eklund MW, Poysky FT, Reed SM, Smith CA (1971) Bacteriophage and the toxigenicity of Clostridium botulinum type C. Science 172: 480–482
Eklund MW, Poysky FT, Meyers JA, Pelroy GA (1974) Interspecies conversion of Clostridium botulinum type C to Clostridium novyi type A by bacteriophage. Science 172: 480–482
Eklund MW, Peterson ME, Poysky FT, Peck FT, Conrad LF (1982) Botulism in juvenile Coho salmon (Onchorhyncushus kisutch) in the United States. Aquaculture 27: 1–11
Eklund MW, Poysky FT, Oguma K, Iida H, Inoue K (1987) Relationship of bacteriophages to toxin and hemagglutinin production by Clostridium botulinum types C and D and its significance in avian botulism outbreaks. In: Eklund MW, Dowell VR (eds) Avian botulism. Thomas, Springfield, pp 191–222
Eklund MW, Poysky FT, Mseitif LM, Strom MS (1988) Evidence for plasmid-mediated toxin and bacteriocin production in Clostridium botulinum type G. Appl Environ Microbiol 54: 1405–1408
Eklund MW, Poysky FT, Habig WH (1989) Bacteriophages and plasmids in Clostridium botulinum and Clostridium tetani and their relationship to production of toxins. In: Simpson LL (ed) Botulinum neurotoxin and tetanus toxin. Academic, San Diego, pp 25–51
Faber K (1890) Die Pathogenic des Tetanus. Berl Klin Wochenschr 27: 717–720
Finn CW, Silver RP, Habig WH, Hardegree MC, Zon G, Garon CF (1984) The structural gene for tetanus neurotoxin is on a plasmid. Science 224: 881–884
George WL, Finegold SM (1985) Clostridia in the human gastrointestinal flora. In: Borriello SP (ed) Clostridia in gastrointestinal disease. CRC, Boca Raton, pp 1–37
Ghanem FM, Ridpath AC, Moore WEC, Moore LVH (1991) Identification of Clostridium botulinum, Clostridium argentinense, and related organisms by cellular fatty acid analysis. J Clin Microbiol 29: 1114–1124
Gimenez DF, Ciccarelli A (1970) Another type of Clostridium botulinum. Zentralbl Bakteriol Parasitenkd Infektionskr Hyg (Abt 1 A) 215: 221–224
Gourreau JM, Guillou JP, Louzis C, Moutou F (1986) Botulisme au Bois de Boulogne. Med Mal Infect 8/9: 524–526
Gurtler V, Wilson VA, Mayall BC (1991) Classification of medically important clostridia using restriction endonuclease site differences of PCR-amplified 16s rDNA. J Gen Microbiol 137: 2673–2679

Haagsma J, Haesebrouck F, Devriese L, Bertels G (1990) An outbreak of botulism type B In horses. Vet Rec 127: 206
Haque A, Sugimoto N, Horiguchi Y, Okabe T, Miyata T, Iwanaga S, Matsuda M (1992) Production, purification, and characterization of Botunolysin, a thio-activated hemolysin of Clostridium botulinum. Infect Immun 60: 71–78
Hatheway CL (1989) Bacterial sources of clostridial neurotoxins. In Simpson LL (ed) Botulinum neurotoxin and tetanus toxin. Academic, San Diego, pp 3–24
Hatheway CL (1990) Toxigenic bacteria. Clin Microbiol Rev 3: 66–98
Hatheway CL (1993a) Bacteriology and pathology of neurotoxigenic Clostridia. In: DasGupta BR (ed) Botulinum and tetanus neurotoxins. Plenum, New York, pp 491–502
Hatheway CL (1993b) Clostridium botulinum and other clostridia that produce botulinum neurotoxin. In: Hauschild AHW, Dodds KL (eds) Clostridium botulinum: ecology and control in foods. Dekker, New York, pp 3–20
Hauschild AHW (1989) Clostridium botulium. In: Doyle MP (ed) Foodborne bacterial pathogens. Dekker, New York, pp 111–189
Hauschild AHW (1993) Epidemiology of human foodborne botulism. In: Hauschild AHW, Dodds KL (eds) Clostridium botulinum: ecology and control in foods. Dekker, New York, pp 69–104
Hauser D, Eklund MW, Kurazono H, Binz T, Niemann H, Gill DM, Boquet P, Popoff MR (1990) Nucleotide sequence of Clostridium botulinum C1 neurotoxin. Nucleic Acid Res 18: 4924
Hauser D, Gibert M, Boquet P, Popoff MR (1992) Plasmid localization of a type E botulinal neurotoxin gene homologue in toxigenic Clostridium butyricum strains, and absence of this gene in non-toxigenic C. butyricum strains. FEMS Microbiol Lett 99: 251–256
Hauser D, Gibert M, Eklund MW, Boquet P, Popoff MR (1993) Comparative analysis of C3 and botulinal neurotoxin genes and their environment in Clostridium botulinum C and D. J Bacteriol 175: 7260–7268
Hauser D, Gibert M, Eklund MW, Popoff MR (1994) Botulinal neurotoxin C1 complex genes, clostridial neurotoxin homology and genetic transfer in Clostridium botulinum. Toxicon (in press)
Hippe H, Andreesen JR, Gottschalk G (1992) The genus Clostridium—nonmedical. In: Balows A, Trüper HG, Dworkin M, Harder W, Scheifer KH (eds) The prokaryotes. Springer, Berlin Heidelberg New York, pp 1800–1866
Hutson RA, Thompson DE, Collins MD (1993) Genetic interrelationships of saccharolytic Clostridium botulinum types B, E, and F and related clostridia as revealed by small-subunit rRNA gene sequences. FEMS Microbiol Lett 108: 103–110
Johnson JL, Francis BS (1975) Taxonomy of the clostridia: ribosomal ribonucleic acid homologies among the species. J Gen Microbiol 88: 229–244
Kim J, Foegeding PM (1993) Principles of control. In: Hauschild AHW, Dodds KL (ed) Clostridium botulinum: ecology and control in foods. Dekker, New York, pp 121–176
Kitasato S (1889) Über den Tetanusbacillus. Z Hyg 7: 224–234
Knock GG (1952) Survey of soils for spores of Clostridium botulinum (Union of South Africa and South West Africa). J Sci Food Agric 3: 86–90
Lawson PA, Llop-Perez P, Hutson RA, Hippe H, Collins MD (1993) Towards a phylogeny of the clostridia based on 16S rRNA sequences. FEMS Microbiol Lett 113: 87–92
Lee WH, Riemann H (1970a) Correlation of toxic and non-toxic strains of Clostridium botulinum by DNA composition and homology. J Gen Microbiol 60: 117–123
Lee WH, Riemann H (1970b) The genetic relatedness of proteolytic Clostridium botulinum strains. J Gen Microbiol 64: 85–90
Matter K, Dreyer F, Aktories K (1989) Actin involvement in exocytosis from PC12 cells: studies on the influence of botulinum C2 toxin on stimulated noradrenaline release. J Neurochem 52: 370–376
McCroskey LM, Hatheway CL (1988) Laboratory findings in four cases of adult botulism suggest colonization of the intestinal tract. J Clin Microbiol 26: 1052–1054
McCroskey LM, Hatheway CL, Fenicia L, Pasolini B, Aureli P (1986) Characterization of an organism that produces type E botulinal toxin but which resembles Clostridium butyricum from the feces of an infant with type E botulism. J Clin Microbiol 23: 201–202
McKinney MW, Levett PN, Haylock RW (1993) Cloning of a DNA sequence unique to Clostridium botulinum group I by selective hybridization. J Clin Microbiol 31: 1845–1849
Moller V, Scheibel I (1960) Preliminary report on the isolation of an apparently new type of Clostridium botulium. Acta Pathol Microbiol Scand 48: 80
Nakamura S, Okado I, Nakashio S, Nishida S (1977) Clostridium sporogenes isolates and their relationship to C. botulinum based on deoxyribonucleic acid reassociation. J Gen Microbiol 100: 395–401

Nakamura S, Okado I, Abe T, Nishida S (1979) Taxonomy of Clostridium tetani and related species. J Gen Microbiol 113: 29–35
Nakamura S, Kimura I, Yamakawa K, Nishida S (1983) Taxonomic relationships among Clostridium novyi types A and B, Clostridium haemolyticum and Clostridium botulinum type C. J Gen Microbiol 129: 1473–1479
Niemann H (1991) Molecular biology of clostridial neurotoxins. In: Alouf JE, Freer J (eds) Sourcebook of bacterial protein toxins. Academic, New York, pp 299–344
Notermans S, Dufrene J, Oosterom J (1981) Persistence of Clostridium botulinum type B on cattle farm after an outbreak of botulism. Appl Environ Microbiol 41: 179–183
Oguma K, Yamaguchi T, Sudou K, Yokosawa N, Fujikawa Y (1986) Biochemical classification of Clostridium botulinum type C and D strains and their nontoxigenic derivatives. Appl Environ Microbiol 51: 256–260
Ohishi I (1983) Response of mouse intestinal loop to botulinum C2 toxin: enterotoxic activity induced by cooperation of non linked protein components. Infect Immun 40: 691–695
Ohishi I (1987) Activation of botulinum C2 toxin by trypsin. Infect Immun 55: 1461–1465
Ohishi I, Odagiri Y (1984) Histopathological effect of botulinum C2 toxin on mouse intestines. Infect Immun 43: 54–58
Popoff MR (1989) Revue sur l'épidémiologie du botulisme bovin en France et analyse de sa relation avec les élevages de volailles. Rev Sci Techn Off Int Epiz 8: 129–145
Popoff MR (1990) Are anaerobes involved in neonatal necrotizing enterocolitis? In: Borriello SP (ed) Clinical and molecular aspects of anaerobes. Wrighston Biomedical, Petersfield, pp 49–57
Popoff MR, Szylit O, Ravisse P, Dabard J, Ohayon H (1985) Experimental cecitis in gnotoxenic chickens monoassociated with Clostridium butyricum strains isolated from patients with neonatal necrotizing enterocolitis. Infect Immun 47: 697–703
Popoff MR, Legardinier JC, Perrot B, Hamard E, Lhomme P, Leclercq H, Beaumont A, Meheut-Ferron P, Lescure G (1986) Analyse épidémiologique du botulisme bovin dans le département de la Manche. Rev Med Vet 137: 857–865
Poulet S, Hauser D, Quanz M, Niemann H, Popoff MR (1992) Sequences of the botulinal neurotoxin E derived from Clostridium botulinum type E (strain Beluga) and Clostridium butyricum (strains ATCC43181 and ATCC43755). Biochem Biophys Res Commun 183: 107–113
Prevots R, Sutter RW, Strebel PM, Cochi SL, Hadler S (1992) Tetanus surveillance—United States, 1989-1990. MMWR CDC Surv Summ 41: 1–9
Ramon G, Descombey PA (1925) Sur l'immunisation antitétanique et sur la production de l'antitoxine tétanique. CR Soc Biol 93: 508–598
Roberts TA, Smart JL (1976) The occurrence and growth of Clostridium spp. in vacuum-packed bacon with particular reference to Cl. perfringens (welchii) and Cl. botulinum. J Food Technol 11: 229–244
Schiavo G, Benfenati F, Poulain B, Rossetto O, Polverino de Laureto P, DasGupta BR, Montecucco C (1992) Tetanus and botulinum-B neurotoxins block neurotransmitter release by proteolytic cleavage of synaptobrevin. Nature 359: 832–835
Smart JL (1983) Type C botulism in intensively farmed turkeys. Vet Rec 113: 198–200
Smart JL, Roberts TA (1977) An outbreak of type C botulism in broiler chickens. Vet Rec 100: 378–380
Smith GR, Oliphant JC (1983) Diagnosis of botulism in water birds. Vet Rec 112: 457–458
Smith LD (1978) The occurrence of Clostridium botulinum and Clostridium tetani in the soil of the United States. Health Lab Sci 15: 74–80
Smith LD, Williams BL (1984) The pathogenic anaerobic bacteria. Thomas, Springfield
Smith LDS (1975) The pathogenic anaerobic bacteria. Thomas, Springfield
Smith LDS (1992) The genus Clostridium—medical. In: Balows A, Trüper HG, Dworkin M, Harder W, Scheifer KH (eds) The prokaryotes. Springer, Berlin Heidelberg New York, pp 1867–1880
Smith LDS, Sugiyama H (1977) Botulism. The organism, its toxins, the disease. Thomas, Springfield
Smith LDS, Sugiyama H (1988) Botulism. the organism, its toxins, the disease (2nd edn.) Thomas, Springfield
Sonnabend O, Sonnabend W, Heinzle R, Sigrist T, Dirnhofer R, Krech U (1981) Isolation of Clostridium botulinum types G and identification of botulinal toxin in humans: report of five sudden unexpected deaths. J Infect Dis 143: 22–27
Sonnabend WF, Sonnabend OH, Gründler P, Ketz E (1987a) Intestinal toxicoinfection by Clostridium botulinum type F in an adult. Lancet 1: 357–361
Sonnabend WF, Sonnabend UP, Krech T (1987b) Isolation of Clostridium botulinum type G from Swiss soil specimens by using sequential steps in an identification scheme. Appl Environ Microbiol 53: 1880–1884

Suen JC, Hatheway CL, Steigerwalt AG, Brenner DJ (1988a) Clostridium argentinense sp. nov.: a genetically homogeneous group composed of all strains of Clostridium botulinum toxin type G and some nontoxigenic strains previously identified as Clostridium subterminale or Clostridium hastiforme. Int J Syst Bacteriol 38: 375–381

Suen JC, Hatheway CL, Steigerwalt AG, Brenner DJ (1988b) Genetic confirmation of identities of neurotoxigenic Clostridium baratii and Clostridium butyricum implicated as agents of infants botulism. J Clin Microbiol 26: 2191–2192

Sugiyama H (1979) Animal models for the study of infant botulism. Rev Infect Dis 1: 683–688

Tacket CO, Rogawski MA (1989) Botulism. in: Simpson LL (ed) Botulinum neurotoxin and tetanus toxin. Academic, San Diego, pp 351–378

Takeshi K, Fujii N, Kimura K, Yamakawa K, Nakamura S, Kumagai M, Oguma K (1991) Characterization of neurotoxigenic Clostridium butyricum strain by DNA hybridization test and by in vivo and in vitro germination tests of spores. Microbiol Immunol 35: 1095–1103

Thompson DE, Brehm JK, Oultram JD, Swinfield TJ, Shone CC, Atkinson T, Melling J, Minton NP (1990) The complete amino acid sequence of the Clostridium botulinum type A neurotoxin, deduced by nucleotide sequence analysis of the encoding gene. Eur J Biochem 189: 73–81

Tizzoni G, Cattoni G (1890) untersuchungen über das Tetanusgift. Arch Exp Pathol Pharmakol 27: 432–450

Vandekerckhove J, Shering B, Bärmann M, Aktories K (1987) Clostridium perfringens iota toxin ADP-ribosylates skeletal muscle actin in Arg-177. FEBS Lett 225: 48–52

Van Ermengen E (1897) über einem neuen anaeroben Bacillus und seine Beziehungen Zum Botulismus. Z Hyg Infekt Krankh 26: 1–56

Weber JT, Goodpasture HC, Alexander H, Werner SB, Hatheway CL, Tauxe RV (1993) Wound botulism in a patient with tooth abcess: case report and review. Clin Infect Dis 16: 635–639

Weickert MJ, Chambliss GH, Sugiyama H (1986) Production of toxin by Clostridium botulinum type A strains cured of plasmids. Appl Environ Microbiol 51: 52–56

Wells CL, Balish E (1983) Clostridium tetani growth and toxin production in the intestines of germfree rats. Infect Immun 41: 826–828

Whelan SM, Elmore MJ, Bodsworth NJ, Brhem JK, Atkinson T, Minton NP (1992) Molecular cloning of the Clostridium botulinum structural gene encoding the type B neurotoxin and determination of its entire nucleotide sequence. Appl Environ Microbiol 58: 2345–2354

Whitman C, Belgharbi L, Gasse F, Torel C, Mattei V, Zoffmann H (1992) Progress towards the global elimination of neonatal tetanus. World Health Stat Q 45: 248–256

Wilde E, Hippe H, Tosunoglu N, Schallehn G, Herwig K, Gottschalk G (1989) Clostridium tetanomorphum sp. nov., nom rev Int J Sys Bacteriol 39: 127–134

Willems A, East AK, Lawson PA, Collins MD (1993) Sequence of the gene coding for the neurotoxin of Clostridium botulinum type A associated with infant botulism: comparison with other clostridial neurotoxins. Res Microbiol 144: 547–556

Zhou Y, Sugiyama H, Johnson EA (1993) Transfer of neurotoxigenicity from Clostridium butyricum to a nontoxigenic Clostridium botulinum type E-like strain. Appl Environ Microbiol 59: 3825–3831

The Present Status of Tetanus and Tetanus Vaccination

A. GALAZKA and F. GASSE

1 Introduction

Tetanus has been known to mankind for thousands of years. The disease is ubiquitous, but it particularly kills many persons in tropical countries, where the disease is called by various local names (BYTCHENKO 1967). The belief that neonatal tetanus (NT) is linked with divine intervention is reflected in the Biblical story of the death of the son of King David and Bathsheba (2 Samuel, 11 and 12).

World Health Organization, Expanded Programme on Immunization, 1211 Geneva, Switzerland

With the development of tetanus toxoid and the organization of modern immunization services tetanus has become a completely preventable disease. Despite the availability of cheap, safe, and effective vaccine tetanus is a continuing medical concern in many countries. In industrialized countries where immunization programs of infants, children, and adolescents have been implemented for several decades, and where almost all infants are born in hospitals, tetanus is now a rare disease occurring mainly in elderly patients. In developing countries, however, tetanus still poses a serious health burden, and the disease affects all age groups, with special preference for newborns and young persons.

Tetanus has been recognized as a priority problem and has been one of the six target diseases of the WHO Expanded Program on Immunization (EPI) since it began in 1974. The main focus of the EPI is directed towards eliminating NT from developing countries.

2 Epidemiology of Tetanus

2.1 Neonatal Tetanus

2.1.1 Neonatal Tetanus in the Past

In the eighteenth and nineteenth centuries NT was prevalent in the Scottish Islands, Iceland, the Westmann Islands, Faroe Island, and Greenland. On St. Kilda, a Scottish island 40 miles west of the Outer Hebrides, the population was progressively decimated by NT (Collacott 1981; Woody and Ross 1989). In 1764 Rev. Kenneth MacAuley wrote (quoted in Collacott 1981):

> The St. Kilda infants are peculiarly subject to an extraordinary kind of sickness: on the fourth, fifth or sixth night after the birth, many of them give up sucking; on the seventh, their gums are so clenched together, that it is impossible to get anything down their throats. Soon after this symptom appears, they are seized with convulsive fits, and after struggling against excessive torments, till their little strength is exhausted, die generally on the eighth day.

MacAuley's description of neonatal tetanus case is still relevant.

In many industrialized countries, the decline in NT morbidity started well before tetanus toxoid was widely available. Improved obstetrical techniques and neonatal care were responsible for the reduction in NT mortality in the United States from 64 per 100 000 live births in 1900 to about 1 per 100 000 live births in 1961–1964 (LaForce et al. 1969). In the United States, NT accounted for 10% of all tetanus cases in 1965–1966 (LaForce et al. 1969) and 1% in 1982–1984 (CDC 1985).

In Denmark improvements in hygiene resulted in a 45% reduction in NT morbidity between the 1920s and 1940s (Fig. 1). The steady fall in NT morbidity in Denmark after 1950 coincided with an increasing degree of active immunization of the population. The complete elimination of NT in Denmark at the end of the

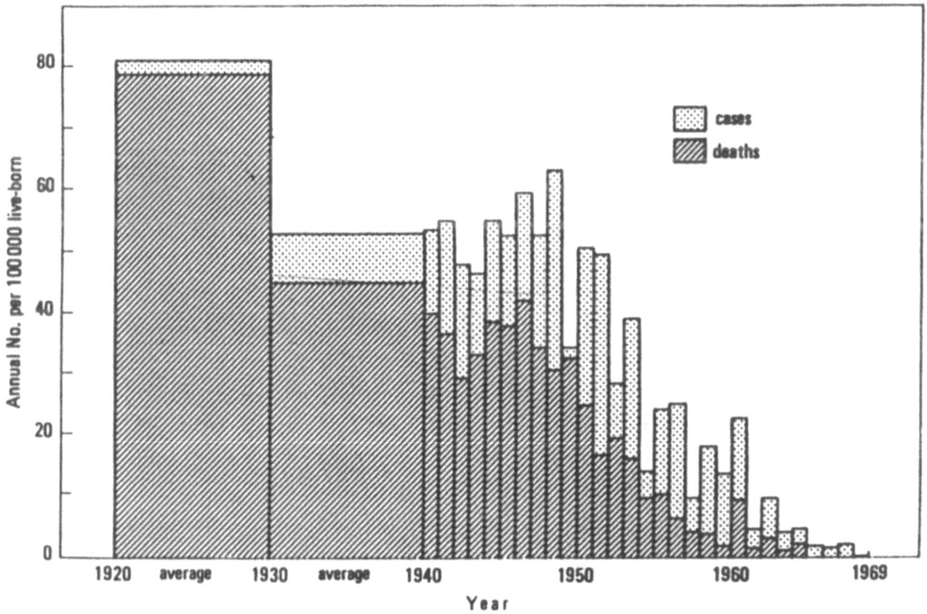

Fig. 1. Number of reported cases and deaths from NT per 100 000 live births in Denmark, 1920–1969. (From CHRISTENSEN (1972)

1960s was probably due to the high tetanus antitoxin levels in pregnant women who had been immunized during their childhood (CHRISTENSEN 1972).

2.1.2 Weakness of Routine Surveillance Systems

NT remains a seriously underreported disease, even in areas with well-developed disease surveillance systems. At the beginning of the 1980s WHO estimated that routine reporting systems identified no more than 5% of tetanus cases occurring worldwide (EPI 1982). Most of unreported cases occur in poor, rural areas where babies are delivered at home by untrained or poorly trained attendants. Other reasons for underreporting are the relative neglect of neonates by health services and traditional attitudes to neonatal deaths in some cultures (STANFIELD and GALAZKA 1984). NT strikes in the first days of life when the child is kept within the home. Stigmas common to many traditional belief systems or due to associations with unclean delivery practices serve as disincentives to seeking care (WHITMAN et al. 1992). The case fatality ratio is high, and the few infants who survive do not have visible disabilities to identify that the disease has occurred.

Surveillance data do not accurately reflect global trends in NT incidence. Although 90% of countries now report NT cases separately from other forms of tetanus, routine reporting systems still identify only a small proportion of the NT cases estimated to occur globally (WHITMAN et al. 1992). Tetanus is not a reportable disease in China, the most populous country in the world. Based on the results of a community study conducted in 1990 it was estimated that about 90 000 deaths due to NT occur each year in China (10 000 in urban areas and 80 000 in rural areas; EPI 1993a).

2.1.3 Magnitude of the Problem

Knowledge of the magnitude of NT has been greatly increased from the results of the community surveys based on a protocol developed in India (Basu and Sokhey 1984; Sokhey 1991). During the 1980s more than 40 developing countries conducted NT mortality surveys. They found NT mortality rates ranging from 1 to 2 per 1000 live births in Jordan and Sri Lanka to 67 deaths per 1000 live births in some areas in India (Galazka et al. 1989). These early studies conducted before immunization and clean delivery programs were widely established revealed that approximately 1 million newborns contract NT each year, of which 800 000 died. By uncovering the magnitude of NT mortality, the community surveys were critically important in changing health policy makers' perceptions and in generating political support for NT control and elimination (Whitman et al. 1992). Studies in developing countries show that NT is the cause of 8%–69% of neonatal mortality. In India a prospective community study conducted in 1983–1984 showed that NT was the second cause (after septicemia) of neonatal mortality, and the NT mortality rate was 9.7 per 1000 live births (Singhai et al. 1990).

Globally, the annual number of NT cases reported to WHO increased considerably between 1974 and 1989, probably reflecting the increasing sensitivity of the reporting systems (Fig. 2). In 1989 there were 29 305 cases of NT reported globally, and NT accounted for 35% of all tetanus cases. Some downward trend has been observed for 1990 and 1991. To estimate NT mortality in developing countries WHO takes into consideration expected mortality rates in the prevaccine era and the most recent percentage of pregnant women immunized with tetanus toxoid. WHO estimates that in 1992 there were 578 000 infant deaths due to NT in developing countries (Table 1). Of these, 210 000 (36%) occurred in Southeast Asia, 152 000 (26%) in Africa, 114 000 (20%) in the Western Pacific region, 90 000 (16%) in the Eastern Mediterranean, 11 000 (2%) in the Americas, and 1000 (0.2%) in Europe.

2.1.4 Risk Factors

Community-based NT mortality surveys have confirmed the importance of risk factors for NT, such as missing immunization with tetanus toxoid during pregnancy, home delivery, unhygienic cutting of the umbilical cord, application of potentially infectious substances to the umbilical stump (Bjerregaard et al. 1993), and a history of NT in a previous child (Hlady et al. 1992). A case-control study in Pakistan showed that application of clarified butter (ghee) during the first few days

Fig. 2. Number of reported NT and NNT cases in developing and industrialized countries, 1974–1992. (From EPI 1993b)

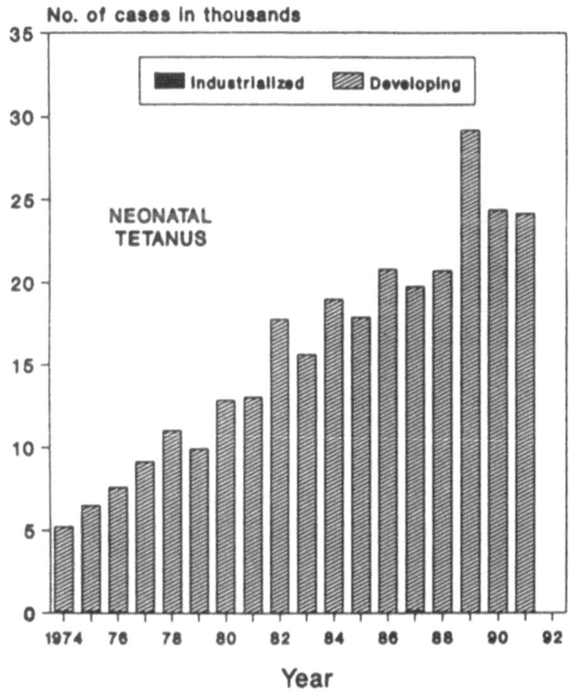
No. of cases in thousands
Industrialized
Developing
NEONATAL
TETANUS
0
5
10
15
20
25
30
35
1974
76
78
80
82
84
86
88
90
92
Year

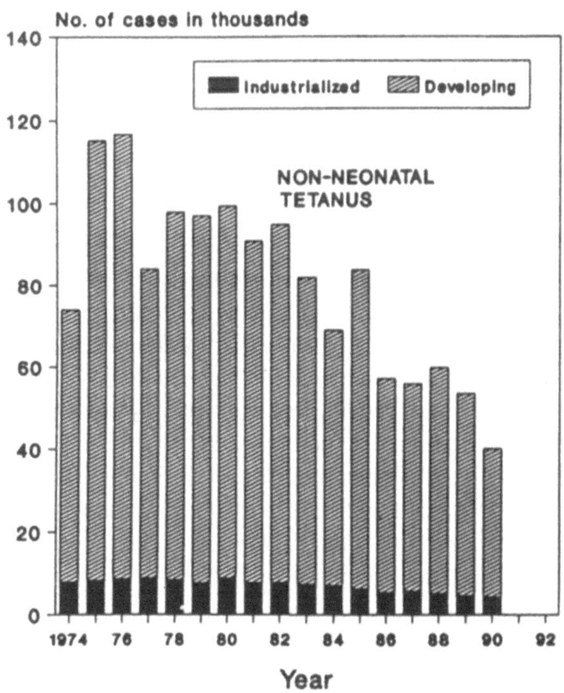
No. of cases in thousands
Industrialized
Developing
NON-NEONATAL
TETANUS
0
20
40
60
80
100
120
140
1974
76
78
80
82
84
86
88
90
92
Year

Table 1. Estimated number of actual and prevented neonatal tetanus deaths in developing countries, 1992 (from EPI 1993b)

	Actual	Prevented
Africa	152 000	102 000
America	11 000	46 000
Eastern Mediterranean	90 000	53 000
Europe	1 000	300
Southeast Asia	210 000	467 000
Western Pacific		
without China	15 000	16 000
with China	114 000	18 000
Total		
without China	479 000	684 000
with China	578 000	686 000

of life was a significant risk factor (TRAVERSO et al. 1991). In one survey a high risk for NT was associated with unwashed hands, suggesting that a possible source of spores may be the hands of the birth attendant (LEROY and GARENNE 1991). The type of tool used to cut the umbilical cord seems not to be an important risk factor (TRAVERSO et al. 1989; LEROY and GARENNE 1991). The main mode of contamination may be unhygienic dressing the umbilical stump (LEROY and GARENNE 1991). Results of all these studies suggest that teaching mothers and birth attendants simple hygienic techniques have a significant impact on NT mortality.

2.2 Nonneonatal Tetanus

2.2.1 Magnitude of the Problem

Tetanus in age groups other than neonates has different morbidity and mortality and different predisposing causes. There are no reliable data showing the real magnitude of nonneonatal tetanus (NNT) in the world. Existing surveillance data grossly underestimate the problem. Studies in industrialized countries show underreporting of NNT morbidity and mortality. Reported cases of tetanus do not represent more than 13% of the actual number of cases in Switzerland (ZUBER et al. 1993). The completeness of reporting tetanus deaths in the United States range between 40% and 76% (SUTTER et al. 1990). The number of NNT cases reported to WHO from developing countries ranged from 90 335 in 1980 to 35 749 in 1990 (Fig. 2). The number of NNT cases reported from industrialized countries during the same period was much lower and ranged from 8 968 cases in 1980 to 4 812 cases in 1990. The clear fall in the reported NNT incidence in developing countries since the mid-1980s coincided with the introduction of immunization programs. It has been estimated that in the 1980s, 300 000–700 000 cases and 120 000–300 000 deaths due to NNT occurred each year in the world, excluding China (REY and TIKHOMIROV 1989). We consider that the upper estimate of 700 000 cases is closer to reality. We believe that in the beginning of the 1990s some 360 000 cases and 140 000 deaths could be attributed to NNT.

2.2.2 Maternal Tetanus

Maternal tetanus is defined as tetanus occurring during pregnancy or within 6 weeks after termination of pregnancy. This includes: postpartum or puerperal tetanus (usually resulting from septic procedures during delivery), postabortion tetanus (following septic maneuvers during induced abortion), and tetanus during pregnancy (FAUVEAU et al. 1993). The incidence of maternal tetanus is unknown. Most information comes from hospital studies, but because only a small proportion of women in developing countries deliver their infants in hospitals, these data are not representative of the population (LISKIN 1992). In recent reviews on maternal morbidity and mortality, tetanus is rarely mentioned as a cause of maternal deaths or disease. In the WHO Factbook only India, Malawi, Nigeria, and Sudan reported between 2% and 10% maternal deaths with tetanus as the cause (WHO 1991). Few cases of tetanus during pregnancy have been reported (DARE and MAKINDE 1989; CZARNECKA et al. 1966; HOLMDAHL and THOREN 1962; JANUSZKIEWICZ et al. 1973; SINGLETON and WITT 1956).

Nevertheless, puerperal of postabortion tetanus may be a significant cause of maternal morbidity and mortality in developing countries due to the unhygienic conditions often associated with home deliveries. In India maternal tetanus constituted 4%–8.5% of all NNT (PATEL and MEHTA 1963; YADAV et al. 1990, 1991). In Upper Volta maternal tetanus was responsible for 23% of 155 NNT cases hospitalized in 1968–1972 in a rural hospital; there were 32 postpartum cases and 3 cases of postabortion tetanus (GENDRON 1972). An analysis in the same hospital for 1973–1975 showed similar results, with 20% of all tetanus attributable to maternal tetanus (COSNARD et al. 1977). In 72 (18.6%) of the tetanus cases hospitalized for NNT in Ibadan, Nigeria, the uterus was the portal of entry. Of these 52 were due to puerperal tetanus, 12 occurred following an abortion, and in 5 women the infection was contracted through douching the vagina with native soap or using dirty rags as tampons. In the three remaining patients tetanus was associated with carcinoma of the cervix, ulcerated prolapsed uterus, or cervical erosion (ADEUJA and OSUNOKUN 1971). In Lagos, Nigeria, 10% of adult tetanus cases were attributed to maternal tetanus, largely following induced abortion (AFONJA et al. 1973). In a recent review it was estimated that between 15 000 and 30 000 cases of maternal tetanus occur each year (FAUVEAU et al. 1993). We consider that the latter estimate is more realistic.

3 The Impact of Mass Immunization Against Tetanus

The introduction of tetanus toxoid into childhood immunization programs has considerably changed the pattern of immunity against this disease. The increase of immunity levels in children has resulted in a reduction in tetanus incidence, especially in younger segments of the population.

3.1 Tetanus Immunity by Age and Sex

Only immunized persons are protected against tetanus; those not immunized are at risk of tetanus from contaminated wounds, including the umbilical stump.

As shown in Fig. 3, tetanus immunity is clearly age related. The level of tetanus antibodies is highest in persons under 30 years of age. Immunity subsequently decreases with increasing age and falls to the lowest levels in those aged over 50 years. Some 50% of adults older than 50 years are nonimmune since they never have been vaccinated or have not received the appropriate booster doses (BEYTOUT et al. 1988). Repeated serological surveys show a continuous increase in the proportion of immune adolescents and young adults

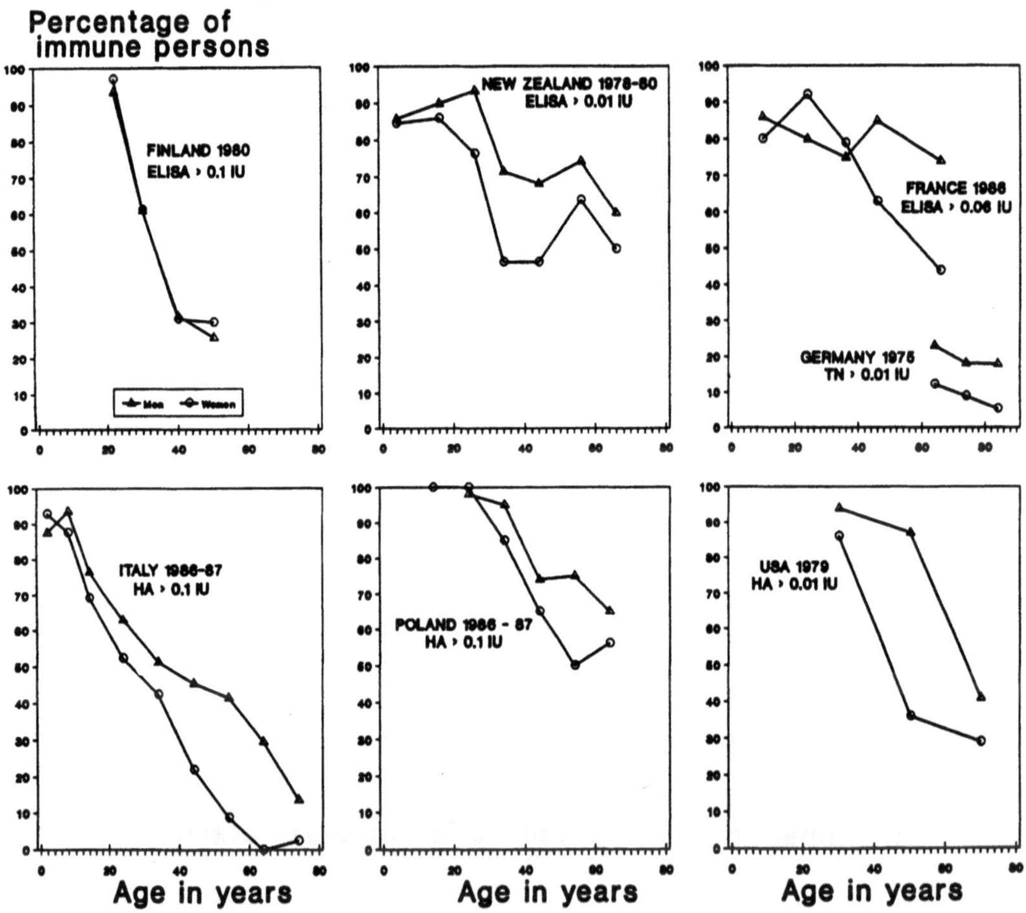

Fig. 3. Immunity against tetanus in men and women in different age groups in several countries. *HA*, Passive hemagglutination test; *TN*, neutralization test in vivo; ELISA, enzyme-linked immunosorbent assay. [From, VILJANEN and NIEMINEN 1980 (Finland); LAU 1987 (New Zealand); GERMAN-FATTAL et al. 1987 (France); FINGER et al. 1975 (Germany); BRUSCOLINI et al. 1991 (Italy); GALAZKA and KARDYMOWICZ 1989 (Poland); CROSLEY et al. 1979 (USA)]

(Fig. 4). In Burma the percentage of tetanus-immune blood donors increased from 17% in 1973–1974 to 56% in 1981–1982 among young adults, but the corresponding proportion did not change in those aged over 40 years (Table 2). There are geographical differences in tetanus immunity levels. In Italy the prevalence of nonimmune persons was higher in the south than in the north (Table 3). This probably reflects a difference in health care delivery (D'AMELIO et al. 1993; ROSMINI et al. 1987).

There is a clear difference between levels of tetanus immunity in men and women. In industrialized countries men are better protected, probably due to additional immunizations given during military service or professional activities (Fig. 5). More than 80% of military recruits had acquired protective levels of tetanus antitoxin through past immunization (CUMBERLAND et al. 1993; D'AMELIO et al. 1993; FAITHFULL-DAVIES 1989). Women in industrialized countries, especially older women, often belong to the most susceptible groups. In India, a country where tetanus toxoid is given to women of child-bearing age, and especially to pregnant women to prevent NT, the immunity status of women is better than that of men (Fig. 5).

Limited serological data from developing countries show the impact of immunization programs on the pattern of tetanus immunity. Studies show an increase in immunity among Indian women (MISRA and RAO 1988) and Burmese

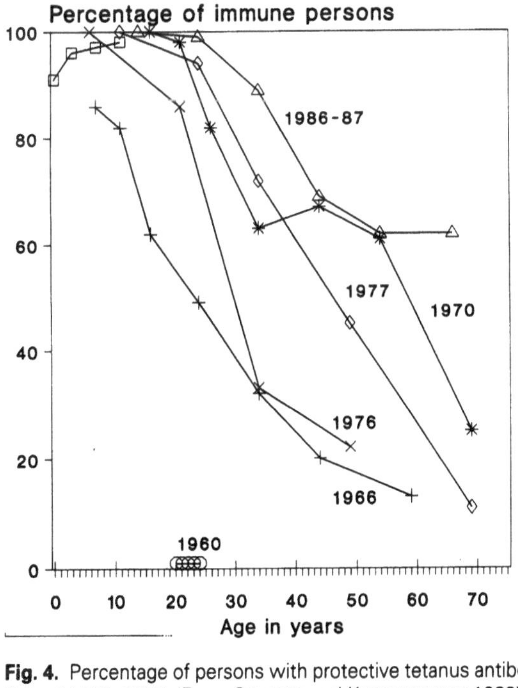

Fig. 4. Percentage of persons with protective tetanus antibody levels, in repeated serological studies, Poland 1960–1987. (From GALAZKA and KARDYMOWICZ 1989)

Table 2. Percentage of blood donors immune to tetanus by age groups, Burma, 1973–1974 and 1981–1981 (from MYA et al. 1985)

Age (years)	1973–1974	1981–1982
18–29	16.8	55.9
30–39	17.0	30.2
40–49	23.2	22.2
>50	0	0
Total	17.2	47.7

Table 3. Percentages of nonimmune military recruits born before 1968 and after 1968, Italy, 1987–1993

	Born before 1968, drafted in 1981[a]	Born after 1968, drafted in 1991[b]
North and Center	17.6	5.5
South and Islands	44.7	15.9

[a] From ROSMIMI et al. (1987)
[b] From D'AMELIO et al. (1993)

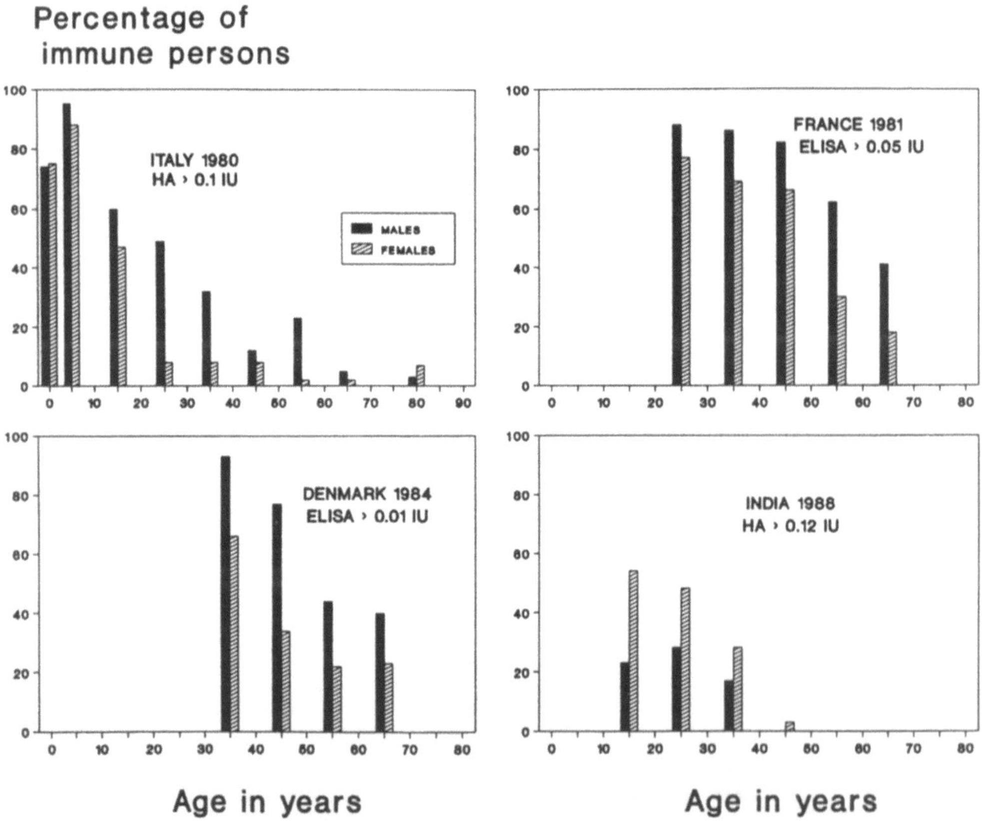

Fig. 5. Percentage of persons with protective tetanus antibody levels, by age and sex, in four countries. [From GASPARINI et al. 1980 (Italy); BOURLEAUD and HUET 1985 (France); KJELDSEN et al. 1988 (Denmark); MISRA and RAO 1988 (India)]

male military staff (Mya et al. 1985). In Peru over 60% of pregnant women had protective antibody levels; the percentage of immune women was similar among women presenting in labor in two capital city hospitals (serving poor and lower middle class populations) and a regional hospital (in the Amazon jungle). That the rate in the jungle hospital was comparable to that in the two urban hospitals probably reflects recent vaccination efforts in the jungle region of Peru, an area traditionally endemic for tetanus. However, the survey was limited to hospitalized women and the immunization coverage among women giving birth outside of hospital may be lower owing to decreased access to health care (Vernacchio et al. 1993). On the Turks and Caicos Islands, where an immunization program for children had been offered for 6 years, some 75% of children 6–15 years old had protective titers of antibodies, while only 18% of those above 20 years of age were protected (Morgan et al. 1981).

3.2 Reduction in Tetanus Incidence

In industrialized countries the introduction of immunization with tetanus toxoid in the 1940s and 1950s led to a significant reduction in tetanus incidence. Today, tetanus is a rare disease in most industrialized countries. In Switzerland an estimated 134 tetanus cases occurred between 1980 and 1989 (Zuber et al. 1993). During 1989–1990 there were 117 cases of tetanus reported in the United States, for an average annual incidence of 0.02 per 100 000 population (Prevost et al. 1992), compared with 0.39 per 100 000 in 1947 when national reporting began. As shown in Fig. 2, the number of cases of NNT reported by developing countries was considerably lower in the second half of 1980s, and this trend is continuing. There are spectacular examples of the impact of immunization on the tetanus incidence in developing countries. Data from a relatively good surveillance system in Sri Lanka show a significant decline in NT and NNT, which is associated with increasing immunization coverage of pregnant women and children with vaccines containing tetanus toxoid (Fig. 6). Data from Egypt show a long-term reduction in the incidence of NNT. In 1980–1986 a steady increase in the reported number of NT cases may reflect improvements in surveillance, while the downward trend in NT incidence in the past 6 years may be attributable to the immunization program (Fig. 7). Immunization coverage of pregnant women in Egypt with two or more doses of tetanus toxoid increased from about 10% in 1987 to about 70% in 1991–1992; the immunization coverage with three doses of diphtheria-pertussis-tetanus (DPT) vaccine in infants exceeded 80% during the past 6 years. The reduction in tetanus morbidity observed in national surveillance data in Egypt is confirmed by hospital study data (El-Sherbini 1991).

High coverage with tetanus toxoid can be remarkably successful in preventing NT. A good example of this is in the city of Maputo, Mozambique. When 95% coverage with two doses of tetanus toxoid was achieved, the annual

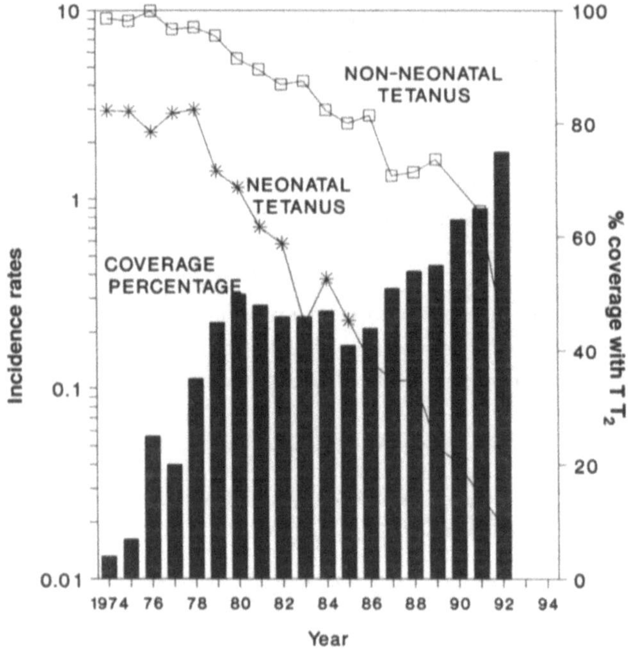

Fig. 6. Neonatal tetanus (per 1000 live births) and non-neonatal tetanus (per 100 000 population) incidence rates and immunization coverage (at least two doses of tetanus toxoid) in pregnant women, Sri Lanka 1974–1992. (From EPI 1993b)

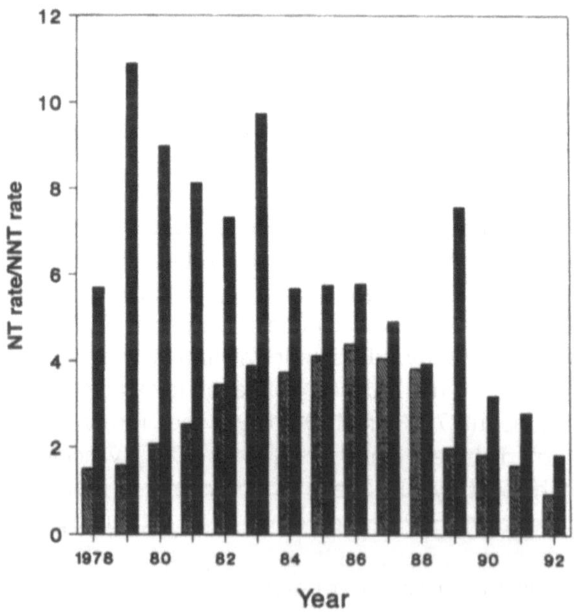

Fig. 7. Incidence rate due to neonatal tetanus (hatched bars; per 1000 live births) and non-neonatal tetanus (*black bars*; per 100 000 population), Egypt 1978–1992. (From EPI 1993b)

number of reported hospital cases of NT dropped from 170–250 cases before 1979 to 0 cases in 1986 (EPI 1988). NT mortality declined dramatically in a district in Aceh Province, Indonesia, between 1984 and 1987. During 1985 a tetanus toxoid immunization crash program was implemented in Aceh Province which resulted in 84% of women 10–45 years of age receiving two doses to tetanus toxoid (YUSUF et al. 1991). Baseline and follow-up data confirmed an 85% reduction in NT mortality during this period, from 32 to 5 per 1000 live births. The neonatal mortality attributable to causes other than tetanus remained unchanged. In Venezuela NT mortality decreased by 93% from 1970–1974 to 1985–1991 (EPI 1993c). Such downward trends in NT morbidity are also reported from Thailand (CHONGSUVIVATWONG et al. 1993), Zambia (MALEK and MUKELABAI 1990), and some states in India (SOKHEY 1991). In other countries, such as Nigeria and Uganda, the impact of immunization programs is yet not visible in the hospital-based data (ANTIA-OBONG et al. 1993; BWIRE and KAWUNA 1992).

3.3 Changes in the Ages of Tetanus Patients

The most dramatic reductions in tetanus morbidity have occurred in age groups covered by routine immunizations. Tetanus has become a rare disease in children and young adults, age groups most frequently affected before introduction of immunization programs. In industrialized countries tetanus has become a disease of the elderly. In France, Poland, Switzerland, and the United States persons below 20 years of age constitute only 2%–6% of all tetanus cases, and 64%–93% of tetanus cases occur in persons over 50–60 years of age (Table 4). Such a trend has been observed in England and Wales and in Australia, conducting immunization programs in childhood; (GALBRAITH et al. 1981; NEWTON-JOHN 1984). In France the median age of tetanus in 1991–1992 was 74 years, and the youngest cases reported were a child aged 4.5 years and an adult aged 46 years (LOMBARD and LEPOUTRE 1993). In developing countries tetanus is a serious health problem in neonates and school-aged children (Fig. 8, Table 4). However, available data on the age distribution in developing countries relate to earlier periods, before the implementation of immunization of children and women of child-bearing age. In some of these countries, where a reduction in the total number of tetanus cases has already been seen, there may also be a change in age distribution of tetanus cases.

3.4 Reduction in Tetanus Mortality

Existing surveillance systems do not provide adequate data on tetanus mortality. WHO estimates that in 1992 immunization or clean delivery practices prevented 686 000 NT deaths in developing countries (Table 1).

Table 4. Age distribution of tetanus cases (percentage of cases below 20 and above 50 years of age) in developing and industrialized countries

	Type of study*	< 20 years	> 50 years	Reference
France				
1969–1979	HS	1[a]	69	LAMISSE et al. (1981)
1990	NR	–	93[b]	PELLETIER and ROURE (1991)
1991–1992	NR	–	93[b]	LOMBARD and LEPOUTRE (1993)
United States				
1982–1984	NR	4	71	CDC (1985)
1985–1986	NR	5	71	CDC (1987)
1987–1988	NR	6	68	CDC (1990)
1989–1990	NR	6	64	PREVOST et al. (1992)
Switzerland				
1968–1989	HS	0	64	JOLLIET et al. (1990)
1980–1989	NR	3	81	ZUBER et al. (1993)
Poland				
1960	NR	46	26	GALAZKA and ABGAROWICZ 1973
1970	NR	15	53	GALAZKA and ABGAROWICZ 1973
1980	NR	3	81	ANUSZ (1982)
1990	NR	2	86	ANUSZ (1992)
England, Wales				
1970	NR	27[c]	50[d]	GALBRAITH et al. (1981)
1979	NR	0[c]	55[d]	GALBRAITH et al. (1981)
Senegal				
1960–1967	HS	68	6	REY and TIKHOMIROV (1989)
1985–1986	HS	65	9	SOW 1987, cited in REY and TIKHO-MIROV (1989)
Upper Volta				
1968–1972	HS	62	4	GENDRON (1972)
1972–1975	HS	69	3	COSNARD et al. (1977)
Nigeria				
1963–1969	HS	28	11	ADEUJA and OSUNTOKUN (1971)
1967–1970	HS	48	8	AFONJA et al. (1973)
Cameroon				
1972	HS	23	18	ANCELLE et al. (1974)
Jamaica				
1980–1986	HS	49	32[d]	FIGUEROA and CLARKE (1988)
Chile				
1975–1976	NR	56[a]	19[d]	DE LA FUENTE et al. (1986)
1982–1983	NR	28[a]	27[d]	DE LA FUENTE et al. (1986)

*NR, Nationwide reporting; HS, hospital study.
[a] < 15 years. [b] > 60 years. [c] < 14 years. [d] > 45 years.

4 Immunization Against Tetanus

4.1 Strategies for Tetanus Control

The elimination of tetanus is an attainable goal. Strategies to control and eliminate tetanus should take into consideration differences between NT and NNT.

NT may be eliminated by combining two approaches: (a) increasing immunization coverage among women of child-bearing age, especially pregnant women, with tetanus toxoid, and (b) improving maternity care, with particular emphasis on

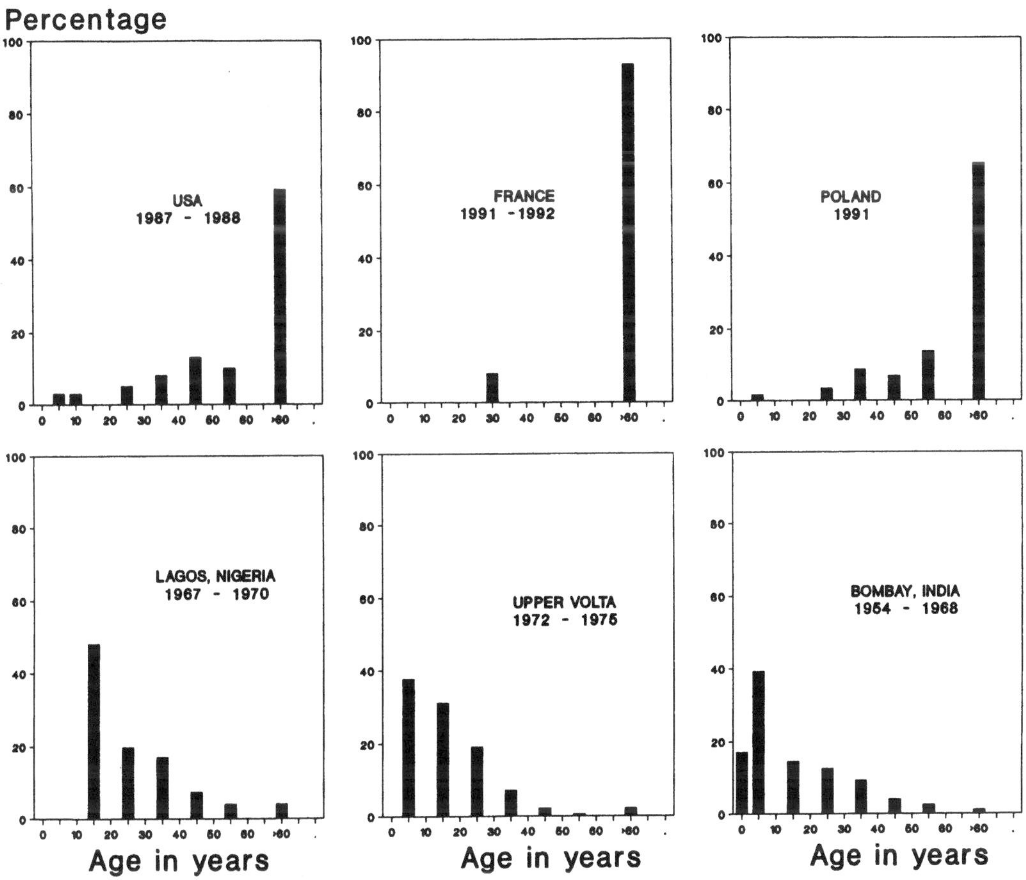

Fig. 8. Age distribution of tetanus patients in several countries, 1954–1992. [From CDC 1990 (USA); LOMBARD and LEPOUTRE 1993 (France); KUSZEWSKI 1993 (Poland); AFONJA et al. 1973 (Nigeria); COSNARD et al. 1977 (Upper Volta); PATEL and MEHTA 1975 (India)]

increasing the proportion of deliveries that are attended by trained persons. In 1989 WHO declared its commitment to the global elimination of NT by the year 1995 (EPI 1990). Elimination of NT is defined as reduction in the case rate to below one per 1000 live births. To achieve and maintain NT elimination the percentage of children protected at birth through tetanus toxoid immunization of mothers or through clean delivery should be above 80%.

For control NNT immunization of the whole population is essential. Other measures, such as prevention of trauma and preventive care of the wounded, are useful but not sufficient to control tetanus. Traumatic wounds remain the main means of acquiring NNT, but in 8%–31% of patients no portal of entry is detected (REY and TIKHOMIROV 1989). In developing countries otorrhea, traditional practices, and intramuscular injections are often identified as risk factors (ANCELLE et al. 1974; AYIVI et al. 1992; PATEL and MEHTA 1963, 1975; REY and TIKHOMIROV 1989; SOW 1987, cited in REY and TIKHOMIROV 1989). There are three main elements in the strategy to control NNT: (a) active immunization of the whole population, with the main emphasis on primary immunization of infants and children and on properly spaced booster doses for adults to maintain immunity through the whole life (b) would care and prophylactic measures in wounded persons, and (c) health education to make the public aware of the risk of tetanus and the importance of its prevention (REY and TIKHOMIROV 1989).

4.2 Immunization Schedules

In most countries the immunization schedule for infants recommends three primary doses of DPT vaccine. The EPI recommends starting the primary series of DPT vaccine early (Table 5) to achieve early protection against pertussis. The primary series should also be completed early as more infants will be immunized when they attend baby clinics before mothers return to work. Three primary doses of DPT vaccine induce an antibody level above the minimum protective threshold (0.01 IU/ml as determined by the in vivo neutralization test), with a mean level between 0.1 and 1 IU/ml (ANDERSON et al. 1988; BARKIN et al. 1984; EDWARDS et al. 1989; KIMURA et al. 1991; PICHICHERO et al. 1986). The primary DPT series will provide tetanus immunity for 1–3 years. Reinforcing infant immunization with a fourth dose of DPT vaccine at 15–24 months of age prolongs tetanus immunity for another 5 years. A fifth dose of tetanus toxoid—given as diphtheria–tetanus vaccine for children (DT) at school entry provides immunity for another 10 years, i.e., until 17 or 18 years of age. An additional dose given as Td vaccine in older children at school leaving or in adults during military service assures sufficient immunity for at least two more decades (BIZZINI et al. 1978; CHRISTENSEN and BOTTIGER 1987; GALAZKA 1993; SIMONSEN et al. 1986).

In most developing countries routine immunization in infancy was implemented in the middle 1980s. The cohorts of women entering child-bearing age in the 1990s were not immunized with tetanus toxoid in their infancy or adolescence, and they represent a large group of susceptible persons. Since 1987 WHO has recommended tetanus toxoid immunization for all women of child-bearing age (EPI 1987). For previously unimmunized women of child-bearing age in developing countries WHO recommends five doses of tetanus toxoid. The

Table 5. The immunization schedule for infants recommended by the WHO Expanded Programme on Immunization

Age	Vaccines	Hepatitis B Vaccine	
		Scheme A[a]	Scheme B[b]
Birth	BCG, OPV0[c]	HB1	
6 weeks	DPT1, OPV1	HB2	HB1
10 weeks	DPT2, OPV2		HB2
14 weeks	DPT3, OPV3	HB3	
9 months	Measles, yellow fever[d]		HB3

BCG, Bacille Calmette-Guérin vaccine; OPV, Oral poliomyelitis vaccine
[a]Where perinatal transmission of hepatitis B virus is common.
[b]Where early transmission of hepatitis B virus is not common.
[c]Where poliomyelitis has not been controlled.
[d]Where yellow fever poses a risk.

schedule should include a first dose given at the first contact (including any time during pregnancy), a second dose at least 4 weeks after the first dose, and a third dose 6–12 months after the second. A third dose of tetanus toxoid induces high and durable immunity, lasting at least 5 years, with mean levels between 1 and 10 IU/ml. A fourth booster dose prolongs immunity for 10 years. Immunity persists for at least 20 years after the fifth dose of TT.

Some developed countries recommend the routine use of Td vaccine every 10 years to maintain tetanus immunity throughout life. However, there are no data on the coverage rate for these additional booster doses. The response may vary individually, and pronounced dispersion in postbooster titers can be expected with time following the previous dose. Some results suggest that revaccination every 20 years may be sufficient to maintain immunity (SIMONSEN et al. 1986, 1987).

Giving a dose of Td vaccine to wounded persons is an important measure to maintain immunity against tetanus in adults. Nevertheless, such "emergency" booster doses are needed no more often than once every 10 years.

4.3 Progress in Immunization Coverage and Need for Sustaining the Achieved Level

When the EPI started in 1974, immunization coverage in children under 1 year of age with childhood vaccines, including DPT vaccine, was estimated to be lower than 5% worldwide. At the beginning of the 1980s immunization coverage rates increased progressively, and by 1990 80% of children received three doses of DPT vaccine by their first birthday (Fig. 9). Global statistics, however, mask disparities among regions, countries, provinces and districts. In 1992, for example, 15 countries in Africa reported immunization coverage below 50% for three doses of DPT vaccine. Much lower coverage (43% in 1992) has been achieved for tetanus toxoid immunization of pregnant women in developing countries to protect their babies from NT. This reflects the reality that in many countries immunization of women is considered a second priority by mothers and by health workers as well.

Major objectives for immunization programs in the 1990s are to maintain a high level immunization coverage with childhood vaccines and to reach 90% coverage of children with DPT vaccine and pregnant women with tetanus toxoid by the year 2000.

5 Research Needs

Research within the EPI is directed at improving disease control strategies (improved vaccines, studies of the acceptability of immunization and studies on the simplification of vaccine administration), strengthening surveillance and program monitoring tools, and providing methods and materials for the cold chain and logical support.

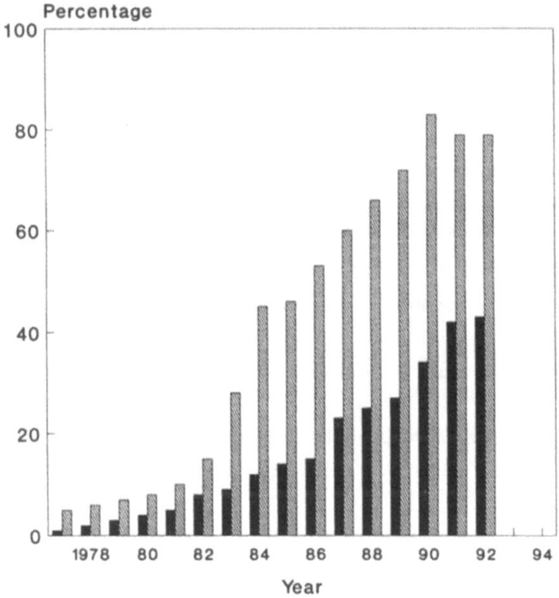

Fig. 9. Global immunization coverage for three doses of DPT vaccine in infants (*hatched bars*) and at least two doses of tetanus toxoid in pregnant women (*black bars*) in developing countries, 1977–1992. (From EPI 1993b; data before 1984 are estimated)

5.1 Slow-Release Tetanus Toxoid

The development of a tetanus toxoid which can ensure solid and long-lasting immunity with one injection administered during the first visit at a health setting could substantially reduce cost, solve the problem of drop-outs between doses, and raise immunization coverage. Work is in progress to develop a slow-release tetanus toxoid encapsulated in biodegradable polylactide/polyglycolide polymers (Aguado and Lambert 1992). Biodegradable polymer microspheres or microcapsules developed over the past decade for reliable, timed release of contraceptive steroids have significant potential for adaptation to antigen release for immunization (Morris et al. 1994). Preliminary experiments in mice have shown that a good primary immune response is produced, and that microencapsulated tetanus toxoid primes the mice more effectively, since the secondary antibody response is higher and persists longer than tetanus toxoid–alum priming (Esparza and Kissel 1992). Controlled-released systems are very promising; however, several issues still need to be resolved. The procedure of preparation may denature the antigen, standardization and regulatory issues may be difficult to establish, and the stability of encapsulated tetanus toxoid at body temperature is unknown.

5.2 Mechanisms of Transplacental Transfer of Tetanus Antibody

The concept of the control of NT is based on the fact that tetanus antitoxin passively transferred from the immunized mother to the fetus provides transient protection of the newborn infant from tetanus. Early studies found that the tetanus antitoxin levels in cord serum and maternal serum were usually equal, although in 20%–30% of cases the cord serum had a lower titer than the maternal serum. Recently it was observed that the cord/maternal ratio of tetanus antibodies is higher in Europe than in Africa (GENDREL et al. 1990). The transfer of tetanus antibodies depends on the quantity and quality of antibodies, and both these parameters may differ depending on immunization schedules used. High immunoglobulin levels in African mothers exposed to multiple antigenic stimuli may play a role in the extent of the placental transfer of antibodies.

Factors influencing the transplacental transfer of tetanus antibody are not known, and mechanisms involved in the transfer of immunity from mother to fetus have not been rigorously investigated. With the development of new or improved vaccines against other common neonatal pathogens, such as group B streptococci and respiratory syncytial virus, there is renewed interest in the potential transfer of protective immunity during pregnancy (ENGLUND et al. 1993).

5.3 A Simple Field Test To Assess Tetanus Immunity Status

There is a need to develop a simple, rapid, and cheap test to assess immunity status against tetanus. Existing tests, such as passive hemagglutination and enzyme-linked immunsorbent assay, require trained persons and equipped laboratories. Ideally, a test which can be used in modest health centers and outpatient departments should use saliva rather than blood sample. The development of such a test could rationalize specific prophylaxis with tetanus-immune globulin in wounded persons, be used to evaluate immunization programs, and help in assessing the immune response in immunocompromised persons.

References

Adeuja AOG, Osuntokun BO (1971) Tetanus in the adult Nigerians. a review of 503 patients. East Afr Med J 48: 683–691

Afonja AO, Jaiyeola BO, Tunwashe OL (1973) Tetanus in Lagos: a review of 228 adult Nigerian patients. J Trop Med Hyg 76: 171–174

Aguado MT, Lambert PH (1992) Controlled-release vaccines—Biodegradable polylactide/polyglycolide (PL/PG) microspheres as antigen vehicles. Immunobiol 184: 113–125

Ancelle JP, Hagbe P, Noutat J, Befidi R (1974) Le tetanos a Yaounde. Med Afr Noire 21: 953–957

Anderson EL, Belshe RB, Bartram J (1988) Differences in reactogenicity and antigenicity of a cellular and standard pertussis vaccines combined with diphtheria and tetanus in infants. J Infect Dis 157: 731–737

Antia-Obong OE, Young MU, Effiong CE (1993) Neonatal tetanus: prevalence before and subsequent to implementation of the Expanded Programme on Immunization. Ann Trop Paediatr 13: 7–12

Anusz Z (1982) Tetanus—1980 (in Polish). Przegl Epidemiol 36: 125–132

Anusz Z (1992) Tetanus—1990 (in Polish). Przegl Epidemiol 46: 111–114

Ayivi B, Zohoun T, Agossou R, Dan V, Hazoume FA (1992) Le tetanos chez l'enfant, aspects epidemiologique, clinique et therapeutique: a propos de 784 cas. Publ Med Afr 25: 54–64

Barkin RM, Samuelson JS, Gotlin LP (1984) DTP reactions and serologic response with reduced dose schedule. J Pediatr 105: 189–194

Basu RN, Sokhey J (1984) Special combined survey for poliomyelitis and neonatal tetanus as supplement to routine surveillance system in India. J Commun Dis 16: 148–153

Beytout D, Nguyen Trung T, Laveran H, Mamouret-Beytout A (1988) Controle de l'etat d'immunite antitetanique dans la population du Puy-de-Dome. Med Mal Infect 12: 897–899

Bizzini B, Champigneux P, Guyotjeannin C, Lebreton P, Tisseuil H, Chevalier J, Blaise F, Thurel C, Turpin A (1978) Etude de l'immunite antitetanique d'une population d'ouvriers des usines Renault. Ann Microbiol 129B: 437–440

Bjerregaard P, Steinglass R, Mutie DM, Kimani G, Mjomba M, Orinda V (1993) Neonatal tetanus mortality in coastal Kenya: a community survey. Int J Epidemiol 22: 163–169

Bourleaud J, Huet M (1985) Contribution a l'etude du tetanos en France. Med Mal Infect 15: 7–16

Bruscolini F, Pianetti A, Baffone W, Albano V, Salvaggio A (1991) Stato dell'immunita' antitetanica in un campione della popolazione di Urbino. Ig San Pub 47: 3–12

Bwire R, Kawuma HJS (1992) Hospital-based study on neonatal tetanus, Buluba Hospital, 1985–1989. J Trop Med Hyg 95: 62–66

Bytchenko B (1967) Tetanus as a world problem. In: Eckman L (ed) Principles on tetanus. Proceedings of the international conference on tetanus, Bern, 15–19 July 1966. Huber, Bern pp 21–41

Chongsuvivatwong V, Bujakorn L, Kanpoy V, Treetrong R (1993) Control of neonatal tetanus in southern Thailand. Int J Epidemiol 22: 931–935

Christensen B, Bottiger B (1987) Epidemiology and immunity to tetanus in Sweden. Scand J Infect Dis 19: 429–435

Christensen PE (1972) Comments on epidemiology and immunology of tetanus in Denmark. In: Proceedings of the Third International Conference on Tetanus, Sao Paolo, 17–22 Aug 1970. Pan American Sanitary Bureau Scientific Publication no 253, Washington, pp 27–30

Centers for Disease Control (CDC) (1985) Tetanus—United States, 1982–1984. Morb Mortal Wkly Rep 34: 602–611

Centers for Disease Control (CDC) (1987) Tetanus—United States, 1985–1986. Morb Mortal Wkly Rep 36: 477–481

Centers for Disease Control (CDC) (1990) Tetanus—United States, 1987–1988. Morb Mortal Wkly Rep 39: 37–41

Collacott RA (1981) Neonatal tetanus in St. Kilda. Scott Med J 26: 224–227

Cosnard G, Joullie M, Davin R (1977) Le tetanos a l'hopital de Ouahigouya (Haute-Volta). Med Trop 37: 45–49

Crossley K, Irvine P, Warren JB, Lee BK, Mead K (1979) Tetanus and diphtheria immunity in urban Minnesota adults. JAMA 242: 2298–2300

Cumberland NS, Kidd AG, Karalliedde L (1993) Immunity to tetanus in United Kingdom populations. J Infect 27: 255–260

Czarnecka E, Kosinska E, Slomko J, Szlapka H, Wolowicka L (1966) A case of tetanus in pregnancy with survival of mother and child (in Polish). Pol Tyg Lek 21: 106–107

D' Amelio R, Stroffolini T, Wirz M, Biselli R, Matricardi PM, Gentili G, Collotti C, Pasquini P (1993) Tetanus immunity among Italians born after the introduction of mandatory vaccination of children. Vaccine 9: 935–936

Dare FO, Makinde OO (1989) Tetanus in pregnancy in a Nigerian women—a case report. Trop Doct 19: 81

de la Fuente M, Gonzates S, Guevara G, Castillo J, Ignatiew J (1986) Epidemiologia dél tetanos en Chile en los ultimos 25 años. Rev Chil Pediatr 57: 273–277

Edwards KM, Bradley RB, Decker MD, Palmer PS, Van Savage J, Taylor JC, Dupont WD, Hager CC, Wright PF (1989) Evaluation of a new highly purified pertussis vaccine in infants and children. J Infect Dis 160: 832–837

El-Sherbini A (1991) Study of tetanus neonatorum in Tanta Fever Hospital, 1988–1989. J Trop Pediatr 37: 262–263

Englund JA, Mbawuike IN, Hammill H, Holleman MC, Baxter BD, Glezen WP (1993) Maternal immunization with influenza or tetanus toxoid for passive antibody protection in young infants. J Infect Dis 168: 647–656

Esparza I, Kissel T (1992) Parameters affecting the immunogenicity of microcapsulated tetanus toxoid. Vaccine 10: 714–720

Expanded Programme on Immunization (EPI) (1982) The use of survey data to supplement disease surveillance. Wkly Epidemiol Rec 57: 361–362

Expanded Programme on Immunization (EPI) (1987) Tetanus control. Wkly Epidemiol Rec 62: 380–383

Expanded Programme on Immunization (EPI) (1988) Neonatal tetanus Update. World Health Organization, Geneva

Expanded Programme on Immunization (EPI) (1990) Global Advisory Group. Part I. Wkly Epidemiol Rec 65: 5–11

Expanded Programme on Immunization (EPI) (1993a) Reassessment of the neonatal tetanus problem, China. Wkly Epidemiol Rec 68: 201–204

Expanded Programme on Immunization (EPI) (1993b) EPI information system. WHO document WHO/EPI/CEIS/93, Geneva

Expanded Programme on Immunization (EPI) (1993c) Neonatal tetanus, 1970–1992, Venezuela Wkly Epidemiol Rec 60: 109–112

Faithfull-Davies DN, Parry ES, Sheffield F, Fieldhouse PM, Boxall MA (1989) Tatanus antitoxin in military personnel. J R Army Med Corps 135: 109–114

Fauveau V, Mamdani M, Steinglass R, Koblinsky M (1993) Maternal tetanus: magnitude, epidemiology and potential control measures. Int J Gynecol Obstet 40: 3–12

Figueroa JP, Clarke R (1988) Under-reporting of tetanus and diphtheria cases in Jamaica (1980–1986). West India Med J 37: 143–147

Finger H, Haberman E, Bracharz K, Hof H (1975) Tetanus-Immunität im Senium. Zentralbl Bakteriol Hyg I Abt Orig B 161: 188–193

Galazka A (1993) Tetanus. In: Expanded Programme of Immunization: Immunological basis for immunization, part 3. World Health Organization, Geneva

Galazka A, Abgarowicz A (1973) Tetanus. In: Kostrzewski J (ed) Infectious diseases in Poland and their control in 1961–1970 (in Polish). Panstwowy Zaklad Wydawnictw Lekarskich, Warsaw, pp 252–265

Galazka A, Kardymowicz B (1989) Tetanus incidence and Immunity in Poland. Eur J Epidemiol 5: 474–480

Galazka A, Gasse F, Henderson RH (1989) Neonatal tetanus in the world and the global Expanded Programme on Immunization In: Nistico G, Rizzini B, Bytchenko B, Triau R (eds) Proceedings of the Eighth International Conference on Tetanus, Leningrad 25–28 Aug 1987. Pythagora, Rome, pp 470–487

Galbraith NS, Forbes P, Tillett H (1981) National surveillance of tetanus in England and Wales 1930–79. J Infect 3: 181–191

Gasparini R, Coppola RC, D'Aste E, Sigari G (1980) Prevalenza dell'antitossina tetanica nella popolazione Genovese in rapporto all'eta ed al sesso. Boll Ist Sieroter Milan 59: 418–423

Gendrel D, Richard-Lenoble D, Massamba MB, Picaud A, Moreno JL, Gendrel C, Baziomo JM, Francoual C, Blot P (1990) Transfert placentaire des anticorps antitétaniques et protection du nouveau-né. Arch Fr Pediatr 47: 725–729

Gendron Y (1972) Le tetanos en milieu rural de Haute Volta. Moyens therapeutiques et resultats. Afr Med 11: 739–746

German-Fattal M, Bizzini B, German A (1987) Immunity to tetanus: tetanus antitoxin and anti-BIIb in human sera. J Biol Standard 15: 223–230

Holmdahl M, Thoren L (1962) Tetanus in pregnancy. Report of a case of severe tetanus with survival of mother and child following tracheostomy and artificial respiration. Am J Obstet Gynecol 84: 339–342

Hlady WG, Bennett JV, Samadi AR, Begum J, Hafez A, Tarafdar AI, Boring JR (1992) Neonatal tetanus in rural Bangladesh: risk factors and toxoid efficacy. Am J Public Health 82: 1365–1369

Januszkiewicz J, Galazka A, Adamczyk J, Sporzynska Z (1973) Severe tetanus in late pregnancy. Scand J Infect Dis 5: 233–235

Jolliet P, Magnenat JL, Kobel T, Chevrolet JC (1990) Aggressive intensive care treatment of very elderly patients with tetanus is justified. Chest 97: 702–705

Kimura M, Kuno-Sakai H, Sato Y, Kamiya H, Nii R, Isomura S, Horiuchi K, Kato T, Deguchi M, Saikusa H, Mortimer EA, Cherry JD (1991) A comparative trial of the reactogenicity and immunogenicity of Takeda acellular pertussis vaccine combined with tetanus and diphtheria toxoids. Outcome in 3- to

8-month-old-infants, 9- to 23-month-old infants and children, and 24- to 30-month-old children. Am J Dis Child 145: 734-740
Kjeldsen K, Simonsen O, Heron I (1988) Immunity against diphtheria and tetanus in the age group 30–70 years. Scand J Infect Dis 20: 177–185
Kuszewski K (1993) Tetanus in 1991 (in Polish). Przegl Epidemiol 47: 141–143
LaForce FM, Young LS, Bennett JV (1969) Tetanus in the United States (1965–1966): epidemiologic and clinical features. New Engl J Med 280: 569–574
Lamisse F (1981) Le tetanos. A propos de cinq cent trois observations. Semin Hopital Paris 57: 1076–1080
Lau RCH (1987) Detection of tetanus toxoid antibodies in human sera in New Zealand by ELISA. Epidemiol Infect 98: 199–202
Leroy O, Garenne M (1991) Risk factors of neonatal tetanus in Senegal. Int J Epidemiol 20: 521–525
Liskin LS (1992) Maternal morbidity in developing countries: a review and comments. Int J Gynecol Obstet 37: 77–87
Lombard L, Lepoutre A (1993) Le tetanos en France en 1991 et 1992. Bull Epidemiol Hebd 28: 125–126
Malek ANA, Mukelabai K (1990) Neonatal tetanus and EPI in Luasaka, Zambia: a ten year retrospective study. Alexandria J Pediatr 4: 593–598
Misra SR, Rao RS (1988) Serological survey of tetanus antibodies in the Pondicherry region. Indian J Med Res 87: 161–165
Morgan P,Koplan JP, Miller GJ, Hull B, Ashcroft MT, Rosenberg M (1981) A serological evaluation of an immunization programme in Salt Cay, Turks and Caicos Islands. West Indian Med J 30: 68–71
Morris W,Steinhoff MC, Russell PK (1994) Potential of polymer microencapsulation technology for vaccine innovation. Vaccine 12: 5–11
Mya MM, To M, Aye KK, Saw TT, Htoo B (1985) Tetanus immunity: trends and/or status in some selected adult Burmese populations. Burmese Med J 31: 149–164
Newton-John HF (1984) Tetanus in Victoria, 1957–1980: review of 106 patients managed in one hospital. Med J Austr 140: 194–200
Patel JC, Mehta BC (1963) Tetanus: a study of 2007 cases. Indian J Med Sci 17: 791–811
Patel JC, Mehta BC (1975) Tetanus: study of 8697 cases. In: Proceedings of the Fourth Conference on Tetanus, Dakar, Senegal, 6–12 April 1975. Fondation Merieux, Lyon pp 233–241
Pelletier A, Roure C (1991) Le tetanos en France en 1990 (a partir des declarations obligatoires). Bull Epidemiol Hebd 31: 127–128
Pichichero ME, Barkin RM, Samuelson JS (1986) Pediatric diphtheria and tetanus toxoid-adsorbed vaccine: immune reponse to the first booster following the diphtheria and tetanus vaccine primary series. Pediatr Infect Dis J 5: 428–430
Prevost R, Sutter RW, Strebel PM, Cochi SL, Hadler S (1992) Tetanus surveillance—United States, 1989–1990. Morb Mortal Wkly Rep 41 [Suppl 8]: 1–9
Rey M, Tikhomirov E (1989) Non neonatal tetanus over the world. In: Nistico G, Bizzini B, Bytchenko B, Triau R (eds) Proceedings of the Eigth International Conference on Tetanus, Leningrad, 25–28 August 1987. Pythagora Press Rome-Milan pp 506–518
Rosmini F, Gentili G, Wirz M, Collotti C, Pana A, Terzi I, Conti S, Pasquini P (1987) Immunity to tetanus among Italians born between 1956 and 1963 Eur J Epidemiol 3: 302–307
Simonsen O, Badsberg JH, Kjeldsen K, Moller-Madsen B, Heron I (1986) The fall-off in serum concentration of tetanus antitoxin after primary and booster vaccination. Acta Pathol Microbiol Immunol Scand 94: 77–82
Simonsen O, Bentzon MW, Kjeldsen K, Venborg HA, Heron I (1987) Evaluation of vaccination requirements to secure continuous antitoxin immunity to tetanus. Vaccine 5: 115–122
Singhai PK, Mathur GP, Mathur S, Singh YD (1990) Neonatal morbidity and mortality in ICDS urban slums. Indian Pediatr 27: 485–488
Singleton AR, Witt RW (1956) Tetanus complicating pregnancy: report of a case treated with chlorpromazine with survival of both mother and infant. Obstet Gynecol 7: 540–541
Sokhey J (1991) Elimination of neonatal tetanus: 1995. J Commun Dis 23: 1–10
Stanfield JP, Galazka A (1984) Neonatal tetanus in the world today. Bull World Health Organ 62: 647–669
Sutter RW, Cochi SL, Brink EW, Sirotkin Bi (1990) Assessment of vital statistics and surveillance data for monitoring tetanus mortality, United States, 1979–1984. Am J Epidemiol 131: 132–142
Traverso HP, Bennet JV, Kahn AJ, Agha SB, Rahim H, Kamil S, Lang MH (1989) Ghee applications to the umbilical cord: a risk factor for neonatal tetanus. Lancet 1: 486–488

Traverso HP, Kamil S, Rahim H, Samadi AR, Boring IR, Bennett JV (1991) A reassessment of risk factors for neonatal tetanus. Bull World Health Organ 69: 573–579

Vernacchio L, Madico G, Verastegui M, Diaz F, Collins TS, Gilman RH (1993) Neonatal tetanus in Peru: risk assessment with modified enzyme-linked immunosorbent assay and toxoid skin test. Am J Public Health 83: 1754–1756

Viljanen MT, Nieminen S (1980) Immunity to tetanus in Finland. Scan J Infect Dis 12: 211–213

Whitman C, Belgharbi L, Gasse F, Torel C, Mattei V, Zoffman H (1992) Progress towards the global elimination of neonatal tetanus. World Health Stat Q 45: 248–256

Woody RC, Ross EM (1989) Neonatal tetanus (St. Kilda, 19th century). Lancet 1989 1: 1339

World Health Organization (WHO) (1991) Maternal mortality. A global factbook. WHO, Geneva

Yadav YR, Kala PC, Yadav S (1990) Study of site and nature of injury as prognostic factors in tetanus. Trop Doct 20: 136–137

Yadav YR, Yadav S, Kala PC (1991) Puerperal tetanus. J Indian Med Asso 89: 336–337

Yusuf B, Solter S, Bertsch D, Arnold RB (1991) Impact of a tetanus toxoid immunization mass compaign on neonatal tetanus mortality in Aceh Province, Indonesia. Southeast Asian J Trop Med Public Health 22: 351–356

Zuber PLF, Schierz A, Arestegui G, Steffen R (1993) Tetanus in Switzerland 1980–1989. Eur J Epidemiol 9: 617–624

Botulism: The Present Status of the Disease

C.L. HATHEWAY

1 Introduction

Botulism is a very serious foodborne illness, often fatal, and characterized by neuroparalytic signs and symptoms. Early observations found an association with consumption of sausages, and thus the name botulism (derived from *botulus*, Latin for sausage) was given to the disease (SMITH and SUGIYAMA 1988). Foodborne botulism is the result of ingestion of food containing neurotoxin produced by *Clostridium botulinum*. The disease may also result from toxin produced in vivo by organisms growing in an infected wound (wound botulism) or colonizing the intestinal tract (infant botulism).

Division of Bacterial and Mycotic Diseases, National Center for Infectious Diseases, Centers for Disease Control and Prevention, Atlanta, GA 3033, USA

Botulism is analogous to tetanus in many respects. Both diseases are caused by protein neurotoxins having a molecular mass of approximately 150kDa, produced by organisms in the genus *Clostridium*. The toxins are immunogenic and their biologic activities are neutralized by specific antitoxins. They cause paralysis by binding to receptors on nerve endings, entering the nerve, and interfering with the release of neurotransmitter. When the action is limited to the peripheral motor nerve endings, as with botulism, interference with the release of acetylcholine results in a flaccid paralysis. Tetanus toxin, after gaining entrance into the nerve ending, is transported through the motor nerve axon and gains access to the spinal cord, where it blocks release of inhibitory neurotransmitter and causes the spastic and convulsive condition recognized as tetanus (Bizzini 1989).

Botulism also occurs in many species other than humans, including mammals, birds, and fish (Smith and Sugiyama 1988). As outbreaks of botulism were investigated in humans and animals over the years, different serologic types of botulism neurotoxin were encountered, each requiring its own specific antitoxin reagent for neutralizing its biologic activity. At present, seven toxin types (A–G) are known. Human botulism is essentially restricted to types A, B, and E, although a few cases of type F have been documented (Hauschild 1992). Type C botulism is common in avian species, often occurring in domestic flocks and in massive outbreaks in wild waterfowl (Eklund and Dowell 1987). Type C outbreaks also occur in domestic animals such as dogs, cattle, and mink (Smith and Sugiyama 1988). Reports of type D botulism are rare and are generally associated with cattle (Seddon 1922). One human outbreak of type C (Prevot et al. 1955) and one of type D (Demarchi et al. 1958) have been described. Type G toxin producing organisms were isolated from samples of soil in Argentina (Giménez and Ciccarelli 1970).

This chapter presents various aspects of the disease of botulism as they apply universally, but relies on personal experience in laboratory investigation and observation of outbreaks and incidents of botulism in the United States. Since a variety of organisms produce botulinum neurotoxin, the presentation begins with the bacteriological aspects of the disease.

2 Organisms that Cause Botulism

The bacterial etiology of botulism was established through a comprehensive investigation of an outbreak affecting 23 persons in Belgium in 1895 (Van Ermengem 1897). It was established that botulism is caused by a toxin produced by an anaerobic spore-forming organism. Van Ermengem named the organism that he isolated *Bacillus botulinus*. As with the toxins that cause tetanus and diphtheria, the botulism toxin is antigenic and can be neutralized by specific anitsera produced against it (Kempner 1897). As noted above, seven serologic types of the neurotoxin have been recognized so far. In general, each type is distinct, with little or no cross-neutralization between types, and with few exceptions, one organism produces one type of toxin.

2.1 Phenotypic Diversity of Neurotoxigenic Organisms

The organisms that can produce botulism neurotoxin are diverse. It was discovered early that organisms isolated from different outbreaks could have different phenotypic characteristics (LEUCHS 1910; BENGTSON 1924). Nevertheless, by convention all organisms capable of producing botulism neurotoxin became classified as *C. botulinum* (PREVOT 1953). *Bergey's Manual of Systematic Bacteriology* (CATO et al. 1986) lists four groups of *C. botulinum*, distinguished on the basis of their cultural and physiologic characteristics. These are commonly referred to as groups I, II, III, and IV (SMITH and SUGIYAMA 1988). The phenotypic and genetic features of the organisms are sufficiently distinct from one group to another to warrant separate species names. Recent findings of toxigenic strains of *C. baratii* (HALL et al. 1985; McCROSKEY et al. 1991) and *C. butyricum* (AURELI et al. 1986; McCROSKEY et al. 1986) that have caused human botulism make it impossible to use a single species designation for *all* toxigenic organisms.

Key characteristics for the four *C. botulinum* groups and the other two species are listed in Table 1. The organisms in group I are often referred to as proteolytic strains, and those in group II as nonproteolytic strains, based on their ability to digest complex proteins such as casein. All strains that produce type A toxin belong to group I. Organisms possessing all the characteristics of group I but lacking toxigenicity are classified as *C. sporogenes*. All organisms (except for the *C. butyricum* strains) that produce type E toxin belong to group II. Analogous nontoxigenic organisms with group II characteristics are encountered, but no name has been established for them. Type B and type F toxins may be produced by strains in either group I or group II. As noted, human botulism is essentially restricted to toxin types A, B, E, and F, and thus is caused only by organisms in groups I and II. Although the organism isolated by VAN ERMENGEM has not been available for comparative studies in recent times, it no doubt belonged to group II and was toxin type B. Its isolation from salted ham, its lower optimal growth temperature, and the serologic differentiation of its toxin from that isolated from an outbreak caused by bean salad in Darmstadt in 1904 (LANDMANN 1904; LEUCHS 1910) support this deduction. The Darmstadt strain was most likely toxin type A, which is found only with group I characteristics.

Organisms that produce type C and type D toxin belong to group III and resemble *C. novyi* type A (CATO et al. 1986). The most readily distinguishable characteristic of these organisms is the production of a neurotoxin that is neutralized with either type C or type D antitoxin. The metabolic acid profile (acetic, propionic, butyric) is unique among the neurotoxigenic organisms. They appear to be more fastidious in growing in laboratory media and to have a higher optimum growth temperature (40°C); they grow well at 37° and 45°C, but poorly

Table 1. Characteristics of clostridia capable of producing botulinum neurotoxin

	C. botulinum group				*C. baratii*	*C. butyricum*
	I	II	III	IV[a]		
Toxin type	A,B,F	B,E,F	C,D	G	F	E
Lipase	+	+	+	−	−	−
Lecithinase	−	−	−	−	+	−
Milk digestion	+	−	±	+	−	−
Gelatin	+	+	+	+	−	−
Glucose	+	+	+	−	+	+
Lactose	−	−	−	−	+	+
Mannose	−	+	+	−	+	+
Growth temperature						
Optimum	35°–40°	18°–25°	40°	37°	30°–37°	30°–45°
Minimum	12°	3.3°	15°			10°
Spore heat resistance						
Temperature	112°	80°	104°	104°		
D value	1.23	0.6–1.25	0.1–0.9	0.8–1.2		
Metabolic acids[b]	A,iB,B,iV PP	A,B	A,P,B	A,iB,B,iV PA	A,B	A,B
Similar atoxic organism	*C. sporogenes*	(No species name assigned)	*C. novyi*	*C. subterminale*	(All typical strains)	(All typical strains)

[a]The name *C. argentinense* has been proposed for this group (SUEN et al. 1988a).
[b]Metabolic acids: A, acetic; P, propionic; B, butyric; iB, isobutyric; iV, isovaleric; PP, phenylpropionic (hydrocinnamic); PA, Phenylacetic.

at 25°C (SMITH and SUGIYAMA 1988). These organisms are carried in the intestinal tract of avian and mammalian species without noticeable effect, but produce very high levels of toxin in the carcass after the animal has died. Thus, types C and D botulism are associated with consumption of carrion. Toxigenicity of these organisms is associated with the presence of a bacteriophage (EKLUND et al. 1987). While toxin is demonstrable in the mixed cultures, the phage is often missing in the organisms in the colony isolated from the plate, and the pure cultures contain only nontoxigenic organisms. There are complex relationships between the bacteria and the phage in the carcass of the affected bird or animal with varying stability of association. In the original milieu, toxigenicity may be enhanced by reinfection of nontoxigenic cells by phages released from other cells (EKLUND et al. 1987). Isolated nontoxigenic cells will remain nontoxigenic, and toxigenic organisms may loose their phage on subsequent passage (BENGTSON 1924). Few toxigenic strains of group III organisms are available. The reason for the stability of toxigenicity in the reference strains is not clear.

The relationship between *C. botulinum* group III and *C. novyi* is demonstrated (in addition to DNA studies) by experiments in which strains of each species were interconverted to the other by infection of phage-cured organisms with phages recovered from strains of the opposite species (EKLUND et al. 1987).

Group IV was created to accommodate an organism isolated from a soil sample in Argentina that was phenotypically radically different from groups I–III and produced a neurotoxin with a novel serologic specificity (type G) that caused a flaccid paralysis in mice indistinguishable from botulism (GIMÉNEZ and CICCARELLI

1970). The organism does not produce lipase and is asaccharolytic and thus resembles *C. subterminale*. A second isolate from an Argentine soil sample was reported several years later (CICCARELLI et al. 1976). Although there are more recent reports on isolation of these organisms from autopsy specimens (SONNABEND et al. 1981), no patients with clinically diagnosed botulism have been documented with this toxin type.

Several cases of type F botulism have been caused by strains of neurotoxigenic *C. baratii*. Two cases of infant botulism and three cases of adult botulism due to this organism have been confirmed (HALL et al. 1985; McCROSKEY et al. 1991; CDC, unpublished). Two cases of infant botulism in Italy were caused by *C. butyricum* strains that produce type E neurotoxin (AURELI et al. 1986; McCROSKEY et al. 1986).

2.2 DNA Homology Studies

The distinctions among groups I, II, III, and IV based on phenotypic characteristics are confirmed by DNA comparsion of representative strains (LEE and RIEMANN 1970a,b; NAKAMURA et al. 1977, 1983; SUEN et al. 1988a). Organisms within each group have a high degree of DNA relatedness and include genetically related nontoxigenic organisms. *C. sporogenes* is related to group I strains; *C. novyi*, as well as nontoxigenic organisms that have apparently lost their phages, is related to the type C and D strains in group III. Nontoxigenic organisms closely related to type E organisms are commonly found, often in the same cultures with toxigenic organisms. These appear to have the same relationship to the group II organisms as *C. sporogenes* has to group I, but no separate species name has been designated. One strain each of toxigenic *C. baratii* and *C. butyricum* was tested and found to be closely related to the type strain (nontoxigenic) of its respective species (SUEN et al. 1988b).

Because the variety of organisms that can produce botulism neurotoxin includes strains identifiable as other legitimate species, we can no longer insist that every toxigenic organism be classified as *C. botulinum*. Since the phenotypic characteristics of the group IV organisms are so different from those of the other three groups, and the organisms appear to play no significant role in human or animal botulism, SUEN et al. (1988a) proposed the name *C. argentinense* for the organisms that produce type G toxin as well as for their nontoxigenic counterparts.

Microbiologists working with botulism frequently discuss the feasibility of establishing separate species names for the organisms in groups I, II, and III. The distinctions are significant because the cultural and physiologic characteristics separating these groups undoubtedly account for differential involvement of the groups in the different forms of the disease, and in different vehicles for foodborne outbreaks. However, the close identification of all of these organisms with botulism necessitates a name that reflects the disease. Therefore, one name, *C. botulinum*, continues to serve for identifying three actual species which must be further identified by their respective group designations.

3 Forms of the Disease

Botulism was recognized only as a foodborne disease for many years. This form of the disease is caused by eating food in which the toxigenic organisms have grown and produced the neurotoxin. Botulism in birds or animals is commonly caused by ingestion of carrion or feeds contaminated with carrion, or of insect larvae feeding on carrion. Other forms of botulism in humans have been recognized more recently as being due to toxigenic organisms growing in a wound (wound botulism; MERSON and DOWELL 1973), or colonizing the intestinal tract (infant botulism; ARNON 1992b).

3.1 Foodborne Botulism

Botulism resulting from ingestion of neurotoxin present in foods remains the most frequent form in humans throughout the world. A recent review of foodborne botulism (HAUSCHILD 1992) lists 3353 outbreaks and 9767 cases recorded in 38 countries between 1951 and 1989. In recent years an average of 449 outbreaks with 930 cases have been recorded annually (Table 2); 72% of the outbreaks and 48% of the cases were reported from Poland. Of the 2622 outbreaks in which the toxin type was determined, 34% were type A, 52% type B, and 12% type E. Only two incidents of type F foodborne botulism were reported during this period. Fourteen incidents showed evidence of two toxin types either because of toxins from two different organisms (12 incidents) or because of a single organism that produces two types of toxin (2 incidents). The

Table 2. Annual reported incidence of foodborne botulism by country[a]

	Per year			Toxin type (%)		
Country	Years	Outbreaks	Cases	A	B	E
Poland	1984–1987	325	448	3	94	3
China	1958–1983	38	168	93	5	1
Soviet Union	1958–1964	14	47	33	38	28
West Germany	1983–1989	16	35		>90	
Italy	1979–1987		34	20	60	7
United States	1975–1992	14	30	60	18	22
France	1978–1989	15	25	0	97	2
12 other countries	1951–1989	27	176	14	27	59
Total		449	963	34	52	13

[a]Data selected from HAUSCHILD (1992); data for the United States updated from CDC records.

estimated case/fatality ratio worldwide is 20%. More than 90% of foodborne botulism is caused by home-prepared or home-preserved foods.

The distribution of foodborne botulism in the United States by state and toxin type between 1975 and 1992 is shown in Table 3. A total of 257 outbreaks occurred in 39 states and the Commonwealth of Puerto Rico. Of these, 48 occurred in California; of those with known toxin types 38 were type A, 6 were type B, and 1 was type F. Type E botulism has not been confirmed in California. Of the 58 outbreaks that occurred in Alaska 48 were type E, 6 were type B, and

Table 3. Foodborne botulism in the United States by state and toxin type, 1975–1992

State	Outbreaks					Cases
	A	B	E	Other[a]	Total	
Alaska	4	6	48	0	58	140
Arkansas	2	0	0	0	2	3
Arizona	1	0	0	0	1	1
California	38	6	0	4	48	70
Connecticut	0	1	0	0	1	1
Colorado	8	1	0	0	9	18
Florida	2	1	0	0	3	6
Georgia	1	0	1	0	2	2
Hawaii	0	5	0	0	5	7
Idaho	2	0	0	0	2	2
Illinois	3	1	0	0	4	31
Kentucky	2	3	0	0	5	5
Louisiana	1	1	0	0	2	2
Massachusetts	0	1	0	0	1	1
Maryland	1	1	0	0	2	2
Michigan	2	2	0	0	4	63
Minnesota	0	0	0	1	1	1
Mississippi	1	0	0	0	1	1
Missouri	1	0	0	0	1	2
Montana	2	0	0	1	3	3
North Carolina	2	2	0	0	4	4
North Dakota	1	0	0	0	1	1
Nebraska	2	0	0	0	2	2
New Jersey	0	0	2	1	3	7
New Mexico	2	0	0	1	3	37
Nevada	1	0	0	0	1	2
New York	1	1	2	0	4	9
Ohio	3	1	0	0	4	4
Oklahoma	2	0	0	1	3	9
Oregon	14	0	0	2	16	24
Pennsylvania	1	1	0	0	2	3
Puerto Rico	1	0	0	0	1	3
Tennessee	2	5	0	0	7	7
Texas	3	2	0	0	5	7
Utah	7	0	0	0	7	7
Virginia	2	0	0	0	2	2
Washington	25	1	0	0	26	43
Wisconsin	4	2	0	1	7	7
West Virginia	0	1	0	0	1	1
Wyoming	3	0	0	0	3	3
Total	147	45	53	12	257	543

[a]Toxin types not determined except for 1 outbreak in California which was type F.

4 were type A. In California the vehicle is most often home-canned vegetable products. In Alaska botulism is invariably caused by fish or marine mammal products preserved or stored by traditional native methods (WAINWRIGHT et al. 1988). The outbreaks occurring in other states were mostly type A or type B in home settings, due to home-canned vegetable products. Washington and Oregon rank third and fourth in outbreaks, respectively, almost exclusively type A. Four large outbreaks occurred in restaurants during this period, due to home-type processed peppers (TERRANOVA et al. 1978), potato salad (SEALS et al. 1981; MANN et al. 1982), and cooked onions (MACDONALD et al. 1985).

In Europe the vehicle is most often a meat product such as sausage or ham (HAUSCHILD 1992; SEBALD and JOUGLARD 1977). These outbreaks are almost always toxin type B, and the causative organisms are thought to be nonproteolytic *C. botulinum* group II. Reports on the characteristics of the organisms are rare. In contrast, the type B outbreaks in the United States are associated with vegetable products and the organisms recovered are proteolytic *C. botulinum* group I (CDC, unpublished data), and except for the serologic properties of the toxin they produce, they are indistinguishable from type A strains.

Why do the nonproteolytic group II organisms cause botulism in proteinaceous foods while the proteolytic group I organisms are usually associated with botulism due to vegetable products? An explanation may be that, contrary to current recommendations, the process sometimes used by home canners for preserving vegetables employs sufficient heat to destroy spores of group II organisms, but not of group I organisms. The meat products involved in the European outbreaks are not heated but are preserved by salting and consumed without cooking. The heat-resistant spores of group I organisms that survive the canning process can germinate and produce toxin in the anaerobic environment of the closed jar. *C. botulinum* type E that causes most of the botulism in Alaska is a group II organism, and the foods associated with outbreaks are also not heated (WAINWRIGHT et al. 1988). The organism responsible for a type B outbreak associated with salted fish in Alaska in 1977 was caused by a group II strain (CDC, unpublished records). Botulism in the United States due to toxin type B group II organisms is rare.

3.2 Wound Botulism

Wound botulism is the form most analogous to tetanus. It occurs when the neurotoxin produced by *C. botulinum* growing in a wound is disseminated throughout the body and poisons the nerve endings causing paralysis. This form of botulism is extremely rare. The first documented case occurred in 1943 and was reported in the literature 8 years later (DAVIS et al. 1951). Forty-seven laboratory-confirmed cases are listed in a review of wound botulism in the United States between 1943 and 1990 (WEBER et al. 1993). Cases are listed by state and toxin type in Table 4. Almost half (47%) occurred in California. They are predominantly toxin type A. One case was confirmed on the basis of serum bioassay and neutralization with polyvalent antitoxin, and the type was therefore not determined. Of the 46 patients whose outcomes are known 7 have died (15% case/fatality ratio). The infected wounds were associated with traumatic injury (30 cases), surgery (5 cases), injection of illicit drugs (7 cases), and sinusitis associated with intranasal cocaine use (2 cases). More than seven additional cases have been documented in the United States since 1992 (CDC 1994, unpublished records), including a case associated with a tooth abscess (WEBER et al. 1993).

The rarity of wound botulism in an unimmunized population in comparison with tetanus indicates that *C. botulinum* is not as infectious in wounds as *C. tetani*. A survey of soils in the United States yielded *C. botulinum* type A or type B in 18.5% of the samples and *C. tetani* in 30% (SMITH 1978; DODDS 1992a). Each year in the United States only one to three cases of wound botulism are observed, while 40–60 cases of tetanus occur despite virtually universal immunization against the latter (CDC 1993).

Outside the United States, reports of one case of wound botulism from France (DEVEZE et al. 1979), two from Italy (CORSI et al. 1979; AURELI et al. 1992), one from Australia (FULLERTON et al. 1980), and one from China (WANG and WANG 1983) can be found in the literature.

3.3 Infant Botulism

A form of botulism occurring in infants as a result of colonization of the intestinal tract by *C. botulinum* was recognized in California as a distinct clinical entity in 1976 (PICKETT et al. 1976; MIDURA and ARNON 1976). ARNON (1992b) has recently described the clinical features, pathogenesis, epidemiology, diagnosis, and

Table 4. Wound botulism in the United States by state and toxin type, 1943–1992

State	Type A	Type B	Other	Total
Arizona	2	0	0	2
California	18	3	1[a]	22
Idaho	2	0	0	2
Indiana	0	1	0	1
Iowa	1	0	0	1
Kentucky	0	1	0	1
Maryland	0	2	0	2
Nebraska	1	0	0	1
New York	0	2	0	2
Oregon	0	1	0	1
Pennsylvania	1	1	0	2
Tennessee	1	1	0	2
Texas	2	1	0	3
Washington	4	0	0	4
Wyoming	0	0	1[b]	1
Total	32	13	2	47

[a]Types A and B [b]Unknown type

treatment of infant botulism. Another review on the worldwide incidence of infant botulism through June 1990 lists 14 countries in which cases have been recognized (DODDS 1992b). Among the total of 993 cases, 932 (94%) have been identified in the United States. Other countries confirming cases are Argentina (23 cases), Australia (11), Japan (10), Canada and Italy (4 each), United Kingdom (2), and Chile, former Czechoslovakia, France, Spain, Switzerland, Sweden, and Taiwan (1 each).

Updating the data through 1992, 1134 cases have been recorded in the United States (Table 5; CDC, unpublished records, 1994), where it has become the most frequently occurring form of botulism. Forty-eight percent of the cases occurred in California.

The toxin types of infant botulism are generally either A or B, and the organisms are group I *C. botulinum*. The few exceptions are: (a) two cases in the United States caused by strains of *C. baratii* that produce a neurotoxin serologically similar to type F (HALL et al. 1985; CDC 1993, laboratory records); (b) two cases in Italy caused by strains of *C. butyricum* that produce type E neurotoxin (AURELI et al. 1986; MCCROSKEY et al. 1986); (c) two cases in the United States (HATHEWAY and MCCROSKEY 1987; HATHEWAY and MCCROSKEY 1989) and one in the United Kingdom (DODDS 1992b) caused by strains that produce a mixture of type B and F toxins (type Bf) ; and (d) one case in the United States caused by a strain that produces a mixture of type B and A toxins (type Ba; HATHEWAY et al. 1981; GIMÉNEZ 1984).

It is surprising that more than 50% of the infant botulism cases in the United States are type B, while the incidence of type A foodborne botulism incidents exceed type B by more than threefold. In California type B botulism accounts for about 39% of the infant cases but only 12% of the foodborne outbreaks and 16% of the wound cases. The toxin types of foodborne and wound botulism appear to reflect the distribution of the type of organisms in the environment (MEYER and DUBOVSKY 1922), while the frequency of type B infant botulism is unexpectedly high.

The primary source of the toxigenic organisms is the environment. Evidence that honey was the vehicle for the spores has been found in 26 cases (ARNON 1992b). No other food vehicle has been implicated. Foods as well as anything else which enters the mouths of infants no doubt can be contaminated by dust particles that contain spores and thus serve as secondary carriers. The case/fatality ratio for infant botulism in the United States is less than 2% (ARNON 1992b).

Table 5. Infant botulism in the United States and in California by toxin type, 1975–1992

	Type A	Type B	Other	Total
United States	517	574	14[a]	1134
California	334	213	1[b]	548

[a]Type not reported, 8 cases; type Bf, 2 cases; type Ba, 1 case; types A and B (2 strains of *C. botulinum*), 1 case; type F due to *C. baratii*, 2 cases.
[b]Type Bf, 1 case.

3.4 Intestinal Colonization of Adults

A rew rare cases of botulism appear to be due to intestinal colonization of humans other than infants (McCroskey and Hatheway 1988). Two patients with clinical signs and symptoms of botulism yielded *C. botulinum* type A in their stool cultures for as long as 119 (McCroskey and Hatheway 1988) and 130 (Griffin et al. 1994) days after onset of illness; there was no evidence of exposure of either patient to a toxic food vehicle. The first patient had had bowel surgery (Billroth I) to remove the pyloric valve 5 weeks before she became ill (Chia et al. 1986). The second patient had Crohn's disease for more than 20 years (Griffin et al. 1994); intestinal lesions may have provided sites for colonization or infection by toxigenic organisms. A third patient who had undergone a jejunal-ileal bypass for obesity 3 years earlier became ill with signs of botulism without any known exposure to contaminated food (English et al. 1981); type A toxin was detected in her serum, and *C. botulinum* type A was isolated from her stool.

A fourth adult, a resident of Iceland, appeared to develop botulism 47 days after exposure to a food that had caused botulism in four other family members (McCroskey and Hatheway 1988). A stool sample obtained 19 days after the presumed exposure, although he was not ill, showed the presence of type B neurotoxin in the enrichment culture, but the organism could not be isolated. Both serum and stool obtained 28 days later, at the time of onset of his symptoms, contained demonstrable type B toxin. The organisms isolated from the stool culture, as well as from the other patients in the outbreak, were identified as group II.

A study of neurotoxemia in botulism in France revealed the persistence of type B toxin (most likely from group II organisms) in the blood for periods as long as 122 days after ingestion of toxic food (Sebald and Saimot 1973). This suggests the possibility of in vivo toxin production by organisms remaining in the patient. It is not possible to follow persistence of toxemia in botulism in the United States because all patients available for study are treated with therapeutic antitoxin immediately after the diagnosis is made. Toxemia can be detected only in blood samples taken before treatment.

One case of type F botulism in the United States caused by toxigenic *C. baratii* is also believed to be due to intestinal colonization (McCroskey et al. 1991). There was no food history suggestive of a contaminated food vehicle, although the patient had had earlier gastric surgery which might have promoted intestinal colonization. Two more cases of adult botulism in the United States due to toxigenic (type F) *C. baratii* have been documented (CDC, unpublished records); no toxic food vehicle has been identified in either case.

3.5 Animal Forms

The clinical features of botulism in animals are essentially the same as those of botulism in humans (Critchley 1991). Animals may have a flaccid paralysis often starting in the hind quarters with weakness, muscle tremors, stumbling, and recumbency. The occurrence of botulism in animals and birds has been reviewed in detail by SMITH and SUGIYAMA (1988).

Botulism in cattle has been referred to as lamsiekte, bulbar paralysis, loin disease, and forage poisoning. The first three terms refer to botulism caused by the animals eating carcasses or chewing on bones contaminated with toxin. The toxin type in these cases is either type C or type D. Forage poisoning caused by eating roughage contaminated with decomposing carcasses of small animals is usually toxin type C. Sometimes type B botulism results from the growth of organisms in the feed itself (DIVERS et al. 1986).

Horses also develop botulism due to forage poisoning. They are especially sensitive to type B toxin (SMITH and SUGIYAMA 1988). A condition known as shaker foal syndrome occurs in young foals and is caused by growth of *C. botulinum*, usually type B, in the intestinal tract (SWERCZEK 1980b). The organisms infect intestinal ulcerative lesions that are brought on when the young animals are subjected to stressful conditions (SWERCZEK 1980a). Foals can be protected by immunizing the mares with type B botulism toxoid before foaling (JOHNSTON and WHITLOCK 1987). An instance of wound botulism in a horse resulting as the aftermath of castration has also been documented (BERNARD et al. 1987).

Type C botulism has been confirmed in dogs that ate dead chickens (BARSANTI et al. 1978; RICHMOND et al. 1978). It often occurs in domestically raised mink fed improperly handled meat products (SMITH and SUGIYAMA 1988). For this reason mink are commonly immunized with type C botulism toxoid.

The most extensive botulism outbreaks are observed in wild water fowl (JENSEN and PRICE 1987). Western duck sickness was established as type C botulism in 1930. It has since been diagnosed in at least 17 countries and on all continents except Antarctica, and has affected at least 117 avian species. Large numbers of birds are lost in these incidents, with occasional estimates into the millions. Type C botulism is sometimes a problem in chickens (SMITH and SUGIYAMA 1988). Toxin is produced in the cecum without affecting the birds. When this toxin is excreted in the feces, it may be ingested from the litter and then cause botulism. The inability of affected birds to hold their heads erect gave rise to the descriptive term limberneck in referring to avian botulism. The spread of the disease commonly results from healthy birds eating fly larvae infesting the carcasses of other birds that have died of botulism.

4 Symptoms and Diagnosis of Human Botulism

The diagnosis of botulism is made on the basis of signs and symptoms of the patients. The diagnosis may be confirmed by laboratory tests; however, confirmatory evidence is not immediately available and may be lacking in about 12% of the laboratory investigations (DOWELL et al. 1977). Patients with botulism typically present with acute onset of weakness in muscles innervated by cranial nerves, resulting in difficulty with speaking and swallowing, and double and blurred vision. Subsequently, they develop progressive symmetric descending weakness or paralysis. In foodborne botulism, these symptoms are frequently preceded by gastrointestinal upset, vomiting, and constipation; diarrhea is less common. Patients may also have autonomic dysfunction resulting in dry mouth, blurred vision, orthostatic hypotension, constipation, and urinary retention. The severity and number of symptoms can be quite variable, but usually patients present with several of these symptoms rather than with a single cranial nerve abnormality. Symptoms reported in a study of foodborne botulism are listed in Table 6 (HUGHES et al. 1981). Since botulinum toxin affects only motor and autonomic nerves, sensory abnormalities are usually absent. Mental functioning is not generally impaired, and patients remain alert unless they develop secondary complications, such as respiratory failure.

The need for intubation of the patient may be used as an indication of severity of illness. A recent comparison of severity of botulism by toxin type found that intubation was required for 67% of type A patients, 52% of type B, and 39% of type E (WOODRUFF et al. 1992). Sporadic cases of botulism were more severely ill than patients involved in multicase outbreaks. This suggests that milder cases of sporadic botulism more often go undiagnosed. Wound botulism causes the same signs and symptoms as the foodborne form, but gastrointestinal features are unlikely. The food history is negative, and a site of infection is necessary.

Infant botulism is characterized by an initial constipation (3 or more days without a bowel movement), lethargy, poor feeding, and increasing weakness (ARNON 1992b). Bulbar palsies are indicated by expressionless face, feeble cry, ptosis, and impaired gag, suck, and swallow reflexes. Poor head control, generalized weakness, and hypotonia are evident. Respiratory failure often ensues, requiring mechanical ventilatory support. Sepsis or failure to thrive is most often the admission diagnosis. Cranial nerve function deficits are the most valid indicators of infant botulism.

Table 6. Frequency of symptoms in 55 patients with type A or type B foodborne botulism[a]

Neurologic symptoms	Percentage with symptom
Dysphagia	96
Dry mouth	93
Diplopia	91
Dysarthria	84
Extremity weakness	73
Blurred vision	65[b]
Dyspnea	60[b]
Paresthesia	14
Gastrointestinal symptoms	
Constipation	73
Nausea	64
Vomiting	59
Abdominal cramps	42
Diarrhea	19

[a]Data from HUGHES et al. (1981).
[b]Incidence in type A patients significantly higher than in type B patients.

Clinicians must consider several other conditions when evaluating noninfant patients with neurologic signs and symptoms consistent with botulism. Patients with Guillain-Barré Syndrome (particularly the Miller-Fischer variant), myasthenia gravis, tick paralysis, drug reactions, stroke, or nervous system infections may have signs and symptoms similar to those caused by botulism. A careful history focusing on consumption of foods likely to cause botulism, physical examination with emphasis on the neurologic exam, and selected tests including imaging of the central nervous system with computed axial tomography or magnetic resonance imaging, lumbar puncture, and electromyography with rapid repetitive stimulation may distinguish these illnesses before confirmatory laboratory tests for botulism can be performed. An elevated spinal fluid protein would indicate Guillain-Barré Syndrome, while an augmented potential registered in the electromyogram strengthens the diagnosis of botulism. A negative tensilon test result does help rule out myasthenia gravis.

5 Laboratory Confirmation of Human Botulism

Botulism is confirmed by detection of toxin in a patient's serum or stool, or in a sample of a food consumed before onset of illness (HATHEWAY 1988). Recovery of *C. botulinum* from stools, gastric samples, or wound specimens from patients with signs and symptoms indicative of botulism is usually sufficient for confirmation. The toxigenic organisms are rarely if ever isolated from living humans in the absence of botulism (DOWELL et al. 1977). Recovery of the organism from a food that does not contain demonstrable toxin is inconclusive.

In suspected foodborne botulism, samples of patient serum and stool and of suspected food should be obtained for laboratory examination; sometimes gastric samples are also examined. For wound botulism, serum and wound specimens should be examined. One or more stools are usually sufficient for infant botulism.

5.1 Conventional Methodology

In practice the mouse bioassay remains the method used for detection and identification of botulinum toxin in botulism investigations (HATHEWAY 1988). Monovalent and polyvalent diagnostic *C. botulinum* antitoxins are available from the Centers for Disease Control and Prevention. Aliquots of serum, feces, or food extract, gastric fluid, or culture supernatant are injected i.p. into mice; the mice are observed for development of signs of botulism, and death over the next 4 days. Other aliquots of the samples to which 1 U of botulinum antitoxin has been added are injected into other mice. If active toxin is present in the test sample, the mouse develops signs of botulism and dies, usually within 24 h, and sometimes in as little as 4–12 h. The type specific and polyvalent antitoxins render the toxin inactive; the heterologous type reagents do not.

Stool, wound, and food samples are cultured in enrichment broth media (HATHEWAY 1988). The enrichment cultures are tested for toxin after at least 4 days incubation. Detection of the toxin indicates the presence of the organism, which is then isolated by streaking the culture on egg yolk agar, and picking the organisms recognized by their characteristic lipase reaction. Toxigenicity of the isolates must be confirmed by bioassay. Sometimes the organism can be isolated from the agar plate streaked directly with a suspension of the feces or food. This is often possible in infant botulism, when the patients are heavily colonized.

5.2 In Vitro Methods for Toxin Detection

Sensitive enzyme-linked immosorbent assays for detecting botulinum neurotoxins have been devised and described in the literature (NOTERMANS and NAGEL 1989). At this time none has been evaluated adequately and made commonly available to laboratories involved in investigating botulism outbreaks. Good-quality reagents and a standardized protocol will be needed for a collaborative evaluation by competent investigative laboratories for establishing satisfactory in vitro tests for detecting and identifying all of the known types of botulinum toxin, produced by diverse strains of each toxin type. Collaborative efforts need to be coordinated at this time to provide a reliable alternative to bioassays. However, mouse lethality in combination with observation of the characteristic signs of the illness and protection of mice by type specific antitoxin essentially defines the botulinum neurotoxins. Thus, the bioassay is not able to be completely eliminated.

5.3 Laboratory Findings

5.3.1 Foodborne Botulism

A recent study of foodborne botulism patients in the United States found toxin in the serum of 39% of the patients and the stool samples of 24% (WOODRUFF et al. 1992). *C. botulinum* was recovered from the stools of 55% of the patients whose samples were cultured. Gastric specimens from relatively few patients were tested by culture, but 46% of these cultured were positive. In this study 67% of 309 patients were confirmed by at least one of the laboratory tests. The overall confirmation was higher since positive findings in foods and in other patients in the same group confirm the diagnosis for patients whose specimens test negative.

In France toxin was detected in the serum of about 72% of patients with type B foodborne botulism (SEBALD and SAIMOT 1973). These cases were presumably due to group II organisms. Toxin was detected in only 16% of patients with type B botulism in the United States due to group I organisms (WOODRUFF et al. 1992). These differences may be due to differences in the characteristics of the organisms or the toxins they produce, or because the French patients could be sampled over a longer period after onset because they were not treated with antitoxin. However, studies in the United States indicate that detection of toxemia diminishes as the interval between onset and collection of samples increases (MANN et al. 1982; WOODRUFF et al. 1992).

5.3.2 Wound Botulism

Wound botulism is confirmed in the laboratory by detecting and identifying botulism neurotoxin in the patient's serum or by isolation of *C. botulinum* from wound specimens. Among 41 patients in the United States (WEBER et al. 1993) for whom laboratory results were available 25 had positive serum tests, 27 had positive cultures, and 11 had both. All of the organisms isolated from these patients belonged to group I, as far as can be documented.

5.3.3 Infant Botulism

Infant botulism is confirmed by detecting botulism neurotoxin in the feces and isolating the toxigenic organism. Among 109 culture-confirmed cases in the United States due to group I organisms, tests for stool toxin were positive in 96, negative in 4, and inconclusive because of interfering toxicity in 9 (HATHEWAY and MCCROSKEY 1987). Stools collected a week or more after onset may be toxin negative but culture positive. However, toxin and organisms may persist for many weeks (MIDURA and ARNON 1976). Serum is usually not tested because early observations suggested that demonstrable levels in infant sera were not achieved. Of 67 infants tested for toxemia between 1975 and 1987, 9 were positive: 8 of 22 type A (36.4%) and 1 of 43 for type B (2.3%); HATHEWAY and MCCROSKEY 1987.

6 Treatment

In the early 1900s 71% of botulism cases in the United States were fatal, but by the 1970s fewer than 16% of patients died (CDC 1979). The greatest improvement in survival of botulism patients has been through improvement of supportive care. However, the proper diagnosis of milder cases in more recent times is undoubtedly also a factor. From 57% to 81% of foodborne botulism patients in the United States require respiratory support (WOODRUFF et al. 1992; MORRIS 1981). Patients are often admitted to intensive care units so that respiratory function can be monitored. Assuring airway patency may require insertion of an endotracheal tube or tracheostomy.

Specific treatment of botulism employs infusion of multivalent botulism antitoxin. In the United States the CDC recommends intravenous infusion of two vials of therapeutic trivalent botulism antitoxin (approximately 10 000 IU each of type A, B, and E antitoxins) as soon as the diagnosis is made, without waiting for laboratory confirmation. This product is made available by CDC through the state health departments during consultations on diagnosis and treatment of patients. More than 80% of adult botulism patients are treated (MORRIS 1981). Circulating antitoxin levels measured after treatment are sufficient to overwhelm levels of toxemia encountered in botulism, and antitoxins persist in the circulation with a half-life of 5–7 days (HATHEWAY et al. 1984). Antitoxin does not neutralize toxin that has already bound to the nerve receptors, but it is effective against toxin before it binds. Therefore clinical symptoms may not improve immediately, but administration of antitoxin may stop progression of symptoms. It is most effective if given early. The efficacy of antitoxin therapy has not been studied in a randomized trial; however, the results of one study suggest that it has a beneficial effect on survival and shortens the course of illness (TACKET et al. 1984). Administration of equine antitoxin may cause hypersensitivity reactions. Among 268 treated patients studied between 1967 and 1977, 2.6% developed urticaria, 1.9% anaphylaxis, and 3.7% serum sickness. The overall incidence of adverse reactions was 9% (BLACK and GUNN 1980).

Treatment of infant botulism has been restricted to supportive care (ARNON 1992b). Use of equine antitoxin has been avoided because of the risk of adverse reactions and of sensitizing the patient against equine serum at such an early age, and noting the favorable prognosis without it. A controlled study on the use of a human-derived antibody product prepared from serum of immunized volunteer donors for neutralizing the neurotoxin in the infants is now in progress in California (ARNON 1992a). If the product is efficacious, it may become the preferred treatment for infants as well as other patients with botulism. Antibiotics for clearing toxigenic organisms from the infants' intestines are not recommended because release of toxin in the gut may worsen neurologic symptoms through bacterial cell lysis (ARNON 1992b).

7 Immunization

Humans and animals can be immunized against botulism by use of toxoids corresponding to each of the serologic types of neurotoxin (CARDELLA 1964). The disease is so rare that it is not practical to vaccinate humans at large, although vaccination of native populations in Alaska and northern Canada, who are at higher risk because of their traditional food preservation methods has been considered. A pentavalent toxoid is available through the CDC Drug Service Office for protecting laboratorians who work with the toxin or the organisms.

8 Summary

The main form of human botulism throughout the world is the classic foodborne intoxication. Wound botulism is very rare, and most of the documented cases have been found in the United States. While infant botulism remains rare throughout the world, it has become the most frequent form of the disease in the United States in recent years. On very rare occasions botulism results from growth and toxin production in humans other than infants. Botulism occurs in animals with much higher frequency. The causative organisms constitute a diverse group of clostridia, resulting in nomenclature problems. Human botulism is largely limited to toxin types A, B, and E, while type C botulism predominates in avian and nonhuman mammalian species. The diagnosis of botulism is made on the basis of the neurologic signs and symptoms that it causes in humans and animals. The diagnosis is confirmed by tests that identify the toxin and toxigenic organisms in patient and food specimens. Treatment includes supportive intensive care and use of therapeutic antitoxin.

References

Arnon SS (1992a) Clinical trial of human botulism immune globulin. In: DasGupta BR (ed) Botulinum and tetanus neurotoxins: neurotransmission and biomedical aspects. Plenum, New York

Arnon SS (1992b) Infant botulism. In: Feigen RD, Cherry JD (eds) Textbook of pediatric infectious diseases, 3rd edn. Saunders, Philadelphia

Aureli P, Fenicia L, Pasolini B, Gianfranceschi M, McCroskey LM, Hatheway CL (1986) Two cases of type E infant botulism caused by neurotoxigenic Clostridium butyricum in Italy. J Infect Dis 154: 201–206

Aureli P, Fenicia L, Ferrini AM, Martini M, Beccia G (1992) Wound botulism: clinical and microbiological findings of an Italian case. Eur J Epidemiol 8: 750–752

Barsanti JA, Walser M, Hatheway CL, Bowen JM, Crowell W (1978) Type C botulism in American foxhounds. J Am Vet Med Assoc 172: 809–813

Bengston IA (1924) Studies on organisms concerned as causative factors in botulism. Hyg Lab Bull 136 136:101

Bernard W, Divers TJ, Whitlock RH, Messick J, Tulleners E (1987) Botulism as a sequel to open castration in a horse. J Am Vet Med Assoc 191: 73–74

Bizzini B (1989) Axoplasmic transport and transynaptic movement of tetanus toxin. In: Simpson LL (ed) Botulinum neurotoxin and tetanus toxin. Academic, San Diego

Black RE, Gunn RA (1980) Hypersensitivity reactions associated with botulinal antitoxin. Am J Med 69: 567–570

Cardella MA (1964) Botulinum toxoids. In: Lewis KH, Cassel K jr (eds) Botulism: proceedings of a symposium. US Department of Health, Education, and Welfare, Public Health Service (PHS Publ no 999 FP-1), Cincinnati

Cato EP, George WL, Finegold SM (1986) Genus Clostridium. In: Sneath PHA, Mair NS, Sharpe ME, Holt JG (eds) Bergey's manual of systematic bacteriology, vol 2. Williams and Wilkins, Baltimore

CDC (1979) Botulism in the United States, 1899–1977. Handbook for epidemiologists, clinicians, and laboratory workers. CDC, Atlanta

CDC(1993) Summary of notifiable diseases, United States 1992. MMWR 41(55): 1–65

Chia JK, Clark JB, Ryan CA, Pollack M (1986) Botulism in an adult associated with food-borne intestinal infection with Clostridium botulinum. N Engl J Med 315: 239–241

Ciccarelli AS, Giulietti AM, Giménez GF (1976) Segunda cepa de Clostridium botulinum tipo G. Primer Congresso y IV Jornadas Argentinas de Microbiologia 21 (abstract)
Corsi FM, DiBattista G, Galgani S, Piazza G (1979) Su un caso di botulismo da ferita. Riv Neurobiol 25: 200–205
Critchley EMR (1991) A comparison of human and animal botulism: a review. J R Soc Med 84: 295–298
Davis JB, Mattman LH, Wiley M (1951) Clostridium botulinum in a fatal wound infection. JAMA 146: 646–648
Demarchi J, Mourgues C, Orio J, Prevot AR (1958) Existence du botulisme humain de type D. Bull Acad Natl Med 142: 580–582
DeVeze JL, Jougland J, LaBastie J, Sans P, Arditi J (1979) Botulisme d'inoculation: quand faut-il y penser? Nouv Presse Med 8: 2755–2756
Divers TJ, Bartholomew RC, Messick JB, Whitlock RH, Sweeney RW (1986) Clostridium botulinum type B toxicosis in a herd of cattle and a group of mules. J Am Vet Med Assoc 188: 382–386
Dodds KL (1992a) Clostridium botulinum in the environment. In: Hauschild AHW, Dodds KL (eds) Clostridium botulinum: ecology and control in foods. Dekker, New York
Dodds KL (1992b) Worldwide incidence and ecology of infant botulism. In: Hauschild AHW, Dodds KL (eds) Clostridium botulinum: ecology and control in foods. Dekker, New York
Dowell VRJ, McCroskey LM, Hatheway CL, Lombard GL, Hughes JM, Merson MH (1977) Copro-examination for botulinal toxin and Clostridium botulinum. A new procedure for laboratory diagnosis of botulism. JAMA 238: 1829–1832
Eklund MW, Dowell VR Jr (eds) (1987) Avian botulism: an international perspective. Thomas, Springfield
Eklund MW, Poysky F, Oguma K, Iida H, Inoue K (1987) Relationship of bacteriophages to toxin and hemagglutinin production and its significance in avian botulism outbreaks. In: Eklund MW, Dowell VR Jr (eds) Avian botulism: an international perspective. Thomas, Springfield
English WL, Williams LP, Bryant RE, Gillies MD (1981) Case 48-1980: botulism; letter to the editor. N Engl J Med 304: 789–790
Fullerton P, Gogna NK, Stoddart R (1980) Wound botulism. Med J Aust 1: 662–663
Giménez DF (1984) Clostridium botulinum subtype Ba. Zentralbl Bakteriol Hyg [A] 257: 68–72
Giménez DF, Ciccarelli AS (1970) Another type of Clostridium botulinum. Zentralbl Bakteriol I Abt Orig A 215: 221–224
Griffin PM, Hatheway CL, Rosenbaum RB, Sokolow R (1994) Type A botulism due to intestinal colonization in an adult with Crohn's disease associated with an immune response to the toxin. Manuscript in preparation
Hall JD, McCroskey LM, Pincomb BJ, Hatheway CL (1985) Isolation of an organism resembling Clostridium barati which produces type F botulinal toxin from an infant with botulism. J clin Microbiol 21: 654–655
Hatheway CL (1988) Botulism. In: Balows A, Hausler WH Jr, Ohashi M, Turano A (eds) Laboratory diagnosis of infectious diseases: principles and practice, vol 1. Springer, Berlin Heidelberg, New York
Hatheway CL, McCroskey LM (1987) Examination of feces and serum for diagnosis of infant botulism in 336 patients. J Clin Microbiol 25: 2334–2338
Hatheway CL, McCroskey LM (1989) Unusual neurotoxigenic clostridia recovered from human fecal specimens in the investigation of botulism. In: Hattori T, Ishida Y, Maruyama Y, Morita RY, Uchida A (eds) Proceedings of the 5th International Symposium on Microbiol Ecology: recent advances in microbiol ecology. Japan Scientific Societies, Tokyo
Hatheway CL, McCroskey LM, Lombard GL, Dowell VRJ (1981) Atypical toxin variant of Clostridium botulinum type B associated with infant botulism. J Clin Microbiol 14: 607–611
Hatheway CL, Snyder JD, Seals JE, Edell TA, Lewis GE Jr (1984) Antitoxin levels in botulism patients treated with trivalent equine botulism antitoxin to toxin types A, B, and E. J Infect Dis 150: 407–412
Hauschild AHW (1992) Epidemiology of human foodborne botulism. In: Hauschild AHW, Dodds KL (eds) Clostridium botulinum: ecology and control in foods. Dekker, New York
Hughes JM, Blumenthal JR, Merson MH, Lombard GL, Dowell VR Jr, Gangarosa EJ (1981) Clinical features of type A and B food-borne botulism. Ann Intern Med 95: 442–445
Jensen WI, Price JI (1987) The global importance of type C botulism in wild birds. In: Eklund MW, Dowell VR Jr (eds) Avian botulism: an international perspective. Thomas, Springfield
Johnston J, Whitlock RH (1987) botulism. In: Robinson NE (ed) Current theraphy in equine medicine, 2nd edn. Saunders, Philadelphia
Kempner W (1897) Weiterer Beitrag zur Lehre von der Fleischvergiftung. Das Antitoxin des Botulismus. Z Hyg Infektionskh 26: 481–500

Landmann G (1904) Ueber die Ursache der Darmstaedter Bohnenvergiftung. Hyg Rundsch 14: 449–452
Lee WH, Riemann H (1970a) Correlation of toxic and nontoxic strains of Clostridium botulinum by DNA composition and homology. J Gen Microbiol 60: 117–123
Lee WH, Riemann H (1970b) The genetic relatedness of proteolytic Clostridium botulinum strains. J Gen Microbiol 64: 85–90
Leuchs J (1910) Beitraege zur Kenntnis des Toxins und Antitoxins des Bacillus botulinus. Z Hyg Infektskh 65: 55–84
MacDonald KL, Spengler RF, Hatheway CL, Hargrett NT, Cohen ML (1985) Type A botulism from sauteed onions. JAMA 253: 1275–1278
Mann JM, Hatheway CL, Gardiner TM (1982) Laboratory diagnosis in a large outbreak of type A botulism. Am J Epidemiol 115: 598–605
McCroskey LM, Hatheway CL (1988) Laboratory findings in four cases of adult botulism suggest colonization of the intestinal tract. J Clin Microbiol 26: 1052–1054
McCroskey LM, Hatheway CL, Fenicia L, Pasolini B, Aureli P (1986) Characterization of an organism that produces type E botulinal toxin but which resembles Clostridium butyricum from the feces of an infant with type E botulism. J Clin Microbiol 23: 201–202
McCroskey LM, Hatheway CL, Woodruff BA, Greenberg JA, Jurgenson P (1991) Type F botulism due to neurotoxigenic Clostridium baratii from an unknown source in an adult. J Clin Microbiol 29: 2618–2620
Merson MH, Dowell VR Jr (1973) Epidemiologic, clinical and laboratory aspects of wound botulism. N Engl J Med 289: 1105–1110
Meyer KF, Dubovsky BJ (1922) The distribution of spores of B. botulinus in the United States. J Infect Dis 31: 559–594
Midura TF, Arnon SS (1976) Infant botulism: identification of Clostridium botulinum and its toxin in faeces. Lancet ii: 934–936
Morris JG (1981) Current trends in theraphy of botulism in the United States. In: Lewis GE jr (ed) Biomedical aspects of botulism. Academic New York
Nakamura S, Okado I, Nakashio S, Nishida S (1977) Clostridium sporogenes isolates and their relationship to C. botulinum based on deoxyribonucleic acid reassociation. J Gen Microbiol 100: 395–401
Nakamura S, Kimura I, Yamakawa K, Nishida S (1983) Taxonomic relationships among Clostridium novyi types A and B, Clostridium haemolyticum and Clostridium botulinum type C. J Gen Microbiol 129: 1473–1479
Notermans S, Nagel J (1989) Assays for botulinum and tetanus toxins. In: Simpson LL (ed) Botulinum neurotoxin and tetanus toxin. Academic San Diego
Pickett J, Berg B, Chaplin E, Brunstetter MA (1976) Syndrome of botulism in infancy: clinical and electrophysiologic study. N Engl J Med 295: 770–772
Prevot AR (1953) Rapport d''iintroduction du President du Sous-Comite Clostridium pour l' unification de la nomenclature des types toxigeniques de C. botulinum. Int Bull Bacteriol Nomenclature 3: 120–123
Prevot AR, Terrasse J, Daumail J, Cavaroc M, Riol J, Sillioc R (1955) Existence en France du botulisme humain de type C. Bull Acad Med (Paris) 139: 355–358
Richmond RN, Hatheway CL, Kaufmann AF (1978) Type C botulism in a dog. J Am Vet Med Assoc 173: 202–203
Seals JE, Snyder JD, Edell TA, Hatheway CL, Johnson CJ, Swanson RC, Hughes JM (1981) Restaurant associated type A botulism: transmission by potato salad. Am J Epidemiol 113: 436–444
Sebald M, Jouglard J (1977) Aspects actuels du botulisme. Rev Prat 27: 173–176
Sebald M, Saimot G (1973) Toxémie botulique: intérêt de sa mise en évidence dans le diagnostic du botulisme humain de type B. Ann Microbiol (Inst Pasteur) 124A: 61–69
Seddon HR (1922) Bulbar paralysis in cattle due to the action of a toxicogenic bacillus with a discussion of the relationship of the condition to forage poisoning (botulism). J Comp Pathol Ther 35: 147–190
Smith LDS (1978) The occurrence of Clostridium botulinum and Clostridium tetani in the soil of the United States. Health Lab Sci 15: 74–80
Smith LDS, Sugiyama H (1988) Botulism: the organism, its toxins, the disease, 2nd edn. Thomas, Springfield
Sonnabend O, Sonnabend W, Heinzle R, Sigrist T, Dirnhofer R, Krech U (1981) Isolation of Clostridium botulinum type G and identification of type G botulinal toxin in humans: report of five sudden unexpected deaths. J Infect Dis 143: 22–27
Suen JC, Hatheway CL, Steigerwalt AG, Brenner DJ (1988a) Clostridium argentinense, sp. nov: a genetically homogenous group composed of all strains of Clostridium botulinum toxin type G and some nontoxigenic strains previously identified as Clostridium subterminale or Clostridium hastiforme. Int J Syst Bacteriol 38: 375–381

Suen JC, Hatheway CL, Steigerwalt AG, Brenner DJ (1988b) Genetic confirmation of identities of neurotoxigenic Clostridium baratii and Clostridium butyricum implicated as agents of infant botulism. J Clin Microbiol 26: 2191–2192

Swerczek TW (1980a) Experimentally induced toxicoinfectious botulism in horses and foals. Am J Vet Res 41: 348–350

Swerczek TW (1980b) Toxicoinfectious botulism in foals and adult horses. J Am Vet Med Assoc 176: 217

Tacket CO, Shandera WX, Mann JM, Hargrett NT, Blake PA (1984) Equine antitoxin use and other factors that predict outcome in type A foodborne botulism. Am J Med 76: 794–798

Terranova W, Breman JG, Locey RP, Speck S (1978) Botulism type B: epidemiologic aspects of an extensive outbreak. Am J Epidemiol 108: 150–156

Van Ermengem E (1897) Ueber einen neuen anaeroben Bacillus and seine Beziehungen zum Botulismus. Z Hyg Infektionskh 26: 1–56

Wainwright RB, Heyward WL, Middaugh JP, Hatheway CL, Harpster AP, Bender TR (1988) Foodborne botulism in Alaska, 1947–1985: epidemiology and clinical findings. J Infect Dis 157: 1158–1162

Wang WJ, Wang LP (1983) Wound botulism: report of a case complicating a head injury. Chin J Med 96: 315–317

Weber JT, Goodpasture HC, Alexander H, Werner SB, Hatheway CL, Tauxe RV (1993) Wound botulism in a patient with a tooth abcess: case report and review. Clin Infect Dis 16: 635–639

Woodruff BA, Griffin PM, McCroskey LM, Smart JF, Wainwright RB, Bryant RG, Hutwagner LC, Hatheway CL (1992) Clinical and laboratory comparison of botulism from toxin types A, B, and E in the United States, 1975–1988. J Infect Dis 166: 1281–1286

Cellular Immunology of Tetanus Toxoid

G. CORRADIN[1] and C. WATTS[2]

1 Introduction

Studies on tetanus neurotoxin (TeNT) as an antigen have been informative at a basic level and are continuing to expand our knowledge of what makes an effective immunogen, and how we might be able to exploit the biochemistry and physiology of antigen-presenting cells (APC) to make yet more effective immunogens. This review deals with two related aspects of TeNT as an antigen: (a) immunological responses at the T and B cell level to tetanus in humans and the possibility of exploiting tetanus in the wider context of immunotherapy, and (b) some aspects of the cell biology and biochemistry of TeNT processing in human B lymphoblastoid cells.

2 Immunological Responses to Tetanus Neurotoxin

The detailed analysis of human T and B cell responses to TeNT was undertaken several years ago together with A. Lanzavecchia with a specific set of questions

[1]Institute de Biochimie, Universite de Lausanne, 1066 Epalinges, Switzerland
[2]Department of Biochemistry, Medical Sciences Institute, University of Dundee, Dundee, DD1 4HN, UK

and desired goals in mind: (a) what T cell epitopes are recognized by the human population? (b) what is the specificity of the B cell response and its influence on the activation of specific T cells? (c) how is TeNT processed by human APC? (d) what is the survival time of a specific T cell clone? (e) which defined T cell epitopes can be proposed as T cell carriers for human vaccination? While not all of these questions have been fully answered, the knowledge we have gained from this model system is undoubtedly of importance to an understanding of the immune response to protein antigens in humans. In addition, practical aspects of this research are reflected in the use of TeNT T cell epitopes as carriers for human vaccination and possibly for antitumor therapy.

2.1 Mapping of T Cell Epitopes

The vast availability of TeNT vaccinated donors has certainly facilitated our studies. The TeNT vaccine is administered in a detoxified form incorporated in aluminium hydroxide by coprecipitation and entrapment. The vaccine is prepared by partial purification of TeNT from a clostridial extract, and detoxification is obtained by the addition of formaldehyde which cross-links lysine and tyrosine residues. The detoxified material maintains the overall structure of the parent protein such that antibodies to TeNT cross-react with the native protein. The percentage of TeNT present in these preparations varies but in general does not exceed 50%. The TeNT is adsorbed and entrapped in the aluminium hydroxide. Given the presence of other components in the vaccine preparation, it was not surprising to find that some of the T cell clones obtained were not specific for the purified protein or fragments thereof (Demotz et al. 1989a).

Three epitopes—830–843 (P2), 947–967 (P30), and 1273–1284 (P4)—together with their restriction element were initially identified by a variety of approaches including use of recombinant TeNT fragments, chemical or enzymatic TeNT products, and finally formally confirmed by the corresponding synthetic peptides (Demotz et al. 1989a, b). When these epitopes were used to screen peripheral blood lymphocytes (PBL) from a large panel of donors, a surprising result was obtained. While P4 was recognized only by PBL of individuals carrying major histocompatibility complex (MHC) class II molecules DRB*3.01 or 02 or 03, P2 and P30 could stimulate PBL of all donors tested. P2 could associate with many different DRB*1 alleles and the DRB*3.02 allele while P30 could interact not only with several DR but also with DP molecules (Panina-Bordignon et al. 1989). The reason for this phenomenon was thought to reside in the fact that the α-chain of the DR molecule is almost invariant in the human population, thus rendering more probable the interaction of one peptide with different DR proteins.

At this point the question arose as to whether sequences of these two universal T cell epitopes contain multiple overlapping epitopes specific for each restriction element or a unique epitope. Using different truncated synthetic analogs, it was possible to demonstrate that P2 contains a single epitope while several distinct overlapping epitopes are present in P30 (PANINA-BORDIGNON et al. 1989) as observed for the universal T cell epitope defined in the circumsporozoite protein of *Plasmodium falciparum* (SINIGAGLIA et al. 1988). It was later realized that individuals carrying the allele DR11.04, which presents a valine instead of glycine residue at position 86 of the β-chain, exhibit a greatly reduced or no response to P30 in conjunction with this restriction element. This observation was related to the lower binding of P30 related naturally processed peptides to DR11.04 molecules compared to that observed for DR11.01 protein (DEMOTZ et al. 1993). In addition, it has recently been shown that a differential degradation of TeNT occurs in the cell lines carrying the two alleles (BARBEY et al. 1995). Thus, it is possible that the lack of response is due in part to the Gly/ Val interchange and in part to differential processing of TeNT.

In recent years, several other epitopes have been defined by various groups (see Table 1). In particular, the use of overlapping peptides covering the entire sequence or a large fragment of TeNT which were tested on PBL of different donors not only confirmed the results previously obtained but revealed the existence of additional epitopes (DEMOTZ et al. 1993; REECE et al. 1993). Table 1 summarizes the epitopes thus far defined and the MHC class II molecule association if known.

It was interesting at this point to investigate the chemical properties of p2 and p30 that enabled them to bind to many different class II alleles. For this purpose single alanine substitution was performed for both peptides. Binding to DR11.01

Table 1. TeNT epitopes and their restriction element

Sequence	MHC class II	Reference
73–93	n.d.	ETLINGER et al. 1990
141–171	n.d.	REECE et al. 1993
257–268	n.d.	REECE et al. 1993
580–599	Promiscous	HO et al. 1990
616–631	n.d.	REECE et al. 1993
640–651	n.d.	REECE et al. 1993
652–663	n.d.	REECE et al. 1993
830–843	DR1, DR4, DW4 DR11(5), DR9 DR13(6), DEW52b	DEMOTZ et al. 1989a PANINA-BORDIGNON et al. 1989
916–932	Promiscous	HO et al. 1990
947–967	DR11(5), DR7 DR9, DPW2 DPW4	DEMOTZ et al. 1989a PANINA-BORDIGNON et al. 1989
1084–1099	DR11(5)	DEMOTZ et al. 1993
1124–1139	DR11(5)	DEMOTZ et al. 1993
1174–1189	DR11(9)	DEMOTZ et al. 1993
1273–1284	DRW52a, 52c	DEMOTZ et al. 1989a PANINA-BORDIGNON et al. 1989

n.d., Not determined

and 11.04 was studied using a functional competition assay (VALMORI et al. 1994a). It was found that binding was not profoundly affected by substitution at any position of the two peptides for the two alleles tested. These results probably reflect the structural characteristics of class II molecules and the nature of the interaction with bound peptides (BROWN et al. 1993; STERN et al. 1994). In class II molecules protein pockets accommodating specific side chains may be more shallow and/or less discriminatory than in class I molecules. It may be argued that the capacity of P2 and P30 and of antigenic peptides in general to bind to MHC class II molecules can be attributed to avoidance of deleterious interactions between peptide side chains and class II molecules rather than to the presence of specific anchor residues as found for class I molecules (BOEHNCKE et al. 1993).

Our findings are in agreement with a recent report suggesting that strict allele-specific binding motifs may not exist for class II molecules (CHICZ et al. 1993). On the other hand, they are at variance with other studies which measured direct binding of purified class II molecules and P2 peptide analogs. This discrepancy in the results obtained by various groups can be attributed to the different assays used, to the use of analogs in which nonconservative substitutions were performed, and/or suboptimal peptide length (O'SULLIVAN et al. 1990, 1991). In particular, direct binding analysis of P2 single alanine substituted analogs with purified DR 6 molecules indicated that Ile 832, Asn 835, and Gly 840 may be important for the interaction. In this case, the Ile side chain may interact directly with a pocket present in DR 6 molecules while the role of Gly 840 is thought to be structural allowing the peptide to adopt the preferred conformation in binding to DR 6 molecules (BOITEL et al. 1995). Finally, analysis of the clonal size and persistence of given clonotypes specific for P2 indicates that the response to this peptide is very small in size and short lived (DELLABONA et al. 1993).

2.2 Use of Tetanus Neurotoxin T Cell Epitopes as T Cell Carriers

It is well established that weak antigens or haptens require the presence of a carrier molecule for efficient stimulation of an antibody response. The carrier molecule provides the T cell helper epitopes necessary for the stimulation of carrier specific T cells which in turn stimulate hapten-specific B cells (MITCHISON 1971). The second advantage in the use of a carrier molecule is the lack of genetic restriction usually encountered for small molecules such as peptides or haptens. On the other hand, carrier molecules generally elicit a vigorous anti-carrier antibody response. In addition, previous priming with a carrier molecule followed by immunization with the same carrier conjugated to a hapten may result in suppression of the antihapten response (HERZENBERG et al. 1980). This suppression is thought to be mediated by carrier specific B cells and/or antibodies. It is therefore clear that this suppression phenomenon may hamper the use of intact TeNT as a carrier for human vaccination and prevents taking full advantage of the fact that a large fraction of the world population is vaccinated with TeNT.

It was of course of considerable practical interest to define TeNT epitopes for their use as T cell carriers. ETLINGER et al. (1990) showed that such an approach can indeed avoid the problem of the suppression mentioned above. Multiple universal T cell epitopes present in TeNT were also extremely attractive for use as universal carriers. For this purpose, a series of synthetic constructs were designed including P2 and P30 together or alone and a B cell epitope derived from the immunodominant repetitive region of different plasmodia (VALMORI et al. 1992). These epitopes were colinearly synthesised and assembled as multiple peptide antigens (MAP) which were shown to be more immunogenic than the linear construct (TAM 1988) In fact, high antibody levels were obtained in all cases in different strains of mice (VALMORI et al. 1992). Some of these constructs conferred a high degree of protection against a subsequent parasite challenge (WANG et al. 1995). A human phase I/II trial is underway to determine the efficacy of TeNT t.t. MAP constructs in the human population. A similar approach was also tested using single-chain linear peptides and P2 as a carrier (KUMAR et al 1992).

TeNT epitopes were also examined for their capacity to increase the cytotoxic T cell response (CTL) to otherwise weak CTL epitopes. The rationale for these experiments was that since CTL require interleukin 2 for activation and proliferation, TeNT epitopes might provide this interleukin 2. This has indeed been confirmed to be correct (WIDMAN et al. 1992; VALMORI et al. 1994b). Coinjection of P2 or P30 with a number of CTL peptides resulted in a dramatic increase in the CTL response specific to the CTL peptides. Clinical studies using these or similar formulations are now underway.

2.3 Use of Tetanus Neurotoxin Epitopes in Cancer Therapy

Another possible and interesting spin-off of this research is the use of universal TeNT T cell epitopes for antitumor therapy. In this case a universal T cell epitope is coupled to B cell lymphoma anti-idiotypic antibody. This antibody, when injected in vivo or used in vitro, targets the TeNT epitopes on the cell lymphoma which then presents the TeNT epitope to specific T cells. Giving the presence of cytotoxic $CD4^{+}$ T cells within the population of TeNT peptide specific T cells, the B cell lymphoma is hopefully killed or prevented from proliferating. The feasibility of this procedure has been tested in vitro in a human system using specific anti-human IgG antibodies to which P2 was coupled (YU et al. 1994). A high degree of killing of B cell lymphoma target cells was observed at a conjugate concentration 100- to 1000-fold lower than that of peptide alone. This approach can be now extended to other tumor-specific markers.

3 Cell Biology of Tetanus Neurotoxin Processing

The phenomenon of antigen processing and presentation on class II MHC has recently attracted the interest of cell biologists interested in cellular protein and membrane traffic pathways. T cells provide a very sensitive assay for processing and transport pathways which may be minor and not detectable using conventional biochemical assays. For example, it is usually considered that material entering cells by endocytosis is either completely degraded in lysosomes or recycled intact to the cell surface. The phenomenon of antigen presentation to $CD4^+$ T cells demonstrates an alternative whereby partially degraded material is returned to the cell surface bound to class II MHC molecules. Different receptors are used for the uptake of intact antigen (membrane Ig or Fc receptors) and for the efflux of processed antigen (class II MHC).

We have been able to follow the fate of antigen by exploiting TeNT-specific human B cell lines immortalized with Epstein-Barr virus (EBV) (LANZAVECCHIA 1985). These cells have allowed an analysis of antigen processing following receptor-mediated uptake of small amounts of highly radiolabeled antigen concentrated on cell surface IgG receptors (WATTS et al. 1989). Similar B cell lines which are both MHC class II positive and membrane Ig positive cannot be easily generated at present in the murine system because of the lack of an EBV equivalent suitable for immortalisation. Consequently the range of antigens to which EBV immortalized lines can be isolated is limited to the normal range of human immunogens. Negligible binding of TeNT to non-specific EBV B cells occurs indicating that the gangliosides responsible for toxin binding to neuronal cells are not expressed on B lymphocytes. The majority of our studies have used TeNT rather than the toxoid to facilitate biochemical analysis of processing. In the subsequent sections "antigen" refers to TeNT unless stated otherwise.

3.1 Antigen Endocytosis and Processing

TeNT-specific human B cells endocytose monovalent antigen rapidly via clathrin-coated pits (WATTS and DAVIDSON 1988; WATTS et al. 1989). In fact uptake of the surface IgG occurs constitutively, i.e., it is not triggered by antigen binding (DAVIDSON et al. 1990) and is therefore similar to the behavior of other cell surface receptors such as those for transferrin and low-density lipoprotein. Recent studies in the murine system also indicate constitutive uptake of membrane Ig molecules although the fraction of molecules found within endosomes at steady state (versus the cell surface) is less in this system (PATEL and NEUBERGER 1993). Some 15–20 min after antigen internalization proteolysis of antigen becomes detectable either as antigen fragments on Tris-tricine gels or as acid-soluble radioactivity returned to the medium (DAVIDSON and WATTS 1989). Indirect data from several laboratories indicates that endocytosis of antigen alone is not sufficient for processing, and that some kind of post endocytic sorting step is also needed to target antigen efficiently to the processing and class II MHC loading compartment. This might, for example, be induced simply by receptor cross-linking which is known to impede Fc receptor recycling (MELLMAN and PLUTNER 1984). However, since monovalent antigen is efficiently processed and presented in both the human and murine B cells, cross-linking is clearly not obligatory and suggests that other mechanisms operate to target antigen to the peptide loading compartment. These as yet uncharacterized postendocytic sorting steps appear more efficient when the IgG is associated with antigen since in EBV cells the lifetime of occupied molecules is some five fold shorter than unoccupied (3 versus 15 h; DAVIDSON et al. 1990).

Analysis of the TeNT fragments generated within human EBV B cells revealed that B cell clones recognizing different epitopes in the TeNT molecule generated distinct fragments. It was shown that this was due to protection of distinct regions of the antigen by the B cell immunoglobulin during processing; in other words, the substrate for processing in B cells expressing high-affinity antibodies is a complex of antigen and antibody and the different substrates arising from different antibody specificities give rise to differential processing patterns (DAVIDSON and WATTS 1989). Recent studies reveal that this can have striking effects on the outcome of processing at the level of T cell epitope presentation (see below).

3.2 Loading and Stability of Class II MHC Molecules

Maturation of class II molecules along the biosynthetic pathway is characterized by loss of the invariant chain and acquisition of a dimeric state resistant to dissociation in sodium dodecyl sulfate (SDS) at 20°C (reviewed in CRESSWELL 1994). Several laboratories have shown that this remarkable stability of the non-covalently linked αβ dimer is due to bound peptide, although instability in SDS does not necessarily signify lack of peptide (LANZAVECCHIA et al. 1992). Acquisition of SDS stability has proven a very useful assay for a key step in the assembly of class II MHC with peptide. In EBV cells specific for TeNT uptake and processing of radiolabeled antigen allowed a direct assay to be developed for the formation of SDS-stable αβ dimers labeled via the peptide moiety rather than by biosynthetic labeling of the α and β chains themselves. We exploited this assay to demonstrate that newly synthesized rather than recycled class II MHC molecules are used to load processed peptides (DAVIDSON et al. 1991). More recently the compartments involved in assembling class II MHC processed peptides have been analyzed using this assay. Two different approaches were used to demonstrate that newly assembled class II MHC carrying processed TeNT peptides are located neither in early or late endosomes nor in conventional lysosomes but apparently in a distinct endosomal population (WEST et al. 1994). Parallel analysis of the trafficking of TeNT gold particles showed that these gained access to a compartment similar if not identical to that observed earlier in non-antigen-specific EBV cells (PETERS et al. 1991). These structures resembled multivesicular bodies and were rich in class II MHC. Taken together with work from other laboratories the current consensus is that assembly of class II MHC probably takes place in a specialized endosome segregated from those which fulfill more general "housekeeping" functions. A specialized site may help resolve the paradox of both proteolytic processing and recycling to the cell surface being accommodated within the endocytic pathway.

MHC/peptide complexes are first detected on the cell surface 45–60 min after antigen uptake, measured using either T cell clones (ROOSNEK et al. 1988) or direct surface biotinylation (S. MOORE, unpublished). Studies using both T cell clones specific for TeNT epitopes and direct biochemical measurements show that, once formed, most peptide/MHC complexes are stable, and that their longevity is determined by the lifetime of the class II MHC molecules themselves (LANZAVECCHIA et al. 1992). On EBV-transformed B cells this is about 36 h. Consequently over several days the potency of TeNT pulsed B cells decreases, but in some cases T cell stimulation is measurable even 10 days later.

3.3 Influence of Antibodies on Antigen Processing

As discussed above, high-affinity antibodies can affect the outcome of processing by protecting the region of antigen around the B cell epitope. Particularly striking are large peptides of up to 16 kDa in size seen in two B cell clones both specific for different conformational epitopes in the TeNT C fragment (DAVIDSON and WATTS 1989; DEMOTZ et al. 1989a). The size of these fragments suggested that antibody binding exerts a stabilizing influence beyond the actual contact surface with antigen, and we asked whether T cell epitopes within these protected peptides are equally well loaded on class II MHC in the presence and absence of the bound antibody, in other words, whether the final outcome of antigen processing is affected by the antibodies which drive antigen capture.

To answer this question we derived in collaboration with A. Lanzavecchia, a large panel of T cell clones specific for various epitopes in TeNT and examined the ability of these clones to be triggered when different TeNT/antibody complexes are the substrates for processing. We found that in some cases very striking effects can be seen which are either suppressive in nature, i.e., antibody binding can prevent the loading of some T cell epitopes (WATTS and LANZAVECCHIA 1993) or in other cases are stimulatory, i.e., antibody binding can boost the loading of T cell epitopes (SIMITSEK et al. 1995). A particularly striking aspect of these experiments is that a single monoclonal anti-C fragment antibody (11.3) simultaneously boosts the presentation of one T cell epitope (1273–1284) while suppressing another (1174–1189). Importantly, the effects of antibodies such as these are dominant in the sense that the suppressive effect on presentation can override the enhancing effects on antigen uptake which antibodies mediate either as surface receptors on B cells or via Fc receptors on other cells (WATTS and

LANZAVECCHIA 1993). A summary of the effects on two different antibodies reactive with the C fragment of tetanus toxin on three different T cell epitopes is shown schematically in Fig. 1. In the light of these results in the TeNT system it seems possible that the frequently observed dominance of a few T cell epitopes in an immune response may partially be due to prevailing antibodies.

Precisely how these boosting and suppressive effects occur during antigen processing is not completely clear since the conformational epitope recognized by the 11.3 antibody (which produces these effects) has not been mapped. However, a protein "footprinting" experiment has begun to shed some light on the mechanism. Digestion in vitro of C fragment/11.3 complexes with trypsin or lysosomal enzymes yields a 17-kDa fragment protected by binding to the 11.3 antibody. The N-terminus of this fragment has been sequenced following the incorporation of radiolabeled amino acids and reveals that the suppressed epitope (1174–1189) falls within at least one of the protected fragments and so is sequestered, thus preventing MHC binding (SIMITSEK et al., submitted). Most likely the boosted epitope (1273–1184) also lies in the protected fragment. Conceivably the increased lifetime of this part of the C fragment results in better presentation of some epitopes, but this can be overridden by steric hindrance of peptide loading by antibody in some cases. Using similar B cell lines, other studies have shown that C3b bound to complement receptor CR1 can engage in

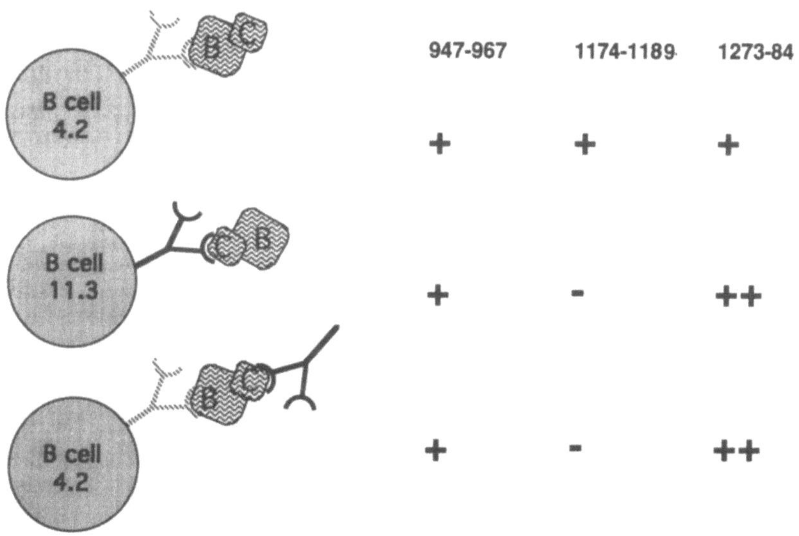

Fig. 1. Antibody effects on presentation of TeNT C fragment epitopes. The effect of membrane immunoglobulin and exogenously added "piggy back" antibody on three different T cell epitopes from the C fragment (865–1315) is shown. B and C fragment domains of TeNT are indicated. Presentation or lack thereof is indicated by +or –. B cell 4.2 is able to present all epitopes, but B cell 11.3 which recognizes the TeNT C fragment cannot present the 1174–1189 epitope except at high antigen concentrations where fluid phase uptake is significant. Conversely, 11.3 cells present the 1273–1284 epitope better than other B cell clones. Both the suppresive and boosting effects can be reproduced by adding the soluble 11.3 antibody as a "piggy back" to antigen bound to 4.2 cells, confirming that is the Ig which controls these effects

disulfide bond formation with TeNT bound to the same cells via membrane Ig (Jacquier et al. 1993). This raises the possibility that complement fragments as well as antibodies might modulate processing of antigens following uptake by antigen presenting cells.

Acknowledgments. Work in the authors laboratories is supported by the Swiss Science Foundation (GC) and by the Wellcome Trust and Medical Research Council (CW).

References

Anderson MD, Fairweather N, Charles IG, Emsley P, Isaacs NW, MacDermott G (1993) Crystallographic characterisation of tetanus toxin fragment C. J Mol Biol 230: 673–674

Barbey C, Watts C, Corradin G (1995) Antigen-processing organelles from DRB1*1101 and DRB1*1104 B cell lines display a differential degradation activity. Eur J Immunol 25: 30–36

Boehncke WH, Takeshita T, Pendleton CD, Houghten RA, Sadegh-Nasseri S, Racoippi L, Berzofsky JA, Germain R (1993) The importance of dominant negative effects of amino acid side chain substitution in peptide-MHC molecule interaction and T cell recognition. J Immunol 150: 331–341

Boitel B, Blank U, Mége D, Corradin G, Sidney J, Sette A, Acuto O (1995) Strong similarity in antigen fine specificity among DRB1*1302 restricted tetanus toxin tt830-843-specific TCRs in spite of highly heterogenous CDR3. J Immunol (in press)

Brown JH, Jardetzky TS, Gorga JC, Stern LJ, Urban RG, Strominger JL, Wyley DC (1993) Three-dimensional structure of the human class II histocompatibility antigen HLA-DR1. Nature 364: 33–39

Chicz RM, Urban RG, Gorga JC, Vignali DAA, Lane WS, Strominger JL (1993) Specificity and promiscuity among processed peptides bound to HLA-DR alleles. J Exp Med 178: 27–47

Cresswell P (1994) Assembly, transport and function of MHC class II MHC molecules. Annu Rev Immunol 12: 259–293

Davidson HW, Watts C (1989) Epitope-directed processing of specific antigen by B lymphocytes. J Cell Biol 109: 85–92

Davidson HW, West MA, Watts C (1990) Endocytosis, intracellular trafficking and processing of membrane IgG and monovalent antigen/membrane IgG complexes in B lymphocytes. J Immunol 144: 4101–4109

Davidson HW, Reid PA, Lanzavecchia A, Watts C (1991) Processed antigen binds to newly synthesised class II MHC molecules in antigen specific B lymphocytes. Cell 67: 105–116

Dellabona P, Casorati G, Friedli B, Angman L, Sallusto F, Tunnacliffe A, Roosneek E, Lanzavecchia A (1993) In vivo persistence of expanded clones specific for bacterial antigens within the human T cell receptor CD4-8-subset. J Exp Med 177: 1763–1771

Demotz S, Lanzavecchia A, Eisel U, Niemann H, Widmann C, Corradin G (1989a) Delineation of several DR-restricted tetanus toxin T cell epitopes. J Immunol 142: 394–402

Demotz S, Matricardi P, Lanzavecchia A, Corradin G (1989b) A novel and simple procedure for determining T cell epitopes in protein antigens. J Immunol Methods 122: 67–72

Demotz S, Barbey C, Corradin G, Amoroso A, Lanzavecchia A (1993) The set of naturally processed peptides displayed by DR molecules is tuned by polymorphism of residue 86. Eur J Immunol 23: 425–432

Etlinger HM, Gillissen D, Lahm HW, Matile H, Sconfeld HJ, Trzeciak A (1990) Use of prior vaccinations for the development of new vaccines. Science 249: 423–425

Herzenberg LA, Tokuhisa T, Herzenberg LA (1980) Carrier-priming leads to hapten-specific suppression. Nature 285: 664–667

Ho PC, Mutch DA, Winkel KD, Saul AJ, Jones GL, Doran TJ, Rzepczyk CM (1990) Identification of two promiscous T cell epitopes from tetanus toxin. Eur J Immunol 20: 477–483

Jacquier MR, Gabert FM, Villiers CL, Colomb MG (1993) Disulphide linkage between C3b and tetanus toxin on tetanus toxin specific EBV-transformed B cells. J Immunol 150: 4253–4260

Kumar A, Arora R, Kaur P, Chauhan VS, Sharma P (1992) "Universal" T helper cell determinants enhance immunogenicity of a Plasmodium falciparum merozoite surface antigen peptide. J Immunol 148: 1499–1505

Lanzavecchia A (1985) Antigen-specific interaction between T and B cells. Nature 314: 537–539

Lanzavecchia A, Reid PA, Watts C (1992) Irreversible association of peptides with class II MHC molecules in living cells. Nature 357: 249–252
Makoff AJ, Ballantine SP, Smallwood A, Fairweather NF (1989) Expression of tetanus toxin fragment C in E. coli: its purification and potential use as a vaccine. Biotechnology 7: 1043–1046
Mellman I, Plutner H (1984) Internalization and degradation of macrophage Fc receptors bound to polyvalent immune complexes. J Cell Biol 98: 1170
Mitchison NA (1971) The carrier effect in the secondary response to hapten-protein conjugates. I. Measurement of the effect with transferred cells and objection to the local environment hypothesis. Eur J Immunol 1: 10–17
O'Sullivan D, Sidney J, Appella E, Walker L, Phillips L, Colon SM, Miles C, Chesnut RW, Sette A (1990) Characterization of the specificity of peptide binding to four DR haplotypes. J Immunol 145: 1799–1808
O'Sullivan D, Arrhenius T, Sidney J, Del Guerico MF, Albertson M, Wall M, Oseroff C, Southwood S, Colon SM, Gaeta F, Sette A (1991) On the interaction of promiscuous antigenic peptides with different DR alleles. J Immunol 147: 2663–2669
Panina-Bordignon P, Tan A, Termijtelen A, Demotz S, Corradin G, Lanzavecchia A (1989) Universally immunogenic T cell epitopes: promiscuous binding to human MHC class II and promiscuous recognition by T cells. Eur J Immunol 19: 2237–2240
Patel KJ, Neuberger MS (1993) Antigen presentation by the B cell antigen receptor is driven by the α/β sheath and occurs independently of its cytoplasmic tyrosine residues. Cell 74: 939
Peters PJ, Neefjes JJ, Oorschot V, Ploegh HL, Geuze HJ (1991) Segregation of MHC class II molecules from MHC class I molecules in the golgi complex for transport to lysosomal compartments. Nature 349: 669–676
Reece JC, Geysen HM, Rodda SJ (1993) Mapping the major human T cell epitopes of tetanus toxin. J Immunol 151: 6175–6184
Roosnek E, Demotz S, Corradin G, Lanzavecchia A (1988) Kinetics of MHC-antigen complex formation on antigen-presenting cells. J Immunol 140: 4079–4082
Simitzek PD, Campbell DG, Lanzavecchia A, Fairweather N, Watts C (1995) Modulation of antigen processing by bound antibodies can boost or suppress Class II MHC presentation of different T celll determinants. J Exp Med (in press)
Sinigaglia F, Guttinger M, Kilgus J, Moran DM, Matile H, Etlinger HM, Trzeciak A, Gillesen D, Pink JR (1988) A malaria T-cell epitope recognized in association with most mouse and human MHC class II molecules. Nature 336: 778–780
Stern LJ, Brown JH, Jardetzsky TS, Gorga JC, Urban RG, Strominger JL, Wyley DC (1994) Crystal structure of the human class II MHC protein complexed with an influenza virus peptide. Nature 368: 215–221
Tam JP (1988) Synthetic peptide vaccine design: synthesis and properties of a high-density multiple antigenic peptide system. Proc Natl Acad Sci USA 85: 5409–5431
Valmori D, Pessi A, Bianchi E, Corradin G (1992) Use of human universally antigenic tetanus toxin T cell epitopes as carriers for human vaccination. J Immunol 149: 717–721
Valmori D, Sabbatini A, Lanzavecchia A, Corradin G, Matricardi PM (1994a) Functional analysis of two universal T cell epitopes in their interaction with DR1101 and DR1104 alleles. J Immunol 152: 2921–2929
Valmori D, Romero J, Men Y, Maryanski JM, Romero P, Corradin G (1994b) Induction of a cytotoxic T cell response by co-injection of a T helper peptide and a cytotoxic T lymphocyte peptide in incomplete Freund's adjuvant: further enhancement by pre-injection of IFA alone. Eur J Immunol 25: 1458–1462
Wang R, Charonenvit Y, Corradin G, Porrozzi R, Hunter RL, Glenn G, Alving CR, Church P, Hoffman SL (1995) Induction of protective polyclonal antibodies by immunization with a *Plasmodium yoelii* circumsporozoite protein multiple antigen peptide vaccine. J Immunol 154: 2784–2792
Watts C, Davidson HW (1988) Endocytosis and recycling of specific antigen in human B cell lines. EMBO J 7: 1937–1945
Watts C, Lanzavecchia A (1993) Suppressive effect of antibody on processing of T cell epitopes. J Exp Med 178: 1459–1463
Watts C, West MA, Reid PA, Davidson HW (1989) Processing of immunoglobulin-associated antigen in B lymphocytes. Cold Spring Harb Symp Quant Biol 1: 345–352
West MA, Lucocq JM, Watts C (1994) Antigen processing and class II MHC peptide loading compartments in human B-lymphoblastoid cells. Nature 369: 147–151
Widmann C, Romero P, Maryanski JL, Corradin G, Valmori D (1992) T helper epitopes enhance the cytotoxic response of mice immunized with MHC class I-restricted malaria peptides. J Immunol Meth 155: 95–99
Yu Z, Healy F, Valmori D, Escobar P, Corradin G, Mach JP (1994) Peptide-antobody conjugates for tumor therapy: a MHC-class II-restricted tetanus toxin peptide coupled to an anti-Ig light chain antibody can induce cytotoxic lysis of a human B-cell lymphoma by specific CD4 T cells. Int J Cancer 56: 244–248

Immunodiagnosis and Immunotherapy of Tetanus and Botulinum Neurotoxins

J.L. Middlebrook and J.E. Brown

1 Introduction

Tetanus and botulism are diseases or poisonings that are highly dramatic in their presentations. They are almost always life threatening and therefore command attention and concern by those around the victim. In earlier times help might have been sought from the shaman or witch doctor, whose spells or potions provided little comfort from the symptoms and eventual death. As modern medicine developed in the last two centuries even well-trained physicians were unable to do much more than a shaman until the advent of immunotherapy in the late 1800s and immunization in the 1920s. Little has changed since then, and we still depend on active immunity or passive immunotherapy to treat both of these highly dangerous poisonings. Immunological techniques and reagents also provide the means of toxin identification, and a timely and definitive diagnosis therefore depends heavily on their use. This chapter reviews the past and current literature on vaccines and various antibody-based products for treatment or prevention of poisonings by both botulinum and tetanus neurotoxins. We also review the techniques employed for their detection and identification.

Toxinology Division, U.S. Army Medical Research Institute of Infectious Diseases, Frederick, MD 21702, USA

2 Tetanus and Botulinum Vaccines

There are few success stories in medical science to match that involving vaccination against tetanus. Tetanus is an age-old disease that is usually fatal.

Vaccination is extremely effective at providing complete and long-lasting immunity, although boosters are required to assure full protection. Tetanus toxoid is one of the earliest of vaccines to have been developed and has been used on a world-wide basis in hundreds of millions of humans and animals. The vaccine is well known as a low cost, safe, and efficacious material. However, as with any drug or biological agent, there is a finite incidence of undesirable side effects and therefore room for improvement of the product by the application of advanced technologies. Moreover, in developing nations even relatively inexpensive vaccines may be unaffordable, and the public health problems of tetanus still reach horrible proportions. Thus, lowering the cost of the vaccine, even without an increase in efficacy, is also a desirable goal.

Botulism is a much more recent disease (poisoning) of concern and became a significant problem only when food processing developed into a widespread human activity. For reasons that are not clear, the incidence of wound botulism, the counterpart of tetanus, is a rare occurrence. Thus, although the causative organisms, *Clostridium botulinum* and *C. tetani,* are both soil organisms and found around the world, it was the latter that presented a far greater natural infection hazard to humans during the early advancement of civilization. As food production moved towards the making of sausage and cured but uncooked meats in the 1800s, more and more outbreaks of botulism poisoning occurred. The problem became more severe as vegetable, fruit, and meat canning, both commercial and home, emerged as popular activities. However, unlike the potential for tetanus, the possibility of contracting botulism could be reduced to negligible levels by adhering to careful canning or preserving techniques, thus avoiding the expense and trouble of immunizing large populations. Therefore, although there is a botulinum toxin vaccine, it has not been used on the scale of tetanus toxoid, and we do not have the vast clinical data set that exists for the latter.

2.1 Tetanus Vaccine

After the identification of a neurotoxin as the cause of clincal tetanus, it became clear that immunization against the toxin should confer protection from the infection. Following the demonstration by RAMON in 1923 that formaldehyde treatment of diphtheria toxin produces a suitable toxoid, RAMON and ZVELLER successfully used formaldehyde-treated tetanus toxin to immunize human volunteers (WIRZ et al. 1990). While it was not immediately apparent how effective this immunization could be, the British and American experience in World War II compared to that of Axis forces presented compelling evidence of efficacy (BOYD 1959). Nearly all of the developed countries in the world now include tetanus vaccination as a routine part of childhood preventative health treatment. Tetanus has ceased to be an appreciable public health problem in those countries, except for the elderly. In a recent evaluation of 117 tetanus cases in the United States 58% of the victims were 60 years of age or older (PREVOTS et al. 1992). Another study based in Europe (DIMACHE et al. 1990) found that 28% of elderly people (> 60 years) who had not received a booster in 10 years were unprotected. Nearly all of these individuals responded to a single immunization and developed a protective titer. These and other studies strongly suggest that persons over the age of 60 should receive a tetanus vaccine booster as a prophylactic measure.

The production of tetanus toxoid is a relatively straightforward process. Using a certified toxigenic strain of *C. tetani*, the organism is cultured in a fermentor under established conditions (WHO Ex Comm 1977). During the late stages of fermentation and growth, the bacteria lyze and release toxin into the medium. At that point it is possible to inactivate the toxin and safely purify it as a nontoxic entity. However, doing so probably results in heterocross-linking and subsequent copurification of bacterial proteins other than neurotoxin. Alternatively, the native (highly toxic) toxin is processed to an acceptable level of purity and then inactivated by chemical treatment. Inactivation is usually accomplished by several weeks' incubation with 0.5% formaldehyde. Most commercial producers of tetanus vaccine bottle the material after adsorption onto the adjuvant aluminum phosphate (alum), although fluid toxoid is also available (Physicians Desk Reference). Various manufacturers have their own recommendations for administration of the vaccine, but the basic course for human neonates is two or three primary immunizations 4–8 weeks apart followed by a booster after about 1 year. Further boosters are probably advisable over the course of one's life, but individual titers vary so much as to make specific recommendations on the time interval meaningless. Common practice is to give a booster to patients presenting with a wound if they have not been immunized within the past 5–10 years. As mentioned above, it is probably advisable to give tetanus boosters to the elderly as a part of preventative health care.

The current methodology for production of tetanus vaccine is fundamentally unchanged from that developed 50 years ago. Probably the greatest impediment to further advances was the lack of details on the primary structure of tetanus toxin. However, there were some investigations into the relationship of gross structure and immunogenicity. First, it is necessary to recall the basic structure of tetanus toxin. As described in detail elsewhere in this volume, tetanus toxin is synthesized in culture as a single-chain protein. However, it is usually isolated as a dichain molecule as a result of bacterial protease nicking at a trypsin-sensitive site (CRAVEN and DAWSON 1973; MATSUDA and YONEDA 1975). This cleavage produces two fragments, known as the heavy (H) and light (L) chains, with molecular weights of 100 000 and 50 000, respectively. These fragments are held together in the absence of reducing agents by an intrachain disulfide. Following reduction it is possible to purify the L and H chains, although it is particularly difficult to obtain the latter free from contaminating full-sized (and fully lethal) toxin.

HELTING and ZWISLER (1977) found that treatment of tetanus toxin with papain releases a polypeptide of approximately 50 000 molecular weight and correctly deduced that the fragment corresponds to the carboxy terminal half of the H chain; they termed this polypeptide "fragment C" (new nomenclature H_c; NIEMANN 1992). The remainder of the toxin (termed "fragment B"; new nomenclature L-H_N) was composed of the amino terminal half of the H chain and the full-sized L chain, still covalently connected by the disulfide mentioned above. HELTING and ZWISLER (1977) also reported that vaccination of animals with either fragment H_c or L-H_N conferred protection from subsequent challenge by native toxin. Further work using monoclonal antibodies has demonstrated that neutralizing epitopes exist on each of the major structural regions of tetanus toxin, the L, H_N, and H_C chains (KENIMER et al. 1983; VOLK et al. 1984).

Our knowledge of the primary structure of tetanus toxin received an enormous boost with the cloning and sequencing of the structural gene for the toxin (EISEL et al. 1986; FAIRWEATHER et al. 1986; FAIRWEATHER and LYNESS 1986). With this information, a rational approach to the development of new vaccines could be undertaken; such an effort was pursued by a group at Wellcome Biotech Ltd. Initially they studied the expression of large fragments of the toxin in *Escherichia coli*. Since the H_C had been shown to be nontoxic and induce protection after vaccination, early efforts focused on that or related polypeptides. Two recombinant plasmids were constructed, one of which expressed a fusion protein consisting of nearly all of the H_C and the *E. coli* trpE protein, while the other expressed 121 residues of the H_N and all of the H_C (FAIRWEATHER et al. 1987). Both of the products were expressed in low yield and insoluble form, but each conferred protection when used as an immunogen in mice (FAIRWEATHER et al. 1987).

Surmising that the low yield and insolubility was a problem associated with expression of a heterologous protein in *E. coli*, the group moved on to try additional constructs. One (pTETtac1) encoded a methionine residue, the first three residues of interferon-γ and $H_{537-1315}$, or approximately two-thirds of the H_N and all of the H_C fragment. This construct produced a protein of the correct size (90 kDa) in *E. coli* cultures, but the material was largely insoluble. Another (pTETtac2) encoded only the H_C with a methionine on the N-terminus. Expressed in *E. coli*, this material was largely soluble and of the correct size. Production was about 12 mg/l cell culture or approximately 3%–4% of total cell protein. Vaccination of mice one time with as little as 0.2 µg H_C conferred protection from 100 LD_{50} tetanus toxin. Confirming evidence that recombinant H_C of tetanus toxin is a good immunogen was obtained from a second research group as a part of their study to make biologically functional H_C (HALPERN et al. 1990).

In an effort to increase the production levels of H_C synthesis in *E. coli* the Wellcome group moved on to analyze the natural gene and its expression in that heterologous system. They observed that the *C. tetani* derived gene is very A-T rich and contains a large number of codons rarely used in *E. coli* (MAKOFF et al. 1989a,b). In a stepwise fashion the structural gene for the H_C was converted to a form having a lower A-T content and no codons rare to *E. coli*. At various stages in the conversion, the level of transcription/translation was examined, and the data were consistent with the notion that expression of product is limited by the availability of rare tRNAs. The final construct (pTATtac115) led to the production of soluble product at the level of 11%–14% of the total cell protein (MAKOFF et al. 1989a,b).

Other expression systems were studied with the aim of making high levels of soluble H_C material suitable for use in humans. One approach was the synthesis of H_C in insect cells using the baculovirus system (CHARLES et al. 1991). Using the native gene construct of pTETtac2, a fragment encoding the structural gene for H_C, the *trp*D ribosome binding site and the *tac* promoter was cloned into the baculovirus expression vector p36C. After cotransfection of *Spodoptera frugiperda* cells with *Autographa californica* nuclear polyhedrosis virus, several recombinants were obtained, and one, BVFC1 was studied further. BVFC-1-infected *S. frugiperda* cells produced a soluble 50-kDa protein that was recognized in immunoblotting by H_C antibody. The material was produced at the level of 10%–12% of total cell protein. After purification this recombinant protein was used to immunize mice and evaluate protection. Mice immunized once with 0.25 μg survived a challenge of 50 LD_{50} tetanus toxin, whereas mice immunized twice with only 0.06 μg survived a similar challenge dose.

Yeast were also evaluated as expression systems for H_C. Two distinct systems were studied, *Saccharomyces cerevisiae* and *Pichia pastoris*. For work with *S. cerevisiae* several versions of the H_C gene were cloned into the expression vector pWYG7 (ROMANOS et al. 1991). When DNA from the plasmid pTETtac2 was used as the structural gene for H_C, no protein was detected in expression experiments, and no mRNA large enough to code for the full-sized protein was found. When DNA from the plasmid with entirely synthetic sequences (pTETtac15) was used, full-sized mRNA was detected, as was full-sized, intracellular H_C protein. Levels of expression were about 2%–3% of total soluble protein. Apparently the rare codon usage or the high A-T content of the natural gene is unsuitable for expression in *S. cerevisiae*. Examination of the culture supernatants from pWYG7 with the synthetic H_C gene revealed a diffuse high molecular weight band, a major band at 65 kDa, and fainter bands smaller than 65 kDa. Treatment of the preparation with endo-H led to conversion of all species to bands of about 50 kDa. These data suggest that extracellular H_C produced by *S. cerevisiae* had been glycosylated at multiple sites.

Immunization with *S. cerevisiae*, expressed H_C presented interesting results (ROMANOS et al. 1991). Mice were vaccinated once with different doses of H_C and then challenged with 100 LD_{50} tetanus toxin. Intracellular H_C was as effective as and possibly more effective than *E. coli* produced H_C at eliciting protection.

Extracellular H_C, as directly isolated, did not induce antibody levels sufficient to protect from the challenge dose of toxin. However, when the extracellular H_C was deglycosylated with endo-H, protection comparable to the intracellular H_C was obtained. Apparently, glycosylation of H_C somehow prevents immune system recognition or processing as is normally accomplished with the non-posttranslationally modified polypeptide.

Expression of H_C in *P. pastoris* was much higher than that achieved in *S. cerevisiae*. Because of experience with the latter strain of yeast, only the synthetic gene for HC (TET 15) was used in the *P. pastoris* work (CLARE et al. 1991). An advantage of this system is the ability to introduce multiple copies of the desired gene and attain ultrahigh levels of expression. Strains that had single copies of the synthetic H_C gene expressed the encoded protein at about 6%–8% of total cell protein in fermentations, higher than in the *S. cerevisiae* system but not dramatically so. However, one isolate that appeared to have 14 copies of the H_C gene, produced H_C at levels of 27% of cell protein, or about 12 g/l culture. This amount of protein is phenomenally high and, since it was stated that *P. pastoris* produced H_C was as effective as native H_C as an immunogen, indicates this system is a good candidate for production of a new generation of tetanus vaccine.

Administration of protein-based vaccines has proven safe and effective in many cases. One drawback is the necessity to give several vaccinations to attain protective and long-lasting titers. Especially in developing countries, getting patients to return several times to a clinic to complete the full course of vaccinations is a frequent and serious problem. If it were possible to give a vaccine that would confer immunity after one administration, much better preventative health care should be realized. One approach to this problem has been the use of live vector-produced antigens. Multiple systems are currently under development, and it is not yet clear whether they will work for any or all antigens. However, tetanus vaccination is clearly a candidate for such an approach and has been studied by several laboratories.

One of the earliest proposed live vector systems was *Salmonella typhi* (MASKELL et al. 1987). Using pTETtac2, which carries the natural gene for H_C, FAIRWEATHER et al. (1990) engineered an *S. typhi* mutant to express H_C in culture independent of the *tac* promoter. Mice were immunized either orally or intravenously with this mutant. Mice immunized with *E. coli* produced H_C served as positive controls, and the parent *S. typhi* strain as negative controls. One oral dose of *S. typhi* with H_C insert (approx. 10^9 organisms) resulted in eight of ten animals surviving a challenge of 500 LD_{50} tetanus toxin, while two doses provided complete protection from a similar challenge. Vaccination by the intravenous route with the same construct resulted in complete protection after either one or two administrations. In both instances the *S. typhi* controls all died, even from as low a challenge as 5 LD_{50} toxin. Controls of *E. coli* produced H_C (10 μg) also provided complete protection from 500 LD_{50} toxin. Similar doses of heat-killed H_C-producing organisms did not lead to a protective response, an indication that replication is required to produce the observed immunity.

Another laboratory studied the feasibility of using bacille Calmette-Guérin (BCG), a live attenuated tubercle bacillus used to immunize humans against tuberculosis, as a live carrier of the H_C protein (STOVER et al. 1991). The authors of the work pointed out that this organism comes with a vast clinical experience of safety for human use, having been given to over 2 billion persons worldwide. The natural gene for H_C was cloned into an extrachromosomal vector (pmv261) that had been shown to express another reporter gene in BCG. When given to mice, the BCG carrier induced antibody to H_C, as measured by enzyme-linked immunosorbent assay (ELISA). Control vaccinations of BCG without vector or transformed BCG that had been heat killed produced markedly lower (perhaps background) ELISA signals. An important question not examined in this study was the potential neutralizing activity of the antibodies stimulated by this live carrier. Thus, the potential of this carrier as a delivery vehicle for tetanus vaccination is yet to be proven.

Another live carrier system examined for potential in tetanus immunization is *Lactococcus lactis*, a food-grade bacterium (WELLS et al. 1993). Using a plasmid encoding the natural gene for H_C (HALPERN et al. 1990), the polymerase chain reaction was applied to clone the structural gene into a construct where it would be expressed fused to the first 11 amino acids of the T7 bacteriophage gene 10 protein. Transferred to *L. lactis* for expression studies, this vector produced a protein of the correct molecular weight (approx. 50 kDa) on polyacrylamide gel electrophoresis that immunoblotted with antibody to H_C. Mice vaccinated subcutaneously with the transfected strain were protected from challenge with tetanus toxin, while animals receiving the parent strain of *L. lactis* did not survive. The longer term aim of this group was apparently to develop the *L. lactis* system to a form suitable for oral administration.

2.2 Botulinum Vaccine

The botulinum toxin toxoid currently available was developed much later than was tetanus toxoid and has been used in far fewer humans. However, it is still regarded to be a safe vaccine. Its efficacy has been evaluated only by surrogate tests since exposure to the toxin is a rare event, and it would not be feasible to do a prospective study. However, to the authors' knowledge, no immunized person, either research or clinical laboratory worker, has ever exhibited systems of poisoning following accidental (potential) exposure during the course of their work.

The earliest published work on making a botulinum toxoid was in the 1930s by the Russian scientist VELIKANOV (1934). His was apparently the first effort to immunize humans. The earliest efforts in humans the United States were production and testing of a toxoid against serotypes A and B (REAMES et al. 1947). The toxoid was a filtered culture autolysate to which formalin was added. The material lost toxicity after 2–3 weeks and was refiltered and bottled as a fluid toxoid. With slight variations, another lot of fluid toxoid was prepared and adsorbed to alum to make an alum-precipitated toxoid. Several hundred human volunteers were then immunized in various regimens with these two types of toxoids, and their seroresponses were followed as a function of time. Each serum sample was evaluated by a mouse neutralization test, which various dilutions of the test serum were incubated with botulinum toxin and then injected into mice. The results indicated that a minimum of two immunizations were required to obtain a high percentage of responders, and that at least 4 weeks should elapse (perhaps as many as 8 weeks) before the second injection is administered.

Because there were a significant number of undesirable local and systemic reactions to the crude toxin-based toxoid, some effort was spent to devise a simple, large-scale purification scheme for both serotypes A and B (Duff et al. 1957a,b). The schemes were very similar and consisted of acid precipitation of toxin from the culture supernatants, $CaCl_2$ extraction of the precipitate, a second acid precipitation, and a final alcohol precipitation. Since this work was carried out before the availability of such analytical techniques as gel electrophoresis, analytical ultracentrifugation was the only technique available to determine purity. By that criterion the material appeared to be pure and has been referred to many times as "crystalline toxin." Today, we know that "crystalline toxin" is only about 10%–15% pure 150 kDa neurotoxin. However, it is clear by the data provided that the scheme did remove a considerable number of extraneous materials. The conditions necessary to detoxify both crystalline A and B toxins were investigated in detail (Wright et al. 1960). Basically, 2–3 weeks in 0.6% formalin made a stable toxoid that elicited immunity in mice and guinea pigs. This toxoid appeared to be stable for at least 36 months, and it was decided to test these improved toxoids in humans.

Purified toxin-based toxoids were prepared in two large lots by Parke, Davis and Company. The toxoids containing approximately the same amount of protein were adsorbed to aluminum phosphate and mixed to produce a bivalent vaccine. Four vaccination schemes were undertaken; 0–2–4–6, 0–8, 0–2–10, and 0–10 weeks (Fiock et al. 1962). The results for serotype A toxoid indicated that the 0–2–10 week schedule was markedly superior to the others in producing rapid immunity. The 0–2–10 week schedule was also the best for serotype B, but only marginally. Moreover, it was clear that higher titers could be achieved against serotype A than B, despite there being the same amount of antigen in the preparations. In contrast, if single boosts were given 1 and 2 years after the above initial series, the 0–2–4–6 scheme produced a better long-term immunity than did the initial series of 0–2–10. Nevertheless, 100% of the recipients were judged to have a satisfactory response 8 weeks after their 1-year booster regardless of the initial regimen.

Since the bivalent toxoid seemed to work well in humans, and because there were additional toxin serotypes of concern, there was a subsequent effort to produce and test a pentavalent toxoid (Fiock et al. 1963). Using partially purified toxins prepared similarly to the bivalent preparations above, toxin serotypes A–E were toxoided and mixed to make a pentavalent vaccine. The immunization schedule selected was 0–2–10 weeks with a booster at 52 weeks. There was a clear immune response to each serotype at 12 weeks. By 52 weeks titers were very low or undetectable, but after the boost individuals exhibited a robust increase in titers to levels approximately tenfold those observed at 12 weeks. By most measures, the poorest response was to serotype B. Immunization with this antigen elicited the lowest percentage of individuals exhibiting measurable titers and, in those who responded titers were usually lower than with the other serotypes. Serotype E toxoid produced slightly better results than serotype B. Responses to serotypes A, C, and D were good. Following this work large amounts of components for the pentavalent vaccine were prepared by the Michigan State Department of Public Health using technology similar to that above. These preparations have served as the supply of botulinum toxin vaccine in the United States until the present time.

Scientifically speaking, there is even more need to improve the botulinum vaccine than that for tetanus toxin. The vaccine is more reactogenic than tetanus toxoid, the supplies of botulinum vaccine are dwindling, and it is very expensive to prepare. We have undertaken a research effort at our Institute to meet this need. Our approach is patterned after that of the Wellcome group with tetanus toxin. Although the work is far from complete and represents a much greater problem than tetanus (seven serotypes of botulinum toxins vs. one of tetanus toxin), significant progress has been made. Much of the work is as yet unpublished, and it is described only, briefly at this point.

We began by working with the natural gene for botulinum toxin serotype A. Using plasmid pCBA3, which encodes most of the natural gene sequence for botulinum toxin H chain (Thompson et al. 1990), we amplified the structural gene by PCR to code for what we defined as the H_C, residues 861–1296. After gel purification and restriction endonuclease treatment the DNA was cloned into the vector pMAL as a fusion protein. As expressed in *E. coli* cultures, this construct produced ELISA-positive material, but not to a concentration that a band could be identified by sodium dodecyl sulfate (SDS) gel electrophoresis (LaPenotiere et al. 1993). Nevertheless, crude material from cultures was used to vaccinate mice; similar material from pMAL vector without H_C insert expressed in *E. coli* was used as a control. After one vaccination mice receiving crude H_C were protected (5/5) from 3 LD_{50} of native toxin, while all the control animals (3/3) died (LaPenotiere et al. 1993). Challenge with higher doses of toxin resulted in survival of some of the animals. These results encouraged us to move on and attempt synthesis of a synthetic gene for botulinum toxin H_C.

Our approach to construction of a synthetic gene for H_C was to design the gene with single codon preferences optimal for *E. coli*. Then some of the codons were altered to less preferred triplets to reduce the A-T content or to position restriction sites so the entire gene could be divided into five similarly sized segments. Each segment was constructed by annealing synthetic oligomers 60–70 nucleotides in length using overlaps of five or six nucleotides. These segments were then ligated and cloned into vectors, amplified, and cut out for assembly into a vector with the appropriate multiple cloning site. The assembled gene was sequenced, and several errors were detected. The gene was cloned into a vector suitable for in vitro mutagenesis (pMUT, Promega), and corrections to the sequence were made. Following sequencing and confirmation of fidelity, the gene was placed in the vector pBR-322, where expression was under the control of the tac promoter. Expressed in *E. coli*, this construct produced immunoreactive material that was identified as a 50-kDa protein by immunoblotting after SDS gel electrophoresis. The protein, however, was largely insoluble, unlike the experience described above with tetanus H_C. Crude material was used to vaccinate mice at the intervals of 0, 2, and 4 weeks. Mice challenged on week 5 survived 1 000 000 LD_{50} serotype A botulinum toxin, indicating that our construct produces a highly effective immunogen. Further work is under way to design an expression system that will produce soluble H_C in *E. coli*.

Because of the success with tetanus H_C expression in baculovirus, we attempted expression of our synthetic botulinum serotype A H_C gene in that system. We employed a commercially available construct designed to produce an excreted product (pAcGP67 B, PharmMingen). Expression of our H_C gene in baculovirus-infected *S. frugiperda* cell did indeed produce a soluble protein of the correct size, that was recognized in immunoblots with serotype A-specific antisera (M. Clayton and J.L. Middlebrook, unpublished data). This same crude preparation was used to vaccinate mice by same regime described above. As with the insoluble product, soluble H_C produced in the baculovirus expression system protected mice from 1 000 000 LD_{50} toxin. Further experiments designed to maximize yield of this soluble material are now under way.

In the authors' viewpoint, the use of vaccines to protect against botulinum and tetanus toxins is coming to an important crossroads. The feasibility of making a new generation of products with several desirable features has now been demonstrated. Using the research and development techniques described above, it will be possible to make improved vaccines that are safer and cheaper to produce. At least in the case of botulinum toxin, preliminary experiments indicate that the H_C is a much more effective antigen than is the current toxoid, and many more doses could be prepared from the same amount of protein. A decision to move forward and use these products in place of the toxoids is easy to reach from the scientific and cost standpoint.

3 Immunotherapy of Tetanus and Botulism

3.1 Tetanus Toxin Poisoning

Passive antibody treatment is an approach used to control or suppress the symptoms of several diseases. Antibody against tetanus for this purpose has been available for decades. Initially, these products were of bovine or equine origin. Basically, they were prepared by immunizing animals with tetanus toxoid and then collecting the sera for therapeutic use. Severe allergic reactions were commonplace (BARDENWARPER 1962), and attempts were made to minimize their incidence by purification of immunoglobulins, so as to reduce the overall foreign protein load to the patient. Further advances were made by enzymatic removal of the most reactogenic region of the antibody, the Fc fragment. However, nearly all of these problems were eliminated by the production of tetanus immune globulin of human origin, which is the product of choice today. This material is prepared from large pools of plasma obtained from tetanus toxoid immunized individuals. The large batches are titered against tetanus toxin in comparison with a United States standard antitoxin and bottled with a known neutralization capacity measured as an international unit (Physicians Desk Reference).

While both the equine and human products are effective in treating tetanus, the human immune globulin seems to be equally effective at lower doses. NATION et al. (1963) felt that they had obtained results comparable to their previous experience with the equine antisera when treating tetanus patients with lower doses of human immunoglobulin. In a double-blind trial of the human and equine products as a treatment for neonatal tetanus (MCCRACKEN et al. 1971) no significant differences in survival were noted when 1/20 of the human vis a vis equine international units were used. It is still not clear what the minimal or optimal dose for treatment really is. There are probably so many factors involved when the patients present as to make such a determination almost impossible.

If there were to be significant improvements in immunotherapy of tetanus, the most likely advance would be a monoclonal antibody. A significant number of studies have been published on the development of tetanus toxin-specific monoclonal antibody libraries, primarily from two viewpoints. One was use of the mouse-derived libraries to probe structure-function relationships of the toxin, while the other had the avowed goal of developing clinically useful antibodies, usually of human origin. Currently this distinction has become blurred in that the undesirable side effects of a mouse antibody can probably be overcome by the use of molecular genetics for conversion to one of several possible new antibody forms, such as a single-chain antibody or mouse-human graft (HODGSON 1991).

At least two groups have prepared mouse monoclonal antibody libraries against tetanus toxin. In the earlier effort (KENIMER et al. 1983) tetanus toxoid or fragment L-H_N were used as immunogens. Almost 100 isolates were obtained, from which 14 were chosen for further study. Two of the monoclonal antibodies were against the H_C, nine against the L chain, and the remaining three against to the H_N. Only two of these antibodies neutralized the toxin, one specific for the H_C and the other for the H_N. In a separate effort VOLK et al. (1984) also used toxoid or fragments to immunize mice and obtained a total of 57 hybridomas producing antibodies to the toxin. They determined that ten of these antibodies neutralized the toxin, six specific for the H_N and four for the H_C. In total, the results of these two groups suggest that the important neutralizing epitopes reside on the H chain.

A greater number of efforts at tetanus toxin neutralizing monoclonal antibody development have been performed with humans as the B cell donors. The first reported isolation of a neutralizing human monoclonal antibody came from a mouse-human heteromyleloma fusion (GIGLIOTTI and INSEL 1982). The antibody was used to neutralize 1–2 LD_{50} toxin; insufficient data were reported to make a stoichiometric ratio calculation. Another of the earliest efforts used a human lymphoblastoid cell fusion partner and peripheral B cells from tetanus toxoid immunized humans to isolate at least one neutralizing antibody (LARRICK et al. 1983). Making some reasonable assumptions from the data presented, one can calculate that a minimum 50-fold molar excess of monoclonal antibody was required to protect mice from approx. 1000 LD_{50}. ZIEGLER-HEITBROCK et al. (1986) also used a human fusion partner to produce two monoclonal antibodies. Neither of the antibodies could neutralize the toxin alone, but produced a delay in the time to death with mice. When used together, the monoclonal antibodies did provide protection, but only from low doses of tetanus toxin. TRAUBAUD et al. (1989) obtained two human monoclonal antibodies to tetanus toxin using B cells from recently toxoid-boosted humans. One of the two antibodies protected mice from a low challenge (two lethal doses) of toxin while the two in combination were synergistic, protecting against five lethal doses.

SIMPSON et al. (1990) isolated a single monoclonal antibody which had a very high affinity (4.2×10^{-10} mol/l) for tetanus toxin. The monoclonal antibody appeared to neutralize tetanus toxin. However, when animals were observed past the customary 4- to 5-day time period for neutralization studies, some of them died. The deaths occurred in a group of animals given dilute antibody that would have been reported as completely protected in most studies. This important observation makes one wonder about the results of many other laboratories that isolated monoclonal antibodies and observed weak but complete protection after 4–5 days. Likewise, ARUNACHALAM et al. (1992) obtained a monoclonal antibody capable of neutralizing tetanus toxin by itself. This antibody would protect from up to 500 LD_{50} and a Fab derived from the antibody protected as well, although only against low (2 LD_{50}) challenge doses.

Two groups have developed human monoclonal antibodies which appear to be promising as clinically useful products. KAMEI et al. (1990) prepared mouse-human hybrid cell lines, one of which produced a monoclonal antibody with a neutralizing potency of 4.3 IU/100 μg protein. Using a combination of three monoclonal antibodies, KAMEI et al. (1990) were able to achieve a marked synergy in neutralization, reaching a value of 43 IU/100 μg protein. At this level of efficacy only 0.6 mg of protein would be required to reach the recommended 250 IU for patient administration compared to a usual dose of more than 100 mg human immune globulin. LANG et al. (1993) also obtained human monoclonal antibodies expressing high neutralizing titers. For reasons not clear, a majority of their isolates were against the L chain of tetanus toxin. The optimal combination of monoclonal antibodies produced a cocktail with a potency of 43 IU/100 μg protein, similar to that of KAMEI et al. However, LANG et al. (1993) found that most neutralizing antibodies in human antitetanus immune globulin were directed to the L chain. Moreover, all of the neutralizing monoclonal antibodies which they obtained were against the L chain. These findings are in opposition to nearly every other study in the field of both botulinum and tetanus toxin immunology. Many laboratories have isolated H chain directed monoclonal antibodies which neutralize the toxins. Immunization with genetically engineered fragment C from either tetanus or botulinum toxin clearly confers very good protection in animal studies (not yet in humans), demonstrating that antibodies against the L chain are not required for good immunity. The reason for this paradox remains unclear.

The development of monoclonal antibodies to the clostridial neurotoxins has been problematic and mysterious. Basically, after many years of effort by several fine laboratories, we still do not have a good, efficacious neutralizing monoclonal antibody against tetanus or any of the seven serotypes of botulinum toxin. In all cases in which the published data were sufficient to make a calculation, it was found that a large molar excess (20 x in the best cases) of any single monoclonal antibody was required to bring about complete protection from even modest doses of tetanus or botulinum toxins (J.L. MIDDLEBROOK, unpublished calculations). This situation is not inherent with monoclonal antibodies to protein neurotoxins as it is possible to isolate monoclonal antibodies that approach the theoretical limit of 1:2 for neutralization of phospholipase A_2 neurotoxins (KAISER and MIDDLEBROOK 1988; MIDDLEBROOK 1991). Paradoxically, the toxoids used to immunize either the mice or humans serving as the cell source for hybridoma production are the same toxoids used to produce good levels of immunity in humans. Perhaps antibody binding to multiple epitopes is required for highly efficient neutralization. Clearly some of the studies cited above provide some evidence for this possibility. In that event it is highly improbable that any single monoclonal antibody will serve well as a therapeutic product. Alternatively, perhaps there is some complication of immune processing of the (toxoid) antigen used to immunize both humans and animals. The extensive formaldehyde treatment clearly inactivates the biological activity of the toxin. What has happened to the structure and conformation of the molecule, and what are the consequences of those changes as regards immune system recognition and processing? Perhaps the availability of genetically engineered nontoxic fragments of both botulinum and tetanus neurotoxins will be used as immunogens and some of these questions can be addressed.

3.2 Botulinum Toxin Poisoning

The early history of botulinum antitoxins for therapy has been reviewed (COOPER 1964). Antibotulinum sera was first produced by KEMPNER in 1897 and by hyperimmunization of horses by LEUCHS in 1910. Efficacy of the use of equine botulinum immune globulin (BIG) was examined in a retrospective study (TACKET et al. 1984). Of 132 cases of type A foodborne poisoning from 1973 to 1980, patients who had received trivalent equine antitoxin had a lower fatality rate and a shorter illness than those who did not receive antitoxin, even after controlling for age and incubation period. Patients who received antitoxin in the first 24 h after onset had a shorter course than those who received antitoxin later. These results suggest that trivalent antitoxin has a beneficial effect on survival and shortens the course of patients with type A botulism. Antitoxin levels and half-lives have been estimated in patients receiving IM/IV trivalent equine BIG (HATHEWAY et al. 1984). In four patients the average peak serum levels of neutralizing activity were 5.2, 3.1, and 2.2 IU/ml for A, B, and E antitoxins, respectively. In one patient the half-life was estimated as 6.5, 7.6, and 5.3 days for antitoxin types A, B, and E, respectively. For immunization purposes, an protective antiserum level is accepted as 0.025 IU/ml, suggesting that initial levels are much higher than needed to neutralize botulinum toxin measured in serum of patients.

Use of equine BIG has been debated because of the possibility of adverse reactions. From 1967 through 1977 the Centers for Disease Control monitored reactions to bivalent (A, B) or trivalent (A, B, E) equine BIG (BLACK and GUNN 1980). Of 268 patients 9% had adverse reactions, with a 2% rate of anaphylaxis and no fatalities. Serum sickness was observed more frequently in those given more than 40 ml of product ($p < 0.02$), equivalent to al least 30 000 IU type A antitoxin. Several factors may have contributed to the observed rate of adverse reactions. The amount of equine protein per dose was not available, the product was antitoxin extracted from pepsin-treated serum, not purified $F(ab')_2$, and the dose may have been 50-fold greater than that required to achieve protective serum levels. The authors concluded that the rate of anaphylaxis was higher than for most other equine products, but that the rate of serum sickness was comparable to that of other enzyme-treated equine products. Equine BIG prepared in Japan and the United States have been compared because of the claim that no case of serum sickness has been observed with the Japanese antitoxin product (LAYTON et al. 1972). The Japanese material appeared to be less contaminated with other serum proteins, probably because of processing by heat coagulation of the pepsin-treated serum in the presence of ammonium sulfate to remove nonimmunoglobulin serum proteins.

More recently, the incidence of adverse reactions to equine rabies immune globulin (RIG) has been examined in a large number of patients. In a retrospective study of 3156 patients in Thailand, adverse reactions were observed in 1.6%, with only one case of anaphylaxis (WILDE et al. 1989b). There were serum sickness-like reactions in fewer than 0.1% of Thai children under 10 years, who may have had less exposure to horse products. In prospective studies the authors evaluated the safety of equine RIG from several manufacturers. The adverse reaction rate with equine RIG from Pasteur Vaccins was less than 1% in 500 patients (WILDE et al. 1987), that from Swiss Vaccine and Serum Institute was 6.2% in 323 cases with no cases of anaphylaxis (WILDE et al. 1989a), and that from Sclavo was 3.6% in 419 patients with one case of steroid therapy (WILDE and CHUTIVONGSE 1990).

The obvious advantages of a human BIG over a horse product led to an effort to establish a supply of human botulism immune plasma from 80 volunteers who received a single pentavalent toxoid booster 7–8 years after previous immunizations (LEWIS and METZGER 1978, 1980). Initially over 500 l plasma was collected with a total amount of 4500, 250, and 1300 IU against types A, B, and E, respectively (LEWIS and METZGER 1980). An additional 2000 l plasma was collected, from which a pilot lot of 52 l was processed to human BIG by the Michigan Department of Public Health. Final titers were above 250, 10, and 80 IU/ml for types A, B, and E toxins, respectively (LEWIS 1981). In 1982, under Bureau of Biologics IND 1332, the remaining plasma was processed to immune globulin by SiO_2 treatment and anion exchange chromatography at the Minnesota ALG Program Laboratory, University of Minnesota (ARNON 1993). Four lots were obtained with approximately the same titer as in the pilot batch and no loss in potency after 6 years of storage. These lots have been tested in phase I safety and pharmacokinetics studies at our Institute, with no adverse effects (D. Hack, unpublished observations). When administered intravenously, a 10-ml dose gave serum titers above 1 IU/ml at 8 h. All volunteers maintained titers greater than 0.02 IU/ml for at least 35 days.

Guinea pigs were given 20 IU human BIG plasma per kilogram either 4 h before or 4 or 8 h after an oral challenge of type A toxin, and all survived with no clinical signs (LEWIS and METZGER 1980). Moreover, the processed human BIG was completely protective (8/8) when given prophylactically to nonhuman primates subsequently challenged by inhalation with botulinum toxin A (1200–3200 mouse LD_{50}/kg), including an animal whose serum titers had declined to undetectable levels (< 0.02 IU/ml; FRANZ et al. 1993). LEWIS and METZGER (1978) suggested that human BIG would be highly desirable for the treatment of infant botulism. In 1991 Arnon's group began a clinical trial testing the efficacy of the Army human BIG for the treatment of infant botulism (FRANKOVICH and ARNON 1991). The trial was postponed because international developments forced the Army to retain its supply of human BIG (SCHWARZ and ARNON 1992). Thereafter, in collaboration with the Massachusetts Public Health Biologic Laboratories, and under the aegis of the Food and Drug Administration Office of Orphan Drug Products, ARNON (1993) established a lot of human botulinum immune globulin from California volunteers (BB-IND-4283) and initiated a randomized, double-blinded, placebo-controlled study. The initial protocol was designed to produce serum levels of 0.125 IU/ml for type A antitoxin and 0.025 IU/ml for type B antitoxin. This prospective study should help establish the efficacy of immunotherapy for human botulism.

Equine BIG has been shown to be efficacious in animal challenge studies as well as in the retrospective analysis of human cases mentioned above. Guinea pigs treated with 1 IU/kg trivalent botulism antitoxin were completely protected from subcutaneous toxin challenge, although protection decreased when antibody was given postchallenge (LEWIS and METZGER 1980). In the early 1980s the University of Minnesota ALG Program, under a contract from the United States Army, prepared heptavalent (A-G) equine antitoxin from the serum of a thoroughbred horse initially immunized with toxoid, then with multiple boosts of toxin. In mice the purified $F(ab')_2$ product (15 mg/kg) was 100% protective from toxin challenge for 7 days after treatment. It was also 100% protective when given within 2 h after a highly lethal intraperitoneal challenge, but decreased to 10% by 12 h postchallenge (M.A. Crumrine, unpublished data). In other experiments with guinea pigs the efficacy of $F(ab')_2$ therapy also decreased with time postchallenge by either the intravenous or inhalation route (M.A. Crumrine, unpublished data). With nonhuman primates, equine BIG $F(ab')_2$ was highly protective (7/8) when given 24 h after inhalation exposure to toxin doses of 30–300 monkey LD_{50} (D.R. Franz, unpublished data). Moreover, therapeutic efficacy was observed in a subgroup (4/4) given one-tenth the recommended human dose on a per kilogram basis (14 IU/kg). In contrast, no animals (0/8) survived when challenged with 150 monkey LD_{50} and given equine BIG at first signs, although four of the animals received four times the human dose on a weight basis. In an earlier study, three of five animals exposed to 50 monkey LD_{50} survived when given equine BIG at first signs without attempts at supportive care (Franz et al. 1993).

Supportive care does improve efficacy of botulinum antibody therapy in monkeys (Oberst et al. 1967), suggesting the possibility of a broad window of therapeutic opportunity if the dose of toxin is not overwhelming. Under BB-IND-3703 the equine $F(ab')_2$ BIG product was tested in a phase I safety and pharmacokinetic trial at our Institute in five volunteers (Sjogren et al. 1992). Individuals were given a dose containing the following amounts of antitoxin: 10 200 IU type A, 4800 IU type B, 1500 IU type C, 72 000 IU type D, 8000 IU type E, 880 type F, and 3200 IU type G. They were observed in the hospital for 48 h and thereafter for 4 months. No evidence of serum sickness was seen in these volunteers, either clinically or in complement cascade tests. Type A titers were above 2.5 IU/ml within 2 h and remained detectable (> 0.02 IU/ml) for at least 21 days, indicating a serum half-life similar to that of trivalent botulinum antitoxin (Hatheway et al. 1984; Lewis and Metzger 1980). In a 1991 Cairo outbreak of type E botulism (Weber et al. 1993) several vials of this product were requested and reported as safely administered to patients.

In December 1990 the United States Army initiated a program at our Institute to produce a heptavalent (type A-G) equine BIG containing purified $F(ab')_2$. Horses were immunized with a single toxin serotype to allow blending of antitoxins to yield equivalent neutralization titers to all toxin types. The initial series of three immunizations at 0, 3, and 7 weeks used formalin-inactivated toxoid equivalent to 2, 0.5, and 0.5 mg neurotoxin per animal, respectively, injected subcutaneously with at least ten injection sites per immunization (Brown et al. 1991). After 9 weeks median titers were : type A, 1370; type B, 140; type E, 620; and type F, 200 IU/ml. Thereafter, animals were boosted at regular intervals with active toxin (0.1 mg per horse) so that antibodies to formalin-sensitive as well as toxoid epitopes should be produced. This project continues with the goal of minimizing the possibility of adverse reactions. The final product will contain $F(ab')_2$ purified by anion exchange chromatography after pepsin treatment of equine IgG isolated by protein G chromatography. Blending components so that the final mixture contains 10 000 IU/dose for each serotype minimizes excess immunoglobulin of any one serotype.

Decreasing the dose of equine BIG $F(ab')_2$ should be considered as a strategy to minimize the possibility of adverse reactions. The current recommendations call for 10 000 IU per dose (about 140 IU/kg), and several doses are usually administered. More than a decade ago FELDMAN stated that the dose of equine antitoxin given is far in excess of what might be needed and called for measures to assess the efficacy of botulinum antitoxin (comments in METZGER and LEWIS 1979). At that time METZGER and LEWIS (1979) reported that guinea pigs given homologous antisera survived a challenge with botulinum toxin 2 weeks later although they had no detectable titers. More recently FRANZ et al. (1993) noted survival of 3/3 nonhuman primates given 14 IU/kg equine BIG F(ab')2 48 h before a highly lethal aerosol challenge although serum antitoxin titers were less than or equal 0.02 IU/ml at time of challenge. This level was effective therapeutically as well, since all four monkeys given 14 IU/kg equine BIG $F(ab')_2$ 24 h after challenge with 6–43 LD_{50} survived (D.R. FRANZ, unpublished data).

Probably the manner of intoxication should also be considered in establishing a dose of immune globulin. An infectious process (e.g., infant botulism) should be treated with sufficient equine BIG to block toxin absorption for the time needed to clear the intestinal colonization. In contrast, environmental exposures, such as laboratory accidents or biowarfare incidents, would require antitoxin to neutralize toxin only from the acute exposure. For all types except E, 1 IU should neutralize 10 000 mouse IP LD_{50}; therefore 10 000 IU IV would neutralize approximately 0.1–1 mg botulinum toxin or at least 10 ml type A culture supernatant entering the circulation, an acute exposure that would be extremely unlikely, except in the most dire laboratory accident.

4 Tetanus and Botulinum Toxin Immunodiagnosis

4.1 Tetanus Toxin

As with botulinum toxin, immunodiagnostic tests would prove useful for both the detection of toxin in patients in whom tetanus is suspected and for the detection of antitetanus toxin levels in wound patients, in sero-epidemiological studies, hyperimmune globulin production efforts, and vaccine potency testing. Several recent reviews have dealt with tetanus immunity in animals (BIZZINI 1993) and with tetanus immune globulin (HABIG and TANKERSLEY 1991), Reviews are also available on earlier immunoassay procedures, including ELISA (MELVILLE-SMITH 1990), and on a detailed assessment of immunity in the Danish population (SIMONSEN 1989).

Because of the clinical distinctiveness of tetanus, and because of the extremely small amount required for intoxication, few efforts have been devoted to the development of immunodiagnostic tests. As the disease has become less prevalent in the developed world due to universal immunization, physicians may have become less confident with its diagnosis (Bleck 1991). In Finland 22% of patients with tetanus were not clinically recognized on presentation (Luisto 1990). Early diagnosis would improve prospects for effective treatment. ELISA procedures for detection of toxin were reported that had a sensitivity of 100 ng/ml using horse antitetanus antibody (Pal et al. 1990) and 1.2 ng/ml using a combination of monoclonal antibodies (Kiessig et al. 1991). However, detection limits at the picogram level are probably necessary.

Detection of levels of antitetanus toxin antibody in patient sera has become more useful for diagnosis in cases of suspected tetanus as the degree of universal immunization increases. The World Health Assembly has adopted the goal of eliminating neonatal tetanus by 1995. In 1992 there were an estimated 787 000 neonatal tetanus deaths, a 20% reduction from 1988 (WHO 1993). Surveillance of tetanus immune status in populations and rapid assessment of tetanus immunity in wound patients will help to further decrease risks of tetanus. Based on experimental data in guinea pigs and clinical experience in horses, the accepted protective serum level is 0.01 IU/ml (McComb 1964). Immunoassays should be able accurately to detect antitetanus antobody levels to at least that level.

Measurements of tetanus antibody levels has traditionally relied on neutralization of mouse lethality (Council of Europe 1981, 1985; Barile et al. 1970). Immunodiagnostic procedures tested as replacement measures have included indirect hemagglutination (Beytout et al. 1989; Reiss et al. 1991; Rey et al. 1978; Nyerges and Lutter 1980; Peel 1980a), double immunodiffusion (Mouillet et al. 1978), immunoelectrophoresis (Virella et al. 1978; Rubin et al. 1981), and radioimmunoassay (Stevens 1978; Layton 1980). Stiffler-Rosenberg and Fey (1977) reported the first use of ELISA to estimate levels of anti-tetanus antibody, using tetanus toxoid adsorbed to the solid phase, and obtained titers comparable to those from a radioimmunosorbent assay. Layton (1980) measured 0.005 IU/ml in human serum or plasma and obtained a correlation coefficient of $r = 0.92$ compared to a solid-phase radioimmunoassay.

Several authors have compared ELISA with the lethality neutralization assay. Melville-Smith (1990), using toxoid adsorbed to solid phase, analyzed 80 human sera and obtained good correlation in sera with neutralization titers above 0.03 IU/ml; four sera not protective in vivo were positive by ELISA. Gentili et al. (1984) found that ELISA correlated well with neutralization in the analysis of 12 human tetanus immunoglobulin preparations($r = 0.84$) and also for titer measurement of 171 human serum specimens ($t = 0.68$; Gentili et al. 1985). Titers measured by ELISA on sera with titers of 0.18 IU/ml or lower showed a higher dispersion around the geometric mean, and an ELISA threshold of 0.1 IU/ml was suggested as the limit of reliability. The authors called for preparation of an international human tetanus immunoglobulin standard to facilitate adoption of immunoassay procedures (Gentili et al. 1986). A separate study (Simonsen et al. 1986) of sera from individuals with varied immunization histories showed satisfactory correlation with high and intermediate levels of neutralizing antitoxin. However, sera with neutralizing titers below 0.16 IU/ml could not be reliably analyzed by ELISA. ELISA values overestimated the neutralizing antibody content of these low titered sera. More recently tetanus toxoid ELISA was found to be highly correlated ($r = 0.936$) with neutralization in 19 human sera examined (SESARDIC and CORBEL 1992).

Because of the shortcomings in measuring titers below 0.2 IU/ml by ELISA several modifications have been proposed to detect titers at least as low as 0.01 IU/ml. A competition ELISA was developed using toxin in solution to compete with toxoid bound to the solid phase (SIMONSEN et al. 1987). In this assay toxin mixed with serial dilutions of test antiserum blocks the binding of that antibody to the solid phase. For values above 0.004 IU/ml, the correlation between neutralization and competition ELISA was $r = 0.98$ ($n = 53$). For neutralization titers below 0.004 IU/ml, the observed ELISA values were consistently low (≤ 0.005 IU/ml). In the prediction of neutralizing activity at the protective level (0.01 IU/ml), the 95% confidence interval of this competition ELISA was 0.0055–0.22 IU/ml. A disadvantage is that the assay requires parallel serum dilutions (one with antigen in solution and one without) and using active tetanus toxin solutions in the range of 0.021–2.1 µg/ml. Another approach to a competition ELISA, termed the toxin binding inhibition test (ToBI test) uses plates coated with diluted antiserum standard. Inhibition of soluble toxin binding to coated wells by serial dilutions of test antiserum is measured (HENDRIKSEN et al. 1988). This study found good agreement between this assay and neutralization ($r = 0.95$) whereas a conventional capture ELISA tended to overestimate neutralization values of low titered sera ($r = 0.74$). This assay has been applied to the combined measurement of tetanus and diphtheria antitoxin in human sera (HENDRIKSEN et al. 1989) and to the estimation of tetanus toxoid potency by measurement of antibody levels in immunized animals (HENDRIKSEN et al. 1991). With human sera the correlation of the ToBI with tetanus toxin lethality neutralization was $r = 0.96$ ($n = 290$), and for diphtheria toxin, the correlation of the ToBI with Vero cell cytotoxicity neutralization was $r = 0.91$ ($n = 140$). In estimating toxoid potencies in multiple trials of the same reference toxoid or in comparison of nine commercial toxoid preparations the potency estimates were nondistinguishable by ToBI from those by neutralization test. The assay is very conservative of toxin, using only 20–40 ng/ml, and the authors stated that toxoid could be used with similar results.

In ELISA, toxin of the highest possible purity should be used to minimize detection of nontoxin antibodies. Purity of antigen, where stated, ranged from 72 to 320 L_f/mg protein in the above studies. Given that pure toxin has 3000 L_f/mg protein N (HABIG and TANKERSLEY 1991), toxin purity was no more than 67%. SIMONSEN et al. (1987) documented that a portion of the immunoassay activity detected in the human sera surveyed was due to antibodies to the nontoxic clostridial proteins in the toxin preparation. Using pure toxin to confirm the good correlation to neutralizing titers seen by competition immunoassays is warranted. Procedures for rapid purification of toxin to high purity are available (SHEPPARD et al. 1987). Better correlation of ELISA value to neutralization titer might be obtained if antibody response to protein fragments or peptides were examined. In a survey of 175 persons, 75% had antibodies by ELISA to a fragment corresponding to the H_c fragment of the protein (GERMAN et al. 1987b). Just 8 of 13 sera exhibiting only anti-H_c antibodies were neutralizing (GERMAN et al. 1987a).

The correlation of immunoassay to neutralizing titer also may depend heavily on the time point in the immunization series at which the serum specimen is taken. For diphtheria toxoid immunization, sera taken after the first administration showed a poor correlation between ELISA and cytotoxicity neutralization, whereas a good correlation was observed in later course sera (LYNG and HERON 1992). Finally, an International Standard for Human Anti-tetanus Immunoglobulin has been established, which should facilitate development of acceptable in vitro immunoassay procedures (SESARDIC et al. 1994).

Some variations to a conventional toxoid/toxin ELISA have been proposed. Two coated-bead enyzme-linked immunoassays have been described (THIELE et al. 1990; VIRELLA et al. 1985). In addition, a highly sensitive method was developed by incorporating Eu^{3+}, labeled antihuman monoclonal antibodies (SCHRODER and KUHLMANN 1991) or Eu^{3+}-labeled recombinant protein G (MARKELA et al. 1993) in a time-resolved fluorescence immunoassay. In the protein G-based assay, the detection limit was stated as 3×10^{-6} IU/ml. A dot-ELISA has been developed and applied to testing human sera (BIZZINI et al. 1990; MASTROENI et al. 1989).

Numerous surveys have applied immunoassay technology to assess the status of immune protection among different target populations. These studies help identify groups at risk because of low titers. SIMONSEN et al. (1984) found good correlation between neutralization and passive hemagglutination in Danes aged 25–30 years, and hemagglutination techniques have been used recently in Russia (BOLOTOVSKII et al. 1992), and Spain (CILLA et al. 1993). ELISA has been used for population studies in Italy (D'ARGENIO et al. 1992; D'AMELIO et al. 1993; GENTILI et al. 1993), Burkina Faso (RUMKE et al. 1993), Peru (VERNACCHIO et al. 1993), Ecuador (BRUSSOW et al. 1993), and the United Kingdom (HAZLEWOOD et al. 1993) and to assess fetal immune response after maternal immunization (DASTUR et al. 1993; ENGLUND et al. 1993). A competitive type ELISA, based on test sera inhibition by of the binding horse radish peroxidase-labeled equine antibody standard, has been used in Taiwan to assess immunity in school age children (LEE et al. 1987; LEE and TSENG 1991).

ELISA has been used for numerous clinical studies to characterize and exploit the tetanus immune response. The IgG subclass response to tetanus toxoid immunization is predominantly IgG1 and IgG4 (FRENCH and HARRISON 1985). The IgG4 population appears to display lower functional affinity than the IgG1 antibodies (DEVEY et al. 1988). ELISA-based tetanus toxoid responses have been used to monitor immunocompetence in immunocompromised patients (MOEN et al. 1986). ELISA has been used to demonstrate the lack of an early immune response to a booster toxoid response in persons not immunized for at least 10 years (PORTER et al. 1992). Tetanus antibody ELISA has been used in a number of studies to examine the immune response when tetanus toxoid is used as a carrier in conjugate vaccines, for example, *Haemophilus influenzae* type b capsular polysaccharide toxoid (BARINGTON et al. 1994). ELISA has been used for tetanus antitoxin determinations in veterinary medicine (PAUL-MURPHY et al. 1989). This broad spectrum of applications would not be feasible by relying on lethality neutralization assays.

4.2 Botulinum Toxin

In order to be able to compare relative potency of different lots of botulinum antitoxin it was necessary to establish antitoxin standards. In the 1920s an antitoxin standard was proposed (BENGTSON 1921, 1924) following procedures long established for the quantitation of tetanus and diptheria antisera. A calibration dose, prepared from dried, ammonium sulfate precipitated toxin, containing about 100 minimal lethal doses (guinea pig) was mixed with serial dilutions of antiserum. The amount of glycerinated serum of each type, which allowed the death of guinea pigs in 96 was considered to contain 0.1 U. Therefore 1 U standard antitoxin should just fail to neutralize approximately 1000 minimal lethal doses. For routine use a test dose of toxin, denoted as L+/10, is determined against 0.1 U standard serum as that amount which when mixed with 0.1 U and injected into 250 g guinea pigs causes death within 96 h. After determination of the test dose for batches of toxin, the potency of an unknown serum was determined by testing serial dilutions against the test dose of toxin. For therapeutic sera, "the smallest amount of antitoxin which when mixed with the test dose of toxin is sufficient to *save* the life of a 250 g guinea pig is considered to contain one tenth of a unit." This procedure was used to assay therapeutic antitoxin prepared by Lederle, at least up to 1964 (COOPER 1964). These sera, plus a type C and E antiserum, are still available in limited quantities (E. Fitzgerald, Food and Drug Administration, personal communication), but are of limited utility because of their low titers, variability among animals, and the ambiguous nature of the unit.

In 1962 the current antitoxin standard, based on mouse lethality, was proposed (BOWMER 1962). For types A–D the unit was proposed as that amount of antitoxin which when mixed with toxin allows survival of 50% of mice given 10 000 LD_{50} each. For type E the level was set at 1000 LD_{50}, but the toxin was not trypsin-treated before testing. International antisera standards were accepted by the WHO Expert Committee on Biological Standardization as equine antisera prepared at the Microbiological Research Establishment in the 1950s for use as British reference preparations. These standards are distributed by the International Laboratory for Biological Standards, Statens Seruminstitut, Copenhagen, Denmark (BOWMER 1963). An international type F standard from rabbit serum was prepared at Statens Seruminstitut containing 4.1 IU per ampoule (KRAG and BENTZON 1965). Using the international standards A–E for reference, United States reference equine antisera were prepared at the Centers for Disease Control, with addition of a type F standard (HARRELL et al. 1964). In the middle 1980s, a second International Standard for type B was prepared and accepted containing 31 IU per ampoule (WHO Expert Committee on Biological Standardization, 1982; J. Lyng, 1985 unpublished). Unfortunately, all current stocks of antiserum standards are in jeopardy of exhaustion.

These serum standards have been used as internal calibration standards in measurement of human serum titers after toxoid immunization, for quantitation of serum titer during potency testing of toxoids, and for quantitation of therapeutic antisera. In evaluating the human immune response, the National Institutes of Health standard was used in early studies (Reames et al. 1947; Wright et al. 1960; Fiock et al. 1962). In 1963 workers at the Army Biological Laboratory converted to use of the newly established International Antiserum Standards (Fiock et al. 1963; Cardella 1964). First, the standard dilution for a toxin test dose was established by mixing serial dilutions of a toxin preparation with equal volumes of antiserum standard at fixed serum titer, followed by injection of 0.5 ml/animal in groups of eight mice weighing 18–20 g. Levels of the antisera standards were set as follows: type A, 0.02 IU/ml; type B, 0.005 IU/ml; type C, 0.02 IU/ml; type D, 0.16 IU/ml; and type E, 0.0125 IU/ml. Deaths were recorded for 4 days, and the 50% survival was determined by a graphical probit method. Following determination of the toxin dose, serial dilutions of a unknown serum were mixed with equal volumes of toxin at fixed concentration, and 0.5 ml volumes were injected. That dilution showing 50% survival was taken as the neutralization endpoint (Wright et al. 1960; Cardella 1964). Titration of human sera in the United States since that time by either the Centers for Disease Control (Hatheway and Dang 1994) or the United States Army (Seigel 1988, 1989) followed guidelines established at that time. Current titrations are performed using twofold serial dilutions of the unknown serum, injection of 0.2 ml, and determination of the 50% survival by the method of Reed and Muench (1938). For type F antiserum our Institute has used a test level of 0.0125 IU/ml (Siegel 1989), while workers at Centers for Disease Control used measured titers at the 0.0025 IU/ml level (Hatheway and Dang 1994). The goal is to determine the lowest possible neutralization titer in an individual's serum as an indicator of some arbitrary level of immunity. A large database of useful human serological titer information has been accumulated.

Unfortunately, a quantitative interpretation of these data is troublesome because of the low concentrations at which titer measurements are made. As mentioned above, accurate quantitation of serum neutralization potencies is required to produce therapeutic immune globulin, to assess toxoid vaccine potency, and to determine the level of immunity that a vaccinated person develops. An accurate determination of the neutralization potency requires careful consideration of the experimental conditions under which a serum standard was calibrated. This may mean that an accurate quantitation assay would require very different conditions from a screening neutralization assay.

Accurate quantitaion of serum potency requires that its neutralization behavior adhere to the principle of multiple proportions. That is, if a certain quantity of toxin is neutralized by a certain amount of antiserum at one concentration, the ratio must remain constant over the range of use of the antiserum. Such behavior is important for a test serum, but it is critical for a serum standard. KEMPNER (1897) observed that antibotulinum serum can display this behavior, and GLENNY and BARR (1932) defined the dilution ratio as a means to assess avidity for diphtheria toxin antiserum. This phenomenon was discussed extensively for diphtheria toxin-antitoxin interactions by JERNE (1951) and is now understood to reflect the average avidity of the mixed populations of antibodies in serum responding to the law of mass action. At the low concentrations under which the neutralization test is currently performed, the international standards do not all react in a linear fashion (J.E. Brown, unpublished observations). FIOCK et al. (1963) realized early on that, at the conditions defined for human sera screening, the behavior of the antiserum standards was non-ideal. The chosen levels of antisera neutralization standards are each able to neutralize about 30 LD_{50} homologous toxin per milliliter, although by reference to the unitage, the test toxin concentrations should have varied from 12.5–1600 LD_{50}/ml for the respective serotypes.

As shown in Table 1, the international antiserum standards were established by testing at relatively high toxin concentrations. Table 1 also shows other factors useful for performing quantitative antibody assays, such as the dilution factors used to achieve partial lethality at that level of test and indicators of avidity, such

Table 1. Conditions for calibration of international standards

Type	Level of test[a]	Dilution factor	No. of mice per level	Probit slope	Index of efficacy[b]	Dilution ratio[c]	Reference
A	L+/10	1.025	20	53	0.11	0.29	BOWMER (1962), FULTHORPE (1955)
B[d]	L+/10	1.05	20	19–40[d]	0.17[f]		LYNG (1985, unpublished)
C	L+	1.03	20	49	0.98		BOWMER (1962)
D	L+	1.064	20	19	0.23		BOWMER (1962)
E	L+/10	1.025	20	78	1.3		BOWMER (1962)
F	L+/100	1.08	5(n=11)	124		1.04[e]	WHO/BS/750 (KRAG and BENTZON 1965)

[a]Level of test, L+, indicates that amount of toxin that when mixed with 1 U antiserum and injected into a group of mice results in 50% survival. L+/10 is the amount of toxin neutralized by 0.1 U antiserum, not 10% of the L+ dose.

[b]Defined as the ratio of toxin neutralized at the tenfold lower antiserum level to that at the calibration test level, corrected for dilution. Low avidity sera would give a value below 1.

[c]Defined as the ratio of serum required to neutralize at the calibration test dose to that required to neutralize 10% of the calibration test dose, corrected for dilution. Low avidity sera would give a value below 1.

[d]Second international standard, (Lyng 1985, unpublished). Probit and dilution ratio data from laboratory no.4.

[e]Data from laboratory 4, unpublished working document WHO/BS/750 (KRAG and BENTZON 1965).

[f]Data from author's laboratory. From (BOWMER 1962), the index of efficacy L+:L+/10 for the first type B international standard was 0.83, indicating a 20% increase in the value of a unit when the second international standard for type B antiserum was adopted.

as probit slope, efficacy index, and dilution ratio (where available). From the efficacy indices and dilution ratios, it is apparent that the standard sera for types A, B, and D cannot be used as quantative standards at less than the originally defined level of test. This lack of proportional response due to low avidity was recognized at the time, especially for type D serum (BOWMER 1962; personal communication, in FULTHORPE 1955).

Experiments with type A standard serve as a good example of this problem of proportional response. FULTHORPE (1955) examined the behavior of a provisional type A antiserum standard apparently identical to the ultimately accepted international standard. It had the same unitage and origin as the international type A standard. He measured the amount of serum required for 50% neutralization of multiples and fractions of an L+/10 toxin dose to determine whether the standard obeyed the principle of multiple proportions. When expressed as the ratio of antiserum predicted to be needed at each level to that actually required, the ratio was constant at and above the 10 × L+/10 level. Below that level the ratio dropped precipitously from 2.4 at 10(L+/10) test doses to 0.064 at a test dose of toxin of 0.01(L+/10). Thus the original calibration level (L+/10) is in a region of antiserum concentration where the ratio is changing. At 1/100 of an L+/10 dose level the provisional antiserum standard required 15 times more antiserum than predicted. If the serum that he characterized is the international standard, his data indicate that accurate determination of serum potencies relative to the international standard would be possible only at one level, L+/10. This is the level recommended by BOWMER (1963).

On the other hand, types C, E, and F international standards appear to be high avidity sera that could be used accurately at levels at least tenfold below the original level of test. The type E serum appears particularly avid. This apparent avidity difference may reflect the higher toxin concentration at which the type E unit was defined. For type E, 1 IU antiserum was defined as the capacity to neutralize 1000 LD_{50} of native type E toxin, which would equal about 100 000 LD_{50} trypsin-nicked toxin. For the type F antiserum standard the principle of multiple proportions seems to hold down to 0.0025 IU/ml (e.g., L+/1600, the current level of serum screening).

In addition to use of antiserum standards at reliable levels, a number of other experimental factors are important for accurate determination of serum potency, including the number of mice used at each serial dilution, dilution factor, data analysis method, and test for parallelism between standard and test serum. Work in our laboratory suggests that accuracy is possible with eight mice per level using a dilution factor of 1.1 (a 10% increment). Probit analysis should be performed instead of LD_{50} determination as described by REED and MUENCH (1938) because the former method allows the calculation of confidence intervals. More importantly, from the slope of the probit regression line, one may infer parallelism between standard and test serum. Probit analysis is available in software packages such as SAS or SPSS, but a graphical method is described by FINNEY (1971). Determination of an accurate serum titer of an unknown specimen requires a two-step approach: first, a range-finding assay, and then a final assay centered on the initial estimate.

One benefit of the Army's equine immunization program has been the opportunity to prepare new lots of equine antiserum standards. We have prepared secondary standards for each of the seven serotypes A–F (Table 2). As recommended by CARDELLA (1964), these standards have been calibrated against the WHO international standards using the level of test at which we routinely analyze human sera to detect low titers. These levels are not the test levels at which the original standards were calibrated but those at which antiserum standards will be most often used. Neutralizing titers at L+ toxin levels have not been determined. A type G antiserum standard, with preliminary values of 103 ± 29 U/vial (following the convention of 10 000 LD_{50} toxin as equivalent to 1 U antitoxin) has been prepared at our Institute.

Future research should focus on replacement of the mouse lethality assay as a means to assess serum potency. Aside from issues of expense and animal welfare, one might expect increased accuracy and precision if problems such as animal variability, prolonged assay times, and dilution of antigen-antibody complexes in the test animal are avoided. One approach would be to use animals lower on the phylogenetic hierarchy. We have made some attempts to develop such models at our Institute, but to no avail. Insects such as fruit flies, worms such as nematodes, and paramecia were exposed to botulinum toxin, but no apparent symptoms developed (J.L. Middlebrook, D.L. Leatherman and H. LaPenotiere, unpublished observations). Another approach may be replacement of the lethality protection assay with a nonlethal paralysis protection assay. Botulinum toxin can be identified by a characteristic, acute local paralysis following injection into the gastrocnemius muscle of mice (SUGIYAMA et al. 1975). Neutralization by homologous type antiserum could be demonstrated. Factors affecting the reliability of this regional paralysis assay for toxin determination have recently been analyzed (PEARCE et al. 1994a,b). The authors found that the paralysis assay provided a better correlate to potencies of medical grade botulinum toxin preparations than did the mouse lethality assay, even when the

Table 2. Calibration of secondary *C. botulinum* antiserum standards

Type	Toxin[a] test level (per mouse)	*n*	Mean ± SD (IU/vial)	CV (%)
A	L+/200	5	440 ± 142	32
B	L+/800	4	330 ± 46	14
C	L+/200	3	458 ± 112	24
D	L+/25	3	234 ± 34	14
E	L+/320	3	632 ± 86	14
F	L+/320	6	910 ± 230	20

[a]Test levels recommended by CARDELLA (1964) to detect antitoxins titers at 0.02, 0.005, 0.02, 0.16, 0.0125, and 0.0125 IU/ml for toxins A–F, respectively.

latter was performed under conditions of high precision. In neutralization assays performed using this model, the injected dose should remain localized at the site rather than being diluted throughout the animal's plasma volume. Antitoxin titers have been estimated by subcutaneous injection of toxin-antiserum mixtures into the inguinocrual region of mice, followed by graded scoring of the abdominal ptosis and local palsy (TAKAHASHI et al. 1990a, b).

A cell culture system responsive to botulinum toxins would provide an enormous advance in evaluating potency of botulinum antisera. SHONE and MELLING (1992) found that type A botulinum toxin inhibits calcium-dependent release of noradrenaline from PC-12 cells with an ED_{50} of 120 n*M* after overnight incubation. Using nerve growth factor differentiated PC-12 cells in a perfusion system, RAY et al. (1993) was able to show inhibition of depolarization-induced calcium-dependent acetylcholine release with an ED_{50} of about 0.01 n*M* botulinum neurotoxin A after a 2-h incubation. Using primary cultures of mouse spinal cord neurons, WILLIAMSON et al. (1992) have shown that tetanus toxin blocks release of K^+-evoked, calcium-dependent release of both excitatory and inhibitory neurotransmitters. Because of the structural and functional similarities of the botulinum toxins to tetanus toxin, the sensitivity of these cultures to botulinum toxins has been examined (L.C. Williamsom, J.E. Brown, and E.A. Neale, unpublished observations). For botulism neurotoxin type A the ED_{50} was approximately 0.2 p*M* after overnight incubation, and the inhibition was neutralized by type A antiserum. Exquisitely sensitive cell lines may not be necessary for measurement of antiserum titers if, as has been shown with diphtheria antisera, correlation with in vivo neutralization titers requires assay in cell culture at relatively high antiserum concentrations to avoid avidity problems common in early course sera (GUPTA et al. 1994).

Serum potency estimation by immunoassay techniques such as ELISA would offer the greatest flexibility, portability, and economy. Anti-botulinum toxin serum antibodies in naturally resistant wild animals were measured by passive hemagglutination (OHISHI et al. 1979). Using ELISA, antibody responses were detected in convalescent sera from two patients recovering from infant botulism (RUBIN et al. 1982) and in one volunteer given toxoid 15 weeks before appearance of detectable neutralizing antibody (DEZFULIAN et al. 1987). SHONE et al. (1986) compared neutralization titers with ELISA titers for ten human sera and observed a limited correlation at high titer values. In a more extensive study SIEGEL (1988) evaluated the immune responses of 186 people for type A antibody levels and 168 individuals for type B responses. The author concluded that using ELISA results to predict neutralizing antibody levels is unwarranted, although a reanalysis of the data shows that the ELISA was 85% predictive of a population where the neutralization test identified 91% as immunized. For type A the sensitivity of the ELISA test was 88% and the specificity was 50%, whereas for type B these values were 98% and 41%, respectively. Appropriate methodological refinements to eliminate the high rate of false positives are in progress. In veterinary practice the detection of an immune response by antibody

ELISA has been used to help diagnose cattle with suspected botulism (JUBB et al. 1993).

Of increasing concern is the development of an immune response in patients receiving botulinum toxin for focal dystonias. Several studies have found circulating antibodies in treated patients (BRIN et al. 1987; GREENE and FAHN 1993; HAMBLETON et al. 1992; HATHEWAY and DANG 1994; JANKOVIC and SCHWARTZ 1991; SIATKOWSKI et al. 1993; ZUBER et al. 1993). Others have failed to find antibody (BIGLAN et al. 1986; GONNERING 1988). Most studies relied on the mouse protection assay, but several have used ELISA as well (BIGLAN et al. 1986; GONNERING 1988). SIATKOWSKI et al. (1993) used a sensitive sphere-linked immunodiagnostic assay and detected antibody at levels that appeared clinically benign. This observation raises the prospect that a sensitive immunoassay could allow clinicians to detect immune responses and alter treatment before antibody titers became inhibitory.

Recently, an amplified ELISA capable of detecting picogram (0.07 p*M*) quantities of toxin (DOELLGAST et al. 1993) has been modified to detect antibody. First, toxin detection was accomplished by employing toxin-specific enzyme-labeled horse antibody and chicken antibody. Complexes formed with toxin in solution were captured on a solid phase coated with anti-chicken IgG, followed by amplification via an enzyme-linked coagulation assay (ELCA; DOELLGAST et al. 1994a). To detect toxin-specific antibody the assay was then modified to measure the degree to which the test serum reduces complex formation of the labeled antibody reagents. The reduction in signal can be used as a measure of the antibody titer (DOELLGAST et al. 1994b). The ELCA-ELISA was used to show significant increases in titer and apparent avidity of serial antisera from a horse during the course of hyperimmunization (DOELLGAST et al., in preparation).

Acknowledgments. We thank Drs. M. Balady and P. Gibbs for helpful discussions and advice and A. Shafer for excellent secretarial assistance. We are also grateful to several scientists cited in the text for sharing their yet unpublished data with us.

References

Arnon SS (1993) Clinical trial of human botulism immune globulin. In: DasGupta BR (ed) Botulinum and tetanus neurotoxins, neurotransmission and biomedical aspects. Plenum, New York, pp 477–482

Arunachalam B, Ghosh S, Talwar GP, Raghupathy R (1992) A single monoclonal antibody that confers total protection from tetanus. Hybridoma 11: 165–179

Bardenwarper HW (1962) Serum neuritis from tetanus antitoxin. JAMA 179: 763–766

Barile M, Hardegree M, Pitmann (1970) Immunization against neonatal tetanus in New Guinea III. The toxin–neutralization test and the response of guinea pigs to the toxoids as used in the immunization schedules in New Guinea. Bull World Health Organ 43: 453–459

Barington T, Gyhrs A, Kristensen K, Heilmann C (1994) Opposite effects of actively and passively acquired immunity to the carrier on responses of human infants to a Haemophilus influenzae type b conjugate vaccine. Infect Immun 62: 9–14

Bengtson IA (1921) Standardization of botulism antitoxins. Am J Public Health 11: 352–357

Bengtson IA (1924) Studies on organisms concerned as causative factors in botulism. Bull US Hyg Lab 136: 1–101

Beytout J, Lauras H, Cailleba A, Nguyen TT (1989) Rapid evaluation of tetanus immunity by a haemagglutination test in the injured at a hospital emergency unit. Biomed Pharmacother 43: 621–625

Biglan AW, Gonnering R, Lockhart LB, Rabin B, Fuerste FH (1986) Absence of antibody production in patients treated with botulinum A toxin. Am J Ophthalmol 101: 232–235

Bizzini B (1993) Clostridium tetani. In: Gyles C, Thoen C (eds) Pathogenesis of bacterial infections in animals, 2nd edn. Iowa State University Press, Ames, pp 97–105

Bizzini B, Leonardi MS, Zummo S, Mastroeni P (1990) Screening of tetanus antibodies with the aid of dot-enzyme-linked immunosorbent assay (dot-ELISA). Validation of the results using standard ELISA and passive hemagglutination test. Boll Ist Sieroter Milan. 69: 391–393

Black RE, Gunn RA (1980) Hypersensitivity reactions associated with botulinal antitoxin. AM J Med 69: 567–570

Bleck TP (1991) Tetanus: pathophysiology, management, and prophylaxis. Dis Mon 37: 545–603

Bolotovskii VM, Kriukov V, Raupov BE (1992) A comparative evaluation of the effectiveness of methods for serological sampling examinations. Zh Mikrobiol Epidemiol Immunobiol 1992: 23–26

Bowmer EJ (1962) Antitoxins of Clostridium botulinum type A,B,C,D and E. Preparation and assay of proposed international standards. Thesis, University of Liverpool

Bowmer EJ (1963) Preparation and assay of the international standards for Clostridium botulinum types A,B,C,D and E antitoxins. Bull World Health Organ 29: 701–709

Boyd J (1959) Symposium of tetanus. Tetanus in two world wars. Proc R Soc Med 52: 109

Brin MF, Fahn S, Moskowitz C, Friedman A, Shale HM, Greene PE, Blitzer A, List T, Lange D, Lovelace RE, McMahon D (1987) Localized injections of botulinum toxin for the treatment of focal dystonia and hemifacial spasm. Mov Disord 2: 237–254

Brown JE, Taylor KL, Brown WL, Clayton MA, Schmidt JJ, Jaax GP, Franz DR (1991) Hyperimmunization of horses for production of immune globulin to botulinum toxins. Abstracts, 10th world congress on animal, plant, and microbial toxins, Venom and Toxin Research Group, University of Singapore, Singapore, p 95

Brussow H, Sidoti J, Freire WB (1993) Tetanus and diphtheria immunization coverage in Ecuadorian children after a national vaccination campaign. J Infect Dis 168: 479–483

Cardella MA (1964) Botulinum toxoids. In: Lewis KH, Cassel KJ (eds) Botulism. U.S. department of health, education, and welfare, Cincinnati, pp 113–130

Charles IG, Rodgers BC, Makoff AJ, Chatfield SN, Slater DE, Fairweather NF (1991) Synthesis of tetanus toxin fragment C in insect cells by use of a baculovirus expression system. Infect Immun 59: 1627–1632

Chatfield SN, Fairweather N, Charles I, Pickard D, Levine M, Hone D, Posada M, Strugnell RA, Dougan G (1992) Construction of a genetically defined Salmonella typhi Ty2 aroA, aroC mutant for the engineering of a candidate oral typhoid-tetanus vaccine. Vaccine 10: 53–60

Cilla EG, Saenz DJ, Urbieta EM, Carcedo MA, Perez TE (1993) Immunity against tetanus in childhood. An Esp Pediatr 38: 499–502

Clare JJ, Rayment FB, Ballantine SP, Sreekrishna K, Romanos MA (1991) High-level expression of tetanus toxin fragment C in Pichia pastoris strains containing multiple tandem integrations of the gene. Biotechnol 9: 455–460

Cooper MS (1964) Antitoxins to C. botulinum. In: Lewis KH, Cassel KJ (eds) Botulism. U.S. department of health, education, and welfare, Cincinnati, 147–164

Council of Europe (1981, 1985) European Pharmacopoeia, 2nd edn. Maisonneuve, Saint-Ruffinel

Craven CJ, Dawson DJ (1973) The chain composition of tetanus toxin. Biochim Biophys Acta 317: 277–285

D'Amelio R, Stroffolini T, Wirz M, Biselli R, Matricardi PM, Gentili G, Collotti C, Pasquini P (1993) Tetanus immunity among Italians born after the introduction of mandatory vaccination of children. Vaccine 11: 935–936

D'Argenio P, Wirz M, Rosmini F, Adamo B, Gentili G, Puccinelli M, Collotti C, Mele A, Esposito D, Forte P, Pasquini P (1992) Tetanus immunity in a partially vaccinated population of children in a district of Naples (Secondigliano). Eur J Epidemiol 8: 93–97

Dastur FD, Shastry P, Iyer E, Awatramani V, Raut S, Mehta SD, Irani SF (1993) The foetal immune response to maternal tetanus toxoid immunization. J Assoc Physicians India 41: 94–96

Devey M, Bleasdale K, Lee S, Rath S (1988) Determination of the functional affinity of IgG1 and IgG4 antibodies to tetanus toxoid by isotype-specific solid-phase assays. J Immunol Methods 106: 119–125

Dezfulian M, Bitar RA, Bartlett JG (1987) Kinetics study of immunological response to Clostridium botulinum toxin. J Clin Microbiol 25: 1336–1337

Dimache G, Stoean C, Durbaca S, Croitoru M, Ionescu M, Nedelcu IN, Corbu I (1990) Study of specific immune response to unadsorbed concentrated tetanus vaccine administered by intradermal route to non-immunized persons in the last ten years. Arch Roum Pathol Exp Microbiol 49: 51–62

Doellgast GJ, Triscott MX, Beard GA, Bottoms JD, Cheng T, Roh BH, Roman MG, Hall PA, Brown JE (1993) Sensitive enzyme-linked immunosorbent assay for detection of Clostridium botulinum neurotoxins A,B, and E using signal amplification via enzyme-linked coagulation assay. J Clin Microbiol 31: 2402–2409

Doellgast GJ, Beard GA, Bottoms JD, Cheng T, Roh BH, Roman, MG, Hall PA, Triscott MX (1994a) Enzyme-linked immunosorbent assay and enzyme-linked coagulation assay for detection of Clostridium botulinum neurotoxins A,B, and E and solution-phase complexes with dual-label antibodies. J Clin. Microbiol 32: 105–111

Doellgast GJ, Triscott MX, Beard GA, Bottoms JD (1994b) Enzyme-linked immunosorbent assay-enzyme-linked coagulation assay for detection of antibodies to Clostridium botulinum neurotoxins A,B, and E and solution phase complexes. J Clin Microbiol 32: 851–853

Duff JT, Wright GG, Klerer J, Moore DE, Bibler RH (1957a) Studies on immunity to toxins of Clostridium botulinum. I. A simplified procedure for isolation of type A toxin. J Bacteriol 73: 42–47

Duff JT, Klerer J, Bibler RH, Moore DE, Gottfried C, Wright GG (1957b) Studies on immunity to toxins of Clostridium botulinum. II. Production and purification of type B toxin for toxoid. J Bacteriol 73: 597–601

Eisel U, Jarausch W, Goretzki K, Henschen A, Engels J, Weller U, Hudel M, Habermann Nieman H (1986) Tetanus toxin: primary structure, expression in E. coli, and homology with botulinum toxins. EMBO J 10: 2495–2502

Englund JA, Mbawuike IN, Hammill H, Holleman MC, Baxter BD, Glezen WP (1993) Maternal immunization with influenza or tetanus toxoid vaccine for passive antibody protection in young infants. J Infect Dis 168: 647–656

Fairweather NF, Lyness VA (1986) The complete nucleotide sequence of tetanus toxin. Nucleic Acid Res 14: 7809–7812

Fairweather NF, Lyness VA, Pickard DJ, Allen G, Thomson RO (1986) Cloning, nucleotide sequencing, and expression of tetanus toxin fragment C in Escherichia coli. J Bacteriol 165: 21–27

Fairweather NF, Lyness VA, Maskell DJ (1987) Immunization of mice against tetanus with fragments of tetanus toxin synthesized in Escherichia coli. Infect Immun 55: 2541–2545

Fairweather NF, Chatfield SN, Makoff AJ, Strugnell RA, Bester J, Maskell DJ, Dougan G (1990) Oral vaccination of mice against tetanus by use of a live attenuated Salmonella carrier. Infect Immun 58: 1323–1326

Finney DJ (1971) Probit analysis, 3rd edn. University Press Cambridge, Cambridge

Fiock MA, Devine LF, Gearinger NF, Duff JT, Wright GG, Kadull PJ (1962) Studies on immunity to toxins of Clostridium VIII. immunological response of man to purified bivalent AB botulinum toxoid. J Immunol 88:277–283

Fiock MA, Cardella MA, Gearinger NF (1963) Studies on immunity of toxins of Clostridium botulinum. IX. Immunological response of man to purified pentavalent ABCDE botulinum toxoid. J Immunol 90: 697–702

Frankovich TL, Arnon SS (1991) Clinical trial of botulism immune globulin for infant botulism. West J Med 154: 103

Franz DR, Pitt LM, Clayton MA, Hanes MA, Rose KJ (1993) Efficacy of prophylactic and therapeutic administration of antotoxin for inhalation botulism. In: DasGupta BR (ed) Botulinum and tetanus neurotoxins, neurotransmission and biomedical aspects. Plenum, New York, pp 473–476

French M, Harrison G (1985) The IgG subclasses of tetanus toxoid antibody following immunization: studies in healthy adults and patients with chronic or recurrent respiratory tract infections. J Clin Lab Immunol 18: 21–25

Fulthorpe AJ (1955) Clostridium botulinum type A and type B toxin-antitoxin reactions. J Hyg 53: 180–195

Gentili G, Pini C, Collotti C (1984) The determination of the potency of human tetanus immunoglobulins by an enzyme-linked immunosorbent assay. J Biol Stand 12: 167–173

Gentili G, Pini C, Collotti C (1985) The use of immunoenzymatic assay for the estimation of tenanus antitoxin in human sera: a comparison with seroneutralization and indirect haemagglutination. J Biol Stand 13: 53–59

Gentili G, Wirz M, Collotti C (1986) Clostridium tetani toxin antibodies. In: Masseyeff R (ed) Antigens and antibodies in chlamydial and bacterial diseases, XI. 3rd edn VCH Weinheim pp 44–52
Gentili G, D'Amelio R, Wirz M, Matricardi PM, Nisini R, Collotti C, Pasquini P, Stroffolini T (1993) Prevalence of hyperimmunization against tetanus in Italians born after the introduction of mandatory vaccination of children with tetanus toxoid in 1968. Infection 21: 80–82
German FM, Bizzini B, German A (1987a) Immunity to tetanus: tetanus antitoxin and anti-BIIb in human sera. J Biol Stand 15: 223–230
German FM, German A, Bizzini B (1987b) Relation between protective potency and specificity of antibodies in sera of tetanus immunized individuals. Eur J Epidemiol 3: 119–122
Gigliotti F, Insel RA (1982) Protective human hybridoma antibody to tetanus toxin. J Clin Invest 70: 1306–1309
Glenny AT, Barr M (1932) The "dilution ratio" of diphtheria antitoxin as a measure of avidity. J Pathol Bacteriol 35: 91–96
Gonnering RS (1988) Negative antibody response to long-term treatment of facial spasm with botulinum toxin. Am J Ophthalmol 105: 313–315
Greene PE, Fahn S (1993) Use of botulinum toxin type F injections to treat torticollis in patients with immunity to botulinum toxin type A. Mov Disord 8: 479–483
Gupta RJ, Higham S, Gupta CK, Rost BR, Siber GR (1994) Suitability of the vero cell method for titration of diphtheria antitoxin in the United States potency test for diphtheria toxoid. Biologicals 22: 65–72
Habig W, Tankersley D (1991) Tetanus. In: Cryz S (ed) Vaccines and immunotherapy. Pergamon, New York, pp 13–19
Halpern JL, Habig WH, Neale EA, Stibitz S (1990) Cloning and expression of functional fragment C of tetanus toxin. Infect Immun 58: 1004–1009
Hambleton P, Cohen HE, Palmer BJ, Melling J (1992) Antitoxins and botulinum toxin treatment. BMJ 304: 959–960
Harrell WK, Green JH, Winn JF (1964) Preparation, evaluation, and use of C. botulinum antitoxins. In: Lewis KH, Cassel KJ (eds) Botulism. US Department of Health, Education, and Welfare, Cincinnati, pp 165–170
Hatheway CL, Dang C (1994) Immunogenicity of the neurotoxins of Clostridium botulinum. In: Jankovoc J, Hallett M (ed) Therapy with botulinum toxins. Dekker, New York, pp 93–107
Hatheway CH, Snyder JD, Seals JE, Edell TA, Lewis GJ (1984) Antitoxin levels in botulism patients treated with trivalent equine botulism antitoxin to toxin types A,B, and E. J Infect Dis. 150: 407–412
Hazlewood M, Nusrat R, Kumararatne DS, Goodall M, Raykundalia C, Wang DG, Joyce HJ, Milford WA, Forte M, Pahor A (1993) The acquisition of anti-pneumococcal capsular polysaccharide Haemophilus influenzae type b and tetanus toxoid antibodies, with age, in the UK. Clin Exp Immunol 93: 157–164
Helting TB, Zwisler O (1977) Structure of tetanus toxin I. Breakdown of the toxin molecule and discrimination between polypeptide fragments. J Biol Chem 252: 187–193
Hendriksen CFM, vander Gun JW, Nagel J, Kreeftenberg JG (1988) The toxin binding inhibition test as a reliable in vitro alternative to the toxin neutralization test in mice for the estimation of tetanus antitoxin in human sera. J Biol Stand 16: 287–297
Hendriksen CFM, vander Gun JW, Kreeftenberg JG (1989) Combined estimation of tetanus and diphtheria antitoxin in human sera by the in vitro toxin-binding inhibition (ToBI) test. J Biol Stand 17: 191–200
Hendriksen CFM, vander Gun JW, Marsman F, Kreeftenberg JG (1991) The use of the in vitro toxin binding (ToBI) test for the estimation of the potency of tetanus toxoid. Biologicals 19: 23–29
Hodgson J (1991) Making monoclonals in microbes. Bio technol 9: 421–425
Jackson RJ, Fujihashi K, Xu-Amano J, Kiyono H, Elson CO, McGhee JR (1993) Optimizing oral vaccines: induction of systemic and mucosal B-cell and antibody responses to tetanus toxoid by use of cholera toxin as an adjuvant. Infect Immun 61: 4272–4279
Jankovic J, Schwartz KS (1991) Clinical correlates of response to botulinum toxin injections. Arch Neurol 48: 1253–1256
Jerne NK (1951) Avidity. Acta Pathol. Microbiol Scand. [Suppl] 87.
Jubb TF, Ellis TM, Gregory AR (1993) Diagnosis of botulism in cattle using ELISA to detect antibody to botulinum toxins. Aust Vet J 70: 226–227
Kaiser II, Middlebrook JL (1988) Preparation of a crotoxin neutralizing monoclonal antibody. Toxicon 26: 855–865
Kamei M, Hashizume S, Sugimoto N, Ozutsumi K, Matsuda M (1990) Establishment of stable human/ mouse hybrid cell lines producing large amounts of anti-tetanus human monoclonal antibodies with high neutralizing activity. Eur J Epidemiol 6: 386-397

Kempner W(1897) Weitere Beiträge zur Lehre der Fleischvergiftung. Das Antitoxin des Botulismus. Z. Hyg. Infektkr 26: 418–500
Kenimer JG, Habig WH, Hardegree MC (1983) Monoclonal antibodies as probes of tetanus toxin structure and function. Infect Immun 42: 942–948
Kiessig ST, Hentschel C, Jahn S, Mehl M. Starke R, Porstmann T (1991) A solid-phase enzyme immunoassay for the detection of tetanus toxin using human and murine monoclonal antibodies. J Basic Microbiol 31: 135–140
Krag P, Bentzon MW (1965) Report on international collaborative assay of antibotulinus serum type F. WHO/BS 750: 65 (unpublished)
Lang AB, Cryz SJ Jr, Schurch U, Ganss MT, Bruderer U (1993) Immunotherapy with human monoclonal antibodies. Fragment A specificity of polyclonal and monoclonal antibodies is crucial for full protection against tetanus toxin. J Immunol 151: 466–472
LaPenotiere HF, Clayton MA, Brown JE, Middlebrook JL (1993) Development of a molecularly engineered vaccine for C. botulinum neurotoxins. In: DasGupta BR (ed) Botulinum and tetanus neurotoxins, neurotransmission and biomedical aspects. Plenum, New York, pp 463–466
Larrick JW, Truitt KE, Raubitschek AA, Senyk G, Wang JCN (1983) Characterization of human hybridomas secreting antibody to tetanus toxoid. Proc Natl Acad Sci USA 80: 6376–6380
Layton G (1980) A microenzyme-linked immunosorbent assay (ELISA) and radioimmunosorbent technique (RIST) for the detection of immunity to clinical tetanus. Med Lab Sci 37: 323–329
Layton LL, Arimoto L, Lamanna C, Olson R, Sharp D, Kondo H, Sakaguchi G (1972) Immunochemical and physiological comparison of horse botulinal antitoxins. Jpn J Med Sci Biol 25: 309–320
Lee HF, Tseng LR (1991) Improvement of the E1A kit for the detection of tetanus immunoglobulin in human sera. Chung Hua Min Kuo Wei Sheng Wu Chi Mien I Hsueh Tsa Chih 24: 221–228
Lee HF, Chen HM, Hsu ST (1987) Serological survey of immune response to tetanus toxoid using antitoxin enzyme immunoassay. Chung Hua Min Kuo Wei Sheng Wu Chi Mien I Hsueh Tsa Chih 20: 269–278
Leuchs J (1910) Beiträge zur Kenntnis des Toxins and Antitoxins des Bacillus botulinus. Z Hyg Infektkr 65: 55–84
Lewis GE (1981) Approaches to the prophylaxis, immunotherapy, and chemotherapy of botulism. In: Lewis GE Jr (ed) Biomedical aspects of botulism. Academic, New York, pp 261–270
Lewis GE Jr, Metzger JF (1978) Botulism immune plasma. Lancet 2: 634–635
Lewis GE Jr, Metzger JF (1980) Studies on the Prophylaxis and treatment of botulism. In: Eaker D, Wadstrom T (eds) Natural toxins, Pergamon, Oxford, pp 601–606
Luisto M (1990) Tetanus in Finland: diagnostic problems and complications. Ann Med 22: 15–19
Lyng J, Heron I (1992) Quantitative estimation of diphtheria and tetanus toxoids. VI. Use of different antibody titration methods for evaluation of immunogenicity in animals during potency assay of diphtheria toxoid. Biologicals. 20: 101–108
Makoff AJ, Ballantine SP, Smallwood AE, Fairweather NF (1989a) Expression of tetanus toxin fragment C in E. coli: its purification and potential use as a vaccine. Biotechnol 7: 1043–1046
V. Chandrika CTMI Volume 195 Chapter 5 Mss 213 to
Makoff AJ, Oxer MD, Romanos MA, Fairweather NF, Ballantine S (1989b) Expression of tetanus toxin fragment C in E. coli: high level expression by removing rare codons. Nucleic Acids Res 17: 10191–10202
Markela E, Stalhberg TH, Hemmila I (1993) Europium-labeled recombinant protein G. A fast and sensitive universal immunoreagent for time-resolved immunofluorometry. J Immunol Methods 161: 1–6
Maskell DJ, Sweeney KJ, O'Callaghan D, Hormaeche CE, Liew FY, Dougan G (1987) Salmonella typhimuriam aroA mutants as carriers of the Escherichia coli heat-labile enterotoxin to the murine secretory and systemic immune systems. Microb Pathog 2: 211–221
Mastroeni P, Leonardi MS, Gazzara D, Bizzini B (1989) Rapid assessment of the antitetanus immune status of a subject using dot-ELISA. Eur J Epidemiol 5: 97–100
Matsuda M, Yoneda M (1975) Isolation and purification of two antigenically active, "complementary" polypeptide fragments of tetanus toxin. Infect Immun 12: 1147–1153
McComb J (1964) The prophylactic dose of homologous tetanus antitoxin. N Engl J Med 270: 175–178
McCracken GH Jr, Dowell DL, Marshall FN (1971) Double-blind trial of equine antitoxin and human immune globulin in tetanus neonatorum. Lancet 1: 1146–1149
Melville-Smith M (1990) Diphtheria and tetanus antitoxins. In: Wreghitt TG, Morgan-Capner P (eds) ELISA in the clinical microbiology laboratory. Public Health Laboratory Service, London, pp 136–147
Metzger JF, Lewis GE (1979) Human-derived immune globulins for the treatment of botulism. Rev Infect Dis 1: 689–692

Middlebrook JL (1991) Preparation and characterization of monoclonal antibodies against pseudexin. Toxicon 29: 359–370
Moen RC, Oemichen SL, Kiggnes AJ, Hong R (1986) ELISA detection of specific functional antibodies in human sera to Escherichia coli, tetanus toxoid, and diphtheria-tetanus toxoids: normal values for IgG, IgA, and IgM. Diagn Immunol 4: 17–23
Mouillet L, Chaniot S, Crajer M, Netter R (1978) Intérêt de l'immunodiffusion radiale dans le titrage des immunoglobulines tétaniques. J Biol Stand 6: 13–20
Nation NS, Pierce NF, Adler SJ, Chinnock RF, Wehrle PF (1963) Tetanus: use of human hyperimmune globulin in treatment. Calif Med 98: 305–307
Niemann H (1992) Clostridial neurotoxins-proposal of a common nomenclature. Toxicon 30: 223–225
Nyerges G, Lutter J (1980) The influence of the method of preservation of erythrocytes on the correlation between tetanus antitoxin values as measured by passive haemagglutination and by seroneutralization tests in mice. J Biol Stand 8: 311–316
Oberst FW, Crook JW, Cresthull P, House MJ (1967) Evaluation of botulinum antitoxin, supportive therapy, and artificial respiration in monkeys with experimental botulism. Clin Pharmacol Ther 9: 209–214
Ohishi I, Sakaguchi G, Riemann H, Behymer D, Hurvell B (1979) Antibodies to Clostridium botulinum toxins in free-living birds and mammals. J Wild Dis 15: 3–9
Pal A, Kumar R, Jailkhani BL (1990) An ELISA for quantitation of tetanus toxin. Indian J Med Res 91: 124–125
Paul-Murphy J, Gershwin LJ, Thatcher EF, Fowler ME, Habig WH (1989) Immune response of the llama (Lama glama) to tetanus toxoid vaccination. Am J Vet Res 50: 1279–1281
Pearce LB, Borodic GE, Johnson EA, First ER, MacCallum R (1994a) The median paralysis unit: a more pharmacologically relevant unit of biologic activity for botulinum toxin. Toxicon (submitted)
Pearce LB, Borodic GE, First ER, MacCallum R (1994b) Measurement of botulinum toxin activity: evaluation of the lethality assay. Ther Appl Pharmacol (in press)
Peel M (1980a) Measurement of tetanus antitoxin. I. Indirect haemagglutination. J Biol Stand 8: 177–189
Peel M (1980b) Measurement of tetanus antitoxin. II. Toxin neutralization. J Biol Stand 8: 191–207
Porter JD, Perkin MA, Corbel MJ, Farrington CP, Watkins JT, Begg NT (1992) Lack of early antitoxin response to tetanus booster. Vaccine 10: 334–336
Prevots R, Sutter RW, Strebel PM, Cochi SL, Hadler S (1992) Tetanus surveillance—United States, 1989–1990. MMRW CDC Surveill Summ 41: 1–9
Ray P, Berman JD, Middleton W, Brendle J (1993) Botulinum toxin inhibits arachidonic acid release associated with acetylcholine release from PC12 cells. J Biol Chem 268: 11057–11064
Reames HR, Kadull PJ, Housewright RD, Wilson JB (1947) Studies on botulinum toxoids, types A and B. III. Immunization in man. J Bacteriol 55: 309–324
Reed LJ, Muench HM (1938) A simple method for estimating fifty percent endpoints. Am J Hyg 27: 493–497
Reiss J, Zawistowska MI, Grzybowski J (1991) A simplified method of rapid hemagglutination test for determining the levels of tetanus antibodies. Med Dosw Mikrobiol 43: 51–54
Rey M, Fontanges R, Robert D, Taylor J, Relyveld E, Vincent-Falquet J, Durand B (1978) Le test d'hémagglutination passive dans l'évaluation de l'immunité antitétanique. Dev Biol Stand 41: 55–63
Romanos MA, Makoff AJ, Fairweather NF, Beesley KM, Slater DE, Rayment FB, Payne MM, Clare JJ (1991) Expression of tetanus toxin fragment C in yeast: gene synthesis is required to eliminate fortuitous polyadenylation sites in AT-rich DNA. Nucleic Acids Res 19: 1461–1467
Rubin LG, Dezfulian M, Yolken RH (1982) Serum antibody response to Clostridium botulinum toxin in infant botulism. J Clin Microbiol 16: 770–771
Rubin M, Wong E, Bell H, Bowman J, Albritton W (1981) A reversed rocket electrophoresis method for the determination of high tetanus antitoxin levels. J Biol Stand 9: 367–371
Rumke HC, Schlumberger M, Floury B, Nagel J, van Steenis B (1993) Serological evaluation of a simplified immunization schedule using quadruple DPT-polio vaccine in Burkina Faso. Vaccine 11: 1113–1118
Schroder JP, Kuhlmann WD (1991) Detection of tetanus antitoxin using Eu(3+)- labeled anti-human immunoglobulin G monoclonal antibodies in a time-resolved fluorescence immunoassay. J Clin Microbiol 29: 1504–1507
Schwarz PJ, Arnon SS (1992) Botulism immune globulin for infant botulism arrives—one year and a Gulf War later. West J Med 156: 197–198

Sesardic D, Corbel MJ (1992) Testing for neutralising potential of serum antibodies to tetanus and diphtheria toxin (letter). Lancet 340: 737–738
Sesardic D, Wong MY, Gaines Das RE, Corbell MJ (1994) The 1st international standard for antitetanus immunoglobulin, human; pharmaceutical, evaluation and international collaborative study. Biologicals 21: 67–75
Sheppard AJ, Hughes M, Stephen J (1987) Affinity purification of tetanus toxin using polyclonal and monoclonal antibody immunoadsorbents. J Appl Bacteriol 62: 335–348
Shone CC, Melling J (1992) Inhibition of calcium-dependent release of noradrenaline from PC12 cells by botulinum type-A neurotoxin. Long-term effects of the neurotoxin on intact cells. Eur J Biochem 207: 1009–1016
Shone C, Appleton N, Wilton SP, Hambleton P, Modi N, Gatley S, Melling J (1986) In vitro assays for botulinum toxin and antitoxins. Dev Biol Stand 64: 141–145
Siatkowski RM, Tyutyunikov A, Biglan AW, Scalise D, Genovese C, Raikow RB, Kennerdell JS, Feuer WJ (1993) Serum antibody production to botulinum A toxin. Ophthalmology 100: 1861–1866
Siegel LS (1988) Human immune response to botulinum pentavalent (ABCDE) toxoid determined by a neutralization test and by an enzyme–linked immunosorbent assay. J Clin Microbiol 26: 2351–2356
Siegel LS (1989) Evaluation of neutralizing antibodies to type A, B, E, and F botulinum toxins in sera from human recipients of botulinum pentavalent (ABCDE) toxoid. J Clin Microbiol 27: 1906–1908
Simonsen O (1989) Vaccination against tetanus and diphtheria. Evaluation of immunity in the Danish population, guidelines for revaccination, and methods for control of vaccination programs. Dan Med Bull 36: 24–47
Simonsen O, Kjeldsen K, Heron I (1984) Immunity against tetanus and effect of revaccination 25–30 years after primary vaccination. Lancet 2: 1240–1242
Simonsen O, Bentzon MW, Heron I (1986) ELISA for the routine determination of antitoxic immunity to tetanus. J Biol Stand 14: 231–239
Simonsen O, Schou C, Heron I (1987) Modification of the ELISA for the estimation of tetanus antitoxin in human sera. J Biol Stand 15: 143–157
Simpson LL, Lake P, Kozaki S (1990) Isolation and characterization of a novel human monoclonal antibody that neutralizes tetanus toxin. J Pharmacol Exp Ther 254: 98–103
Sjogren MH, McKee KT, Clayton MA, Franz DR, Condie RM, Peters CJ (1993) Safety and kinetics of an equine F(ab)2 heptavalent botulinum antitoxin in human volunteers. In: Kamely D, Bannister KA, Sasmor RM (eds) Army Science: the new frontiers: military and civilian applications. Borg Biomedical, Saratoga, WY, pp 157–158
Stevens RH, Saxon A (1978) Immunoregulation in humans: control of antitetanus toxoid antibody production after booster immunization. J Clin Invest 62: 1154–1160
Stiffler-Rosenberg G, Fey H (1977) Measurement of human tetanus antitoxinusing an enzyme-linked immunosorbent assay. Schweiz Med Wochenschr. 107:
Stover CK, de la Cruz VF, Fuerst TR, Burlein JE, Benson LA, Bennett LT, Bansal GP, Young JF, Lee MH, Hatfull GF, Snapper SB, Barletta RG, Jacobs WR Jr, Bloom BR (1991) New use of BCG for recombinant vaccines. Nature 351: 456–460
Sugiyama H, Brenner SL, DasGupta BR (1975) Detection of Clostridium botulinum toxin by local paralysis elicited with intramuscular challenge. Appl Microbiol 30: 420–423
Tacket CO, Shandera WX, Mann JM, Hargrett NT, Blake PA (1984) Equine antitoxin use and other factors that predict outcome in type A foodborne botulism. Am J Med 76: 794–798
Takahashi M, Komiya T, Kameyama S, Sakaguchi G (1990a) Titration of botulinum antitoxin of low levels by the score method. Jpn J Med Sci Biol 43: 163–170
Takahashi M, Noda H, Takeshita S, Fujiwara T, Nakanoin H, Mizunoya T, Sakaguchi G (1990b) Attempts to quantity Clostridium botulinum type A toxin and antitoxin in serum of two cases of infant botulism in Japan. Jpn J Med Sci Biol 43: 233–237
Thiele GM, Rogers J, Collins M, Yasuda N, Smith D, McDonald TL (1990) An enzyme-linked immunosorbent assay for the detection of antitetanus toxoid antibody using aluminium-absorbed coating antigen. J Clin Lab Anal 4: 126–129
Thompson DE, Brehm JK, Oultram JD, Swinfield T-J, Shone CC, Atkinson T, Melling J, Minton NP (1990) The complete amino acid sequence of the Clostridium botulinum type A neurotoxin, deduced by nucleotide sequence analysis of the encoding gene. Eur J Biochem 189: 73–81
Trabaud M-A, Lery L, Desgranges C (1989) Human monoclonal antibodies with a protective activity against tetanus toxin. Acta Pathol Microbiol Immunol Scand 97: 671–676
Velikanov IM (1934) Experimental immunization of man against botulism. Klin Med 12: 1802–1806

Vernacchio L, Madico G, Verastegui M, Diaz F, Collins TS, Gilman RH (1993) Neonatal tetanus in Peru: risk assessment with modified enzyme-linked immunosorbent assay and toxoid skin test. Am J Public Health 83: 1754–1756

Virella G, Fudenberg H, Kyong K, Pandey J, Galbriath R (1978) A practical method for quantitation of antibody responses to tetanus and diphtheria toxoids in normal young children. Z Immunitätsforsch Immunobiol 155: 80–86

Virella G, Hyman B, Kilpatrick JM (1985) Solid-phase enzymoimmunoassay of antitetanus toxoid antibodies. Diagn Immunol 3: 215–219

Volk WA, Bizzini B, Snyder RM, Bernhard E, Wagner RR (1984) Neutralization of tetanus toxin by distinct monoclonal antibodies binding to multiple epitopes on the toxin molecule. Infect Immun 45: 604–609

Weber JT, Hibbs RJ, Darwish A, Mishu B, Corwin AL, Rakha M, Hatheway CL, El-Sharkawy SS, El-Rahim SA, Al-Hamd FS, Sarn JE, Blake PA, Tauxe RV (1993) A massive outbreak of type E botulism associated with traditional salted fish in Cairo. J Infect Dis 167: 451–454

Wells JM, Wilson PW, Norton PM, Gasson MJ, LePage RWF (1993) Lactococcus lactis: high-level expression of tetanus toxin fragment C and protection against lethal challenge. Mol Microbiol 8: 1155–1162

WHO (1993) Expanded program on immunization. The global elimination of neonatal tetanus: progress to date. Wkly Epidemiol Rec 68: 277–282

WHO (1977) Manual for the production and control of vaccines: tetanus toxoid. BLG/UNDP/77.2 Rev 1. WHO, Geneva

WHO Expert Committee on Biological Standardization (1982) clostridium botulinum Type B Antitoxin: replacement of the international standard progress report. WHO/BS 82. 1352 (unpublished)

Wilde H, Chutivongse S (1990) Equine rabies immune globulin: a product with an undeserved poor reputation. Am J Trop Med Hyg 42: 175–178

Wilde H, Chomchey P, Prakongsri S, Punyaratabandhu P (1987) Safety of equine rabies immune globulin (letter) Lancet 2: 1275

Wilde H, Chomchey P, Prakongsri S, Punyaratabandhu P, Chutivongse S (1989a) Adverse effects of equine rabies immune globulin. Vaccine 7: 10–11

Wilde H, Chomchey P, Punyaratabandhu P, Phanupak P, Chutivongse S (1989b) Purified equine rabies immune globulin: a safe and affordable alternative to human rabies immune globulin. Bull World Health Organ 67: 731–736

Williamson LC, Fitzgerald SC, Neale EA (1992) Differential effects of tetanus toxin on inhibitory and excitatory neurotransmitter release from mammalian spinal cord cultures in culture.J Neurochem 59: 2148–2157

Wirz M, Gentili G, Collotti C (1990) Tetanus vaccine: present status. Adv Biotechnol Processes 13: 35–55

Wright GG, Duff JT, Fiock MA, Devlin HB, Soderstrom RL (1960) Studies on immunity to toxins of Clostridium botulinum. V.Detoxification of purified type A and B toxins, and the antigenicity of univalent and bivalent aluminum phosphate adsorbed toxoids. J Immunol 84: 384–389

Ziegler-Heitbrock HWL, Reiter C, Trenkmann J, Futterer A, Reithmuller G (1986) Protection of mice against tetanus toxin by combination of two human monoclonal antibodies recognizing distinct epitopes on the toxin molecule. Hybridoma 5: 21–31

Zuber M, Sebald M, Bathien N,de Recondo RJ, Rondot P (1993) Botulinum antibodies in dystonic patients treated with type A botulinum toxin: frequency and significance Neurology 43: 1715–1718

Clinical Use of Botulinum Neurotoxins

F. Cardoso[1] and J. Jankovic[2]

1 Introduction

Botulinum neurotoxin (BoNT) is the most potent biologic toxin known to a man (Gill 1982). Its clinical effects have been recognized since the end of the nineteenth century, when van Ermengem (1897) related botulism to a toxin produced by *Clostridium botulinum*, an anaerobic bacterium. The analysis of botulism outbreaks has led to the recognition that human intoxication may be caused by BoNT types A, B, E, and F (Green et al. 1983; Sakaguchi 1983). There are three other antigenically distinct toxins (C, D, and G) (Simpson 1981). These serotypes are produced by distinct strains of *C. botulinum* (Schantz and Johnson 1992). Although outbreaks of botulism have been identified as recently as in 1989 (Critchley et al. 1989), in the past two decades medical interest in BoNT has been fueled by the demonstration of its therapeutic usefulness. In 1973 Scott used BoNT type A for treatment of strabismus in nonhuman primates, and 8 years later he reported the results of his initial experience with this toxin to correct

[1]Department of Neurology, The Federal University of Minas Gerais, Av. Pasteur 89, Belo Horizonte, MG,1107, Brazil, 30150-290
[2]Department of Neurology, Baylor College of Medicine, 6550 Fannin, no. 1801, Houston, TX 77030, USA

strabismus in humans (SCOTT 1981). Since then BoNT injections have been demonstrated to be a safe and efficient therapy for a large number of neurologic and nonneurologic diseases (JANKOVIC and BRIN 1991). In December 1989 the United States Food and Drug Administration approved BoNT type A for the treatment of strabismus, blepharospasm, and hemifacial spasm in patients 12 years of age and older.

This chapter reviews the clinical experience with the therapeutic use of BoNT. Although the discussion is centered around BoNT type A (hereafter BoNT refers to BoNT type A unless otherwise stated), the serotype commercially available for clinical use, preliminary results of BoNT type F use are also mentioned (GREENE and FAHN 1993; LUDLOW et al. 1992). A monograph devoted entirely to the therapeutic use of BoNT was recently published (JANKOVIC and HALLETT 1994).

2 Production and Mechanism of Action

C. botulinum cultures grow for 24–36 h, at which time cells undergo lysis. The toxin liberated during lysis is activated by proteases present in the culture broth that convert protoxin into the highly potent toxin. The thus obtained BoNT is precipitated, extracted from the precipitated, acidified, and crystalized (HAMBLETON 1992; MELLING et al. 1988; SCHANTZ and JOHNSON 1994). The crystalline BoNT (900 kDa) contains nontoxic proteins bound noncovalently to the neurotoxic component of 150 kDa (DASGUPTA 1984). The secondary and tertiary structures of the neurotoxic component upon which the toxicity depends is stabilized by the nontoxic proteins (SCHANTZ and JOHNSON 1992). In England crystalline BoNT is prepared by anion-exchange chromatography and RNase treatment (DASGUPTA 1994; HAMBLETON 1992; MELLING et al. 1988; TSE et al. 1982). BoNT potency is measured by mouse units (U), and 1 U is the amount that kills 50% of 18- to 20-g female Swiss-Webster mice (the LD_{50}; SCHANTZ and JOHNSON 1990; SELLIN 1981). One nanogram of the United States BoNT (BOTOX) contains 2.5 U whereas 1 ng of the British toxin (Dysport) contains 40 U (QUINN and HALLETT 1989). For humans the lethal quantity of BoNT (BOTOX) is estimated to be 30–40 U/kg (HERRERO et al. 1967; GILL 1982; SCOTT and SUZUKI 1988).

BoNT causes weakness of the striate musculature by blockade of acetylcholine release at myoneural junction (GUYTON and MCDONALD 1947; SCHANTZ and JOHNSON 1992). The specificity for cholinergic motor nerve terminals is given by the selective location of BoNT acceptors at the presynaptic membrane of these synapses (BLACK and DOLLY 1987). BoNT binds with high affinity to these acceptors, and it is rapidly internalized by an energy-dependent process (DASGUPTA 1994; HAMBLETON 1992; MELLING et al. 1988; SIMPSON 1981, 1989). It appears that the heavy chain of BoNT neurotoxin moiety is capable of inducing channel formation in lipid membranes (SHONE et al. 1987). The molecular mechanism of BoNT was elucidated only quite recently (HUTTNER 1993). The light chain (50 kDa) of BoNT A acts as a zinc-dependent protease that selectively cleaves SNAP-25 (BLASI et al. 1993a; SCHIAVO et al. 1993a,b), a protein needed for vesicle targeting and fusion (SÖLLNER et al. 1993). VAMP/synaptobrevin are targets of types B, D, F and G (HUTTNER 1993; SCHIAVO et al. 1992, 1993 a,b,c, 1994; YAMASAKI et al. 1994), type C cleaves syntaxin (BLASI et al. 1993b) and type E cleaves SNAP-25 (SCHIAVO e t al. 1993 a,b).

The cleavage of any of these proteins prevents the acetylcholine-containing vesicles from fusing with the presynaptic membrane and releasing the neurotransmitter into the synaptic cleft, leading to a functional denervation. A few hours after intramuscular injection of BoNT there is a nearly complete decline in miniature end-plate potential (KAO et al. 1976; PESTRONK et al. 1976), loss of junctional acetylcholine receptors (AVILA et al. 1989), and development of extrajunctional acetylcholine receptors (BORODIC et al. 1991a). Despite the similarities between the effects of paralysis induced by BoNT and nerve section, SKETELJ et al. (1993) demonstrated that the toxin induces a slower and less intense decrease of the junctional A12 acetylcholinesterase. BORODIC et al. (1990, 1991a,b, 1994) found a denervation gradient up to 30 mm from the site of injection of 2–3 U BoNT. Beyond a 30-mm distance the intensity of acetylcholinesterase staining substantially decreases, consistent with histologic evidence of diminishing muscle atrophy. Although BoNT injection induces muscle changes immediately, there is a delay in the onset of clinical effect, possibly related to the spontaneous release of acetylcholine (STANLEY and DRACHMAN 1983). After exposure to BoNT the muscle fibers become atrophied, but 2–7 days later the axon terminal begins to sprout and the proliferating branches form new synaptic contacts on the adjacent muscle fibers (DUCHEN 1970, 1972; PAMPHLETT 1989; SHONE et al. 1987). The sprouting process is inhibited by α-bungarotoxin, an antagonist of acetylcholine receptors (PESTRONK and DRACHMAN 1978).

Several studies have demonstrated increased jitter and changes in fiber density on electromyography (EMG) in muscles remote from the BoNT injection site (GARNER et al. 1990; LANGE et al. 1987; OLNEY et al. 1988; SANDERS et al. 1986). The clinical relevance and duration of these findings is unclear; no objective weakness has been demonstrated in these patients. These remote effects, first detected 3–13 days after BoNT injections (GARNER et al. 1993), may result from hematogenic dispersion of the BoNT (COMELLA et al. 1992a; GARNER et al. 1993). Alternatively, they may be consequence of a local uptake and retrograde axonal transport via anterior horn motoneurons (COMELLA et al. 1992a). WIEGAND et al. (1976) showed that after intramuscular injection BoNT is transported to the spinal cord by retrograde axonal transport. In the spinal cord, where it may be transferred to the contralateral half-segment, the toxin blocks recurrent inhibition mediated by the Renshaw cells (WIEGAND and WELLHONER 1977).

3 Clinical Applications of Botulinum Neurotoxin

3.1 Strabismus and Other Disorders of Ocular Motility

BoNT was initially introduced as an alternative to surgery to realign the eyes of selected patients with congenital or acquired strabismus (American Academy of Ophthalmology 1989; Biglan et al. 1989; Magoon 1984, 1989; National Institutes of Health Consensus Development Conference Statement on Clinical Use of Botulinum Toxin 1991; Scott 1980; Scott et al. 1973, 1990). The toxin may also be used as an adjunct to surgery (National Institutes of Health Consensus Development Conference Statement on Clinical Use of Botulinum Toxin 1991). The injections are performed under electromyographic guidance, using a Teflon-coated needle to localize the extraocular muscle (American Academy of Ophthalmology 1989). The procedure is usually carried out in the office under local anesthesia although sedation with ketamine is employed in infants and children (American Academy of Ophthalmology 1989). BoNT injections are specially useful for horizontal strabismus with angles under 50 prism diopters (American Academy of Ophthalmology 1989; Magoon 1984, 1989; Scott 1980; Scott et al. 1990). After a follow-up of at least 6 months 33%–72% of patients with horizontal strabismus experienced a substantial improvement (a deviation of 10 prism diopters or less; American Academy of Opthalmology 1989; Biglan et al. 1989; Scott et al. 1990). Treatment of vertical strabismus by injecting the inferior rectus is effective but the results are modest in comparison to horizontal strabismus (Magoon and Dakoske 1985). This is due to the technical difficulties to inject other vertically acting muscles (American Academy of Ophthalmology 1989). In 6th nerve palsy and other forms of paralytic strabismus BoNT injections are useful for treatment of the antagonist muscle when paralysis is incomplete and as an adjunct to muscle surgery (American Academy of Ophthalmology 1989; Fitzsimons et al. 1989; National Institutes of Health Consensus Development Conference Statement on Clinical Use of Botulinum Toxin 1991; Rosenbaum et al. 1989; Scott and Kraft 1985). In these cases diplopia and late contractures may be prevented during the recovery phase (National Institutes of Health Consensus Development Conference Statement on Clinical Use of Botulinum Toxin 1991). BoNT injections have also been used to treat comitant strabismus (Biglan et al. 1989; Scott 1980), thyroid ophthalmopathy (Dunn et al. 1986), nystagmus (Helveston and Pogrebiniak 1988), and diplopia caused by internuclear ophthalmoplegia and skew deviation (Newman and Lambert 1992). Despite these reports, there are no controlled studies comparing BoNT injections and surgery for extraocular muscular dysfunction.

Disadvantages and side effects of BoNT therapy for strabismus are need for repeated injections (50%), blepharoptosis (17%–50%), vertical strabismus (17%), diplopia, spatial disorientation, and unstable alignment (American Academy of Ophthalmology 1989; National Institutes of Health Consensus Development Conference Statement on Clinical Use of Botulinum Toxin 1991). Although rare, globe perforation and retrobulbar hemorrhage may occur (National Institutes of Health Consensus Development Conference Statement on Clinical Use of Botulinum Toxin 1991). The side effects are usually transient and do not cause amblyopia.

3.2 Dystonia

Dystonia is a syndrome consisting of sustained and patterned contractions of muscles producing abnormal postures or repetitive twisting (e.g., torticollis) or squeezing (e.g., blepharospasm) movements (Stacy et al. 1993). Specific causes are found in about one-third of patients with dystonia (secondary dystonia), and in the remaining patients no specific etiology can be identified (primary or idiopathic dystonia; Jankovic and Fahn 1993). Included among secondary forms of dystonia is Wilson's disease, an autosomal recessive disease characterized by copper deposition in several organs, including the brain (Brewer and Yuzbasiyan-Gurkan 1992). Tardive dystonia results from exposure to dopamine receptor block drugs (Stacy et al. 1993). There are many other causes of secondary dystonias, the discussion of which is beyond the scope of this review (Jankovic and Fahn 1993).

Dystonia can be classified according to its distribution in focal dystonia, segmental dystonia, multifocal dystonia, generalized dystonia, and unilateral dystonia (hemidystonia; Jankovic and Fahn 1993). A marker for childhood-onset idiopathic generalized autosomal dominant dystonia in Jewish and non-Jewish patients was identified in the q32–q34 region of chromosome 9 (Ozelius et al. 1989, 1992). One study (Waddy et al. 1991) suggests that the same gene is responsible for adult-onset focal idiopathic dystonias but this awaits confirmation. The majority of patients with dystonia fail to improve with conventional pharmacologic medications (Jankovic and Brin 1991). BoNT injections are now regarded as the treatment of choice in focal and segmental dystonias and are indicated for the treatment of severe focal abnormalities in generalized forms (National Institutes of Health Consensus Development Conference Statement on Clinical Use of Botulinum Toxin 1991).

3.2.1 Blepharospasm

Blepharospasm is a form of focal dystonia characterized by intermittent or sustained closure of the eyes due to involuntary contractions of the orbicularis oculi (Jankovic and Fahn 1993). The clinical spectrum of this condition ranges from an increased blinking rate to functional blindness due to complete inability to open the eyelids. Approximately two-thirds of patients with blepharospasm also display involvement of other muscles of the face, oromandibular region, pharynx, larynx, and neck (Gandras et al. 1988; Jankovic and Fahn 1993). Clonazepam and trihexyphenidyl are the most useful drugs used to treat this condition, but they benefit fewer than 30% of patients, often at the expense of undesirable side effects (Clarke 1992; Gandras et al. 1988; Greene et al. 1988; Jankovic and Fahn 1993). Surgical procedures such as orbicularis oculi myectomy, facial nerve sectioning, and brow lift are effective initially, but the dystonia usually recurs within months (Clarke 1992; Talbot et al. 1982).

One controlled (Jankovic and Orman 1987) and a large number of open studies (Brin et al. 1987; Cohen et al. 1986; Dutton and Buckley 1988; Elston 1994; Jankovic and Schwartz 1993a; Jankovic et al. 1990b; Kraft and Lang 1988; Mauriello et al. 1987; Scott et al. 1985) have demonstrated the efficacy and safety of BoNT injections as treatment for blepharospasm. Despite different methods employed in the different studies, most of them show a moderate to marked improvement in 70%–100% (mean 90%) of treated patients (Elston 1994). The improvement typically starts 3–4 days after the injections and lasts an average of 3.5 months. Ptosis and diplopia, observed in up to 44% and 11% of injected patients, respectively, are the most common side effects (Clarke 1992). Other adverse effects include exposure keratitis, pain at the site of injection, entropion, conjunctivitis, increased tearing, and dry eye (Elston 1994). All these adverse experiences are self-limited, and none causes significant disability (Frueh et al. 1988; Jankovic and Brin 1991). Because of its proven safety and efficacy (National Institutes of Health Consensus Development Conference Statement on Clinical Use of Botulinum Toxin 1991) BoNT is now considered a primary form of therapy for blepharospasm (Report of the Therapeutics and Technology Assessment Subcommittee of the American Academy of Neurology 1990).

3.2.2 Cervical Dystonia

Cervical dystonia or spasmodic torticollis, the most common focal dystonia seen in a movement disorder clinic, is characterized by involuntary contractions of the neck muscles, causing patterned, repetitive, clonic (spasmodic) head movements or tonic (sustained) abnormal postures of the head as a result of twisting (torticollis), tilting toward one shoulder (laterocollis), flexing (anterocollis), or extending (retrocollis) the neck (Jankovic and Fahn 1993). Although torticollis, present in 70%–80% of patients, is the posture most commonly observed, the majority of patients display a combination of dystonic head deviations (Chan et al. 1991; Deuschl et al. 1992; Jankovic et al. 1991; Rondot et al. 1991). One-third of patients develop dystonia also in the oromandibular area, shoulder, arm, and trunk (Jankovic and Fahn 1993). Around 70% of patients with cervical dystonia complain of neck pain (Chan et al. 1991; Jankovic et al. 1991). Up to 38% of patients may develop spontaneous remission, but in most cases cervical dystonia is a persistent, life-long disorder (Jahanshshi et al. 1990; Jayne et al. 1984).

Trihexyphenidyl or other anticholinergics, benzodiazepines, dopamine depletors such as reserpine or tetrabenazine, carbamazepine, or the combination of these improve the cervical dystonia in up to 50% of patients (Jankovic and Brin 1991; Jankovic and Fahn 1993). Most patients, however, experience adverse effects that at times may be more disabling than the underlying disorder. Selective peripheral denervation and other surgical techniques may be a useful alternative in selected patients, but the results vary considerably. The benefits are moderate at best, and there is a high relapse and complication rate (BERTRAND and MOLINA-NEGRO 1988; CLARKE 1992).

The usefulness of BoNT injections as treatment for cervical dystonia has been demonstrated in several controlled and open studies (BLACKIE and LEES 1990; BRIN et al. 1987; GELB et al. 1989; GREENE et al. 1990; JANKOVIC et al. 1990b; JANKOVIC and SCHWARTZ 1990, 1993a; LEES et al. 1992; LORENTZ et al. 1991; POEWE et al. 1992; TSUI et al. 1986). In most reports (LEES et al. 1992; JANKOVIC and SCHWARTZ 1990; POEWE et al. 1992), 1 week after injections 80%–90% of patients experience significant improvement in posture and pain lasting an average of 4 months. In 14% of patients BoNT injections fail to provide benefit after one or more treatments, but only fewer than 7% of patients show no response to repeated treatment (LEES et al. 1992; JANKOVIC and SCHWARTZ 1990; POEWE et al. 1992). Complications and adverse effects of BoNT for cervical dystonia, observed in about one-quarter of injected patients (JANKOVIC and SCHWARTZ 1991a; LEES et al. 1992;LORENTZ et al. 1991; POEWE et al. 1992), include swallowing difficulties, neck weakness, pain at the site of injection, increased pain or strain in neighboring muscles, fatigue, malaise, hoarseness of the voice, enhanced head tremor, brachial plexopathy, and development of antibodies against BoNT (LEES et al. 1992; GLANZMAN et al. 1990; POEWE et al. 1992; REPORT OF THE THERAPEUTICS AND TECHNOLOGY ASSESSMENT SUBCOMMITTEE OF THE AMERICAN ACADEMY OF NEUROLOGY 1990). Dysphagia, the most common side effect of BoNT injections into the cervical muscles, is reported by an average of 14%–24% of injected patients (range 0%–90%) and is twice as common in female patients (BLACKIE and LEES 1990; BORODIC et al. 1990; BRIN et al. 1987; COMELLA et al. 1992b; GELB et al. 1989; GREENE et al. 1990; JANKOVIC and SCHWARTZ 1990; KOAY and ALUN-JONES 1989; LEES et al. 1992; LORENTZ et al. 1991; POEWE et al. 1992; TSUI et al. 1987b). This complication is self-limited and seldom severe enough to require a change to soft or liquid diet or even nasogastric feeding. The most likely explanation for the occurrence of dysphagia after BoNT injections is diffusion of the toxin into pharyngeal muscles (COMELLA et al. 1992b; JANKOVIC and FAHN 1993). Other mechanisms, however, such as BoNT remote effects and CNS action, may play a role in the genesis of dysphagia (COMELLA et al. 1992b). Longitudinal analyses of repeated injections of BoNT showed that chronic treatment of cervical dystonia with BoNT is not associated with any decline in benefit (JANKOVIC and SCHWARTZ 1990).

The most important determinants of a favorable response to BoNT injections for cervical dystonia are proper selection of the involved muscles and the toxin dose (JANKOVIC and BRIN 1991). Although one study showed slightly increased magnitude of benefit in cervical dystonia patients injected with EMG assistance (COMELLA et al. 1992a), most physicians define the muscles to be injected by careful clinical examination of the patients (JANKOVIC and FAHN 1993). EMG guidance is employed in rare circumstances when the target muscle cannot be palpated and in patients unresponsive to previous injections (Dubinski et al. 1991; Poewe and Wissel 1994). Borodic and colleagues (1992) showed that multiple-point injections into cervical muscles with BoNT are more efficient than single-point injections with respect to pain and posture deformity improvement.

3.2.3 Oromandibular Dystonia

Oromandibular dystonia is focal dystonia involving the masticatory, lower facial, and tongue muscles, causing trismus, bruxism, opening or lateral deviation of the jaw, and involuntary tongue movement (Jankovic 1988; Jankovic and Fahn 1993). Most patients fail to improve with pharmacotherapy, although baclofen, anticholinergic, and tetrabenazine may be helpful (Greene et al. 1988; Jankovic and Brin 1991; Jankovic and Fahn 1993).

Among patients with jaw-closing dystonia 70% of those treated with BoNT improve in chewing and speech whereas about 10% do not respond (Blitzer et al. 1989; Jankovic et al. 1990b). A few patients develop transient excessive weakness of jaw closure. The treatment of jaw-opening dystonia and lateral-deviation dystonia is more challenging, requiring thorough knowledge of the local anatomy and management of complications. But in experienced hands 50% of patients benefit from injections although many develop swallowing difficulties (Jankovic and Brin 1991; Jankovic et al. 1990b). The treatment of lingual dystonia is even more difficult because about 50% of injected patients have dysphagia; selected patients, however, obtain some benefit (Jankovic et al. 1990b).

3.2.4 Laryngeal Dystonia (Spasmodic Dysphonia)

Dystonia of the larynx, spasmodic dysphonia, is characterized by either adduction or abduction of the vocal cords (Blitzer et al. 1988b; Jankovic and Fahn 1993). The more common adductor spasmodic dysphonia is characterized by a strained, strangled voice which is often interrupted by voiceless pauses. The abductor form of spasmodic dysphonia is less common and is characterized by a breathy, whispering voice. A few patients display a combination of both forms (Pool et al. 1991). Pharmacotherapy, speech therapy, and psychotherapy do not improve spasmodic dysphonia (Jankovic and Brin 1991). Surgical treatment consists of sectioning of the recurrent laryngeal nerve (Dedo 1976). However, 3 years after the procedure only 36% of patients had improved and 48% were worse than before surgery (Aronson and De Santo 1983).

BoNT injections into the thyroarythenoid vocalis muscle complex is regarded as the treatment of choice for adductor spasmodic dysphonia, significantly improving the voice quality in 80%–100% of patients (American Academy of Otolaryngology 1990; Blitzer and Brin 1991; Blitzer et al. 1986, 1988a; Brin et al. 1987, 1989, 1992a; Ford et al. 1990; Gacek 1987; Jankovic et al. 1990b; Lees et al. 1992; Ludlow 1990; Ludlow et al. 1988, 1990; Miller et al. 1987; Report of the Therapeutics and Technology Assessment Subcommittee of the American Academy of Neurology 1990; Whurr et al. 1993; Zwirner et al. 1991). Complications are observed in 25% of injected patients and the most common ones are pain at the injection site and mild, transient dysphagia (WHURR et al. 1993). Unilateral or bilateral injections, EMG-guided percutaneous injections or injections under direct visualization have produced similar results (FORD et al. 1990; JANKOVIC and BRIN 1991; WHURR et al. 1993). The treatment of abductor spasmodic dysphonia is more challenging, but up to 75% of patients experience an 80% improvement (BLITZER and BRIN 1991; BRIN et al. 1990).

3.2.5 Limb Dystonia

Writer's cramp and other occupational cramps, such as pianist's and golfer's cramp, were considered psychogenic disorders in the past, but there is now compelling evidence that these conditions represent task-specific dystonias (JANKOVIC and SHALE 1989; MARSDEN and SHEEHY 1990; ROSENBAUM and JANKOVIC 1988; SACHDEV 1992). Despite treatment with muscle-relaxation techniques, physical and occupational therapy, pharmacotherapy, and surgery, most patients remain disabled (JANKOVIC and FAHN 1993). Several open trials (COHEN et al. 1989; JANKOVIC et al. 1990b; JANKOVIC and SCHWARTZ 1993b; KARP et al. 1994; LEES et al. 1992; RHOAD and STERN 1993; RIVEST et al. 1991; TSUI et al. 1987a) have demonstrated that BoNT injections, performed with or without EMG guidance, provide moderate to significant benefit in 58%–84% of patients. In a double-blind study TSUI and colleagues (1993) showed that 60% of patients with writer's cramp improved after BoNT injections. In one study of 30 patients with adequate follow-up at least two-thirds experienced moderate to marked improvement with the overall average duration of 9.3 weeks (JANKOVIC and SCHWARTZ 1993b). In another study, however, 24 of 37 patients (65%) followed over 2 years discontinued the treatment for various reasons, such as inadequate or loss of response, inaccessibility of a treatment provided, and high cost (KARP et al. 1994). Transient, usually not disabling, hand weakness has been the only reported side effect in all studies. BoNT injections may also benefit patients with other occupational hand movement disorders (JANKOVIC and BRIN 1991). Dystonia of the lower limbs associated with idiopathic torsion dystonia, parkinsonism, levodopa use and others produces pain and gait difficulties (JANKOVIC and FAHN 1993). Both symptoms have been shown to improve with BoNT injections (JANKOVIC and BRIN 1991; TSUI et al. 1990).

3.3 Hemifacial Spasm and Other Facial Dyskinesias

Hemifacial spasm, characterized by involuntary, paroxysmal clonic twitches or tonic contractions of the muscles supplied by the facial nerve, is thought to result from a compression of the 7th cranial nerve by an aberrant artery as it exits from the brainstem (CARDOSO and JANKOVIC 1995). Treatment with anticonvulsant drugs such as carbamazepine, phenytoin, and clonazepam results in mild improvement in only a few patients. Surgical decompression of the facial nerve in the posterior fossa often cures the movement disorder, but permanent facial paralysis, deafness, stroke, other complications, and recurrences are not uncommon (JANETTA et al. 1977). Several studies have demonstrated that injections of BoNT into the affected musculature completely abolishes the involuntary movements for about 5 months in almost all patients (BRIN et al. 1987; DUTTON and BUCKLEY 1988; ELSTON 1986, 1992; JANKOVIC et al. 1990b; KRAFT and LANG 1988; MAURIELLO 1985; MAURIELLO et al. 1987; TOLOSA et al. 1988). The United States Food and Drug Administration approved BoNT injections as a therapy for hemifacial spasm in 1989. Because of its high efficacy and safety, BoNT is considered the treatment of choice for this condition (BORODIC 1994; JANKOVIC and BRIN 1991).

Following peripheral facial paralysis patients may develop aberrant reinnervation which is phenomenologically similar to hemifacial spasm (BORODIC et al. 1993a). As with the latter, pharmacologic treatment is disappointing. BoNT injections have been shown to be an effective therapy for this condition (BORODIC et al. 1993a; DRESSLER and SCHONLE 1991; PUTTERMAN 1990). Levodopa-induced blepharospasm in parkinsonian patients, eyelid freezing triggered by blepharospasm, eyelid entropion, intractable orbicularis myokymia, tardive dystonia and stereotypies are other facial dyskinesias which have been treated effectively with BoNT (CLARKE and SPALTON 1988; JANKOVIC and BRIN 1991; JANKOVIC et al. 1990a; JORDAN et al. 1989; KATZ and ROSENBERG 1987).

3.4 Other Indications

Tremor of the hands and head often improves with medications, but in patients with high-amplitude or task-specific tremors pharmacotherapy is disappointing (JANKOVIC and BRIN 1991). Surgical treatment, thalamotomy, or thalamic stimulation, has proven an effective treatment in 75%–85% of patients with disabling tremors (BENABID et al. 1991; GROSSMAN and HAMILTON 1993). One study (JANKOVIC and SCHWARTZ 1991b) demonstrated that BoNT injections may represent a useful therapy for severe head and hand tremor. In this open trial reduction in the amplitude of the tremor was observed in 67% of patients. Although weakness was reported by 60% of patients, they preferred this complication to having disabling tremor. Another study (TROSCH and PULLMANN 1994) confirmed that BoNT injections may improve tremor in some patients. Voice tremor, a condition usually resistant to conventional pharmacologic treatment, has been successfully treated with BoNT (LUDLOW et al. 1989). In some patients stuttering improves after laryngeal BoNT injections (BLITZER and BRIN 1991; BRIN et al. 1992).

Tics are relatively brief and intermittent movements (motor tics) or sounds (vocal tics) that occur abruptly and, although they may be coordinated and repetitive, are not continuous, and the patients can usually suppress them (STACY et al. 1993). Although in most patients antidopaminergic drugs control tics effectively, some patients with disabling eyelid and cervical tics clearly benefit from BoNT injections (JANKOVIC 1994a,b). In addition to controlling the motor component of the tic, BoNT appears to also diminish or eliminate the premonitory sensory phenomenon experienced by patients with dystonic tics (JANKOVIC 1994b).

Controlled and open studies demonstrated the usefulness of BoNT injections to treat spasticity (DAS and PARK 1989; SNOW et al. 1990; TSUI and O'BRIEN 1993). Increased tone in Stiff-Person syndrome, an autoimmune disorder associated with antibodies against glutamic acid decarboxylase (BLUM and JANKOVIC 1991), has been reported to improve after BoNT treatment (DAVIS and JABBARI 1993). Bladder dysfunction in patients with spinal-cord injury has been treated successfully with BoNT injections into the detrusor and sphincter muscles (DYKSTRA 1994; DYKSTRA and SIDI 1990; DYKSTRA et al. 1988). Local injections of BoNT may improve anismus, intractable constipation caused by spasm of the rectal sphincter (HALLAN et al. 1988; PASRICHA and KALOO 1994). Patients with segmental myoclonus and other hyperkinetic disorders may also benefit from BoNT injections (JANKOVIC and BRIN 1991; POLO and JABBARI 1994). BoNT has also been used for cosmetic reasons to treat glabelar frown lines and other facial wrinkles (CARRUTHERS and CARRUTHERS 1994). Because of its ability to block the activity of the neuromuscular transmission, BoNT may play a role in the management of all conditions characterized by hyperactivity of the lower motor neuron.

Absolute contraindications to the use of BoNT are allergy to the drug and infection or inflammation at the proposed injection site whereas myasthenia gravis, Eaton-Lambert syndrome, motor neuron disease, and coagulopathy are relative contraindications (NATIONAL INSTITUTES OF HEALTH CONSENSUS DEVELOPMENT CONFERENCE STATEMENT ON CLINICAL USE OF BOTULINUM TOXIN 1991; REPORT OF THE THERAPEUTICS AND TECHNOLOGY ASSESSMENT SUBCOMMITTEE OF THE AMERICAN ACADEMY OF NEUROLOGY 1990). Safety for use during pregnancy and lactation has not been firmly established (NATIONAL INSTITUTES OF HEALTH CONSENSUS DEVELOPMENT CONFERENCE STATEMENT ON CLINICAL USE OF BOTULINUM TOXIN 1991).

4 Clinical Significance of Botulinum Toxin Antibodies

Antibodies against BoNT may be present in the serum of patients exposed to toxin for therapeutic reasons (JANKOVIC and BRIN 1991; NATIONAL INSTITUTES OF HEALTH CONSENSUS DEVELOPMENT CONFERENCE STATEMENT ON CLINICAL USE OF BOTULINUM TOXIN 1991). An in vivo neutralization assay is the most reliable method to demonstrate these antibodies (HATHEWAY et al. 1984) although enzyme-linked immunosorbent assays have been developed (DEZFULIAN and BARTLETT 1984; TSUI et al. 1988). Only those antibodies detected by the mouse bioassay have been found to correlate with the absence of clinical response to subsequent injections. These antibodies have been identified in 35%–60% of patients with cervical dystonia who failed to respond to BoNT after repeated injections (HAMBLETON et al. 1992; JANKOVIC and SCHWARTZ 1991a). The lack of improvement in antibody-negative patients may be related to injection technique problems, such as insufficient injected dose and inappropriate muscle selection. Among unselected patients with cervical dystonia treated with BoNT 3%–14% of them have been shown to have circulating antibodies to the toxin (TSUI et al. 1988; ZUBER et al. 1993). The observation that patients with circulating antibodies to BoNT detected by assays other than the mouse bioassay remain responsive to repeated injections (DEZFULIAN and BARTLETT 1984; TSUI et al. 1988) suggests that these antibodies are not biologically active and may be directed to other components of the BoNT complex. Other explanations include low antibody titer, insufficient to completely inactivate the toxin, or a placebo effect of the injections (ZUBER et al. 1993). Patients who receive higher cumulative doses of BoNT are more likely to develop antibodies (GREENE et al. 1994; HAMBLETON et al. 1992; JANKOVIC and SCHWARTZ 1991a; NATIONAL INSTITUTES OF HEALTH CONSENSUS DEVELOPMENT CONFERENCE STATEMENT ON CLINICAL USE OF BOTULINUM TOXIN 1991; ZUBER et al. 1993). Accordingly, in disorders treated with lower BoNT doses, such as strabismus, blepharospasm, and hemifacial spasm, the occurrence of BoNT antibodies is extremely rare (BIGLAN et al. 1986; BORODIC et al. 1993b; GONNERING 1988). Short-interval between successive injections has been suggested as a potential risk factor for the development of BoNT antibodies (GREENE et al. 1994; NATIONAL INSITITUTES OF HEALTH CONSENSUS DEVELOPMENT CONFERENCE STATEMENT ON CLINICAL USE OF BOTULINUM TOXIN 1991; TSUI et al. 1988; ZUBER et al. 1993).

Patients who became nonresponsive to BoNT because of the development of antibodies have been successfully treated with toxin type F (GREENE and FAHN 1993; LUDLOW et al. 1992). However, the length of benefit was reduced as compared with type A toxin. Recently, an experimental study demonstrated that botulinum toxin type B produces regional denervation similar to type A toxin (BORODIC et al. 1993c). It remains to be proven whether this serotype may be clinically useful.

References

American Academy of Ophthalmology (1989) Botulinum toxin therapy of eye muscle disorders: safety and effectiveness. Ophthalmology 96(2): 37–41

American Academy of Otolaryngology (1990) Position statement on the clinical usefulness of botulinum toxin in the treatment of spasmodic dysphonia. Arch Otolaryngol Head Neck Surg Bull 9: 8

Aronson AE, De Santo LW (1983) Adductor spastic dysphonia: three years after recurrent laryngeal nerve section. Laryngoscope 93: 1–8

Avila OL, Drachman DB, Pestronk A (1989) Neurotransmission regulates stability of acetylcholine receptors at the neuromuscular junction. J Neurosci 9: 2902–2906

Benabid AL, Pollak P, Gervason C, Hoffman D, Gao DM, Hommel M, Perret SE, de Rougemont S (1991) Long-term suppression of tremor by chronic stimulation of the ventral intermediate thalamic nucleus. Lancet 337: 403–406

Bertrand CM, Molina-Negro P (1988) Selective peripheral denervation in 111 cases of spasmodic torticollis: rationale and results. Adv Neurol 50: 637–643

Biglan AW, Gonnering R, Lockhart LB, Rabin B, Fuerste FH (1986) Absence of antibody production in patients treated with botulinum A toxin. Am J Opthalmol 101: 232–235

Biglan AW, Burnstine RA, Rogers GL, Saunders RA (1989) Management of strabismus with botulinum toxin A. Opthalmology 96: 935–943

Black JD, Dolly JO (1987) Selective location of acceptors for botulinum neurotoxin A in the central and peripheral nervous systems. Neuroscience 23: 767–779

Blackie JD, Lees AJ (1990) Botulinum toxin treatment in spasmodic torticollis. J Neurol Neurosurg Psychiatry 53: 640–643
Blasi J, Chapman ER, Link E, Binz T, Yamasaki S, DeCamilli P, Südhof T, Niemann H, Jahn R (1993a) Botulinum neurotoxin A selectively cleaves the synaptic protein SNAP-25. Nature 365: 160–163
Blasi J, Chapman ER, Yamasaki S, Binz T, Niemann H, Jahn R (1993b) Botulinum neurotoxin C blocks neurotransmitter release by means of cleaving HPC-1/syntaxin. EMBO J 12: 4821–4828
Blitzer A, Brin MF (1991) Laryngeal dystonia: a series with botulinum toxin therapy. Ann Otol Rhinol Laryngol 100: 85–89
Blitzer A, Brin MF, Fahn S, Lovelace RE (1986) Botulinum toxin for the treatment of spastic dysphonia. Laryngoscope 96: 1300–1301
Blitzer A, Brin MF, Fahn S, Lovelace RE (1988a) Localized injections of botulinum toxin for the treatment of focal laryngeal dystonia (spastic dysphonia). Laryngoscope 98: 193–197
Blitzer A, Brin MF, Fahn S, Lovelace RE (1988b) Clinical and laboratory characteristics of focal laryngeal dystonia: study of 110 cases. Laryngoscope 98: 636–640
Blitzer A, Brin MF, Fahn S (1989) Botulinum toxin injection for the treatment of oromandibular dystonia. Ann Otol Rhinol Laryngol 98: 93–97
Blum P, Jankovic J (1991) Stiff-person syndrome: an autoimmune disease. Mov Disord 6: 12–20
Borodic GE (1994) Hemifacial spasm: evaluation and management, with emphasis on botulinum toxin therapy. In: Jankovic J, Hallet M (eds) Therapy with botulinum toxin. Dekker, New York, pp 331–351
Borodic GE, Joseph M, Fay L, Cozzolino D, Ferrante R (1990) Botulinum A toxin for the treatment of spasmodic torticollis: dysphagia and regional toxin spread. Head Neck 12: 392–398
Borodic GE, Cozzolino D, Ferrante R, Wegner AW, Young RR (1991a) Innervation zone of orbicularis oculi muscle and implications for botulinum A therapy. Ophthalmic Plast Reconstr Surg 7: 54–60
Borodic GE, Pearce LB, Johnson EA, Schantz EJ (1991b) Clinical and scientific aspects of botulinum A toxin. Ophthalmol Clin North Am 4: 491–503
Borodic GE, Pearce LB, Smith K, Joseph M (1992) Botulinum A toxin for spasmodic torticollis: multiple vs single injection points per muscle. Head Neck 14: 33–37
Borodic GE, Pearce LB, Cheney M, Metson R, Brownstone D, Townsed D, McKenna M (1993a) Botulinum A toxin for treatment of aberrant facial nerve regeneration. Plast Reconstr Surg 91: 1042–1045
Borodic GE, Pearce LB, Ferrante R (1993b) Therapeutic botulinum toxin: histologic effects and diffusion properties. In: DasGupta BR (ed) Botulinum and tetanus neurotoxins. Plenum, New York, pp 623–645
Borodic GE, Pearce LB, Smith KL et al (1993c) Botulinum B toxin as an alternative to botulinum A toxin: a histologic study. Ophthalmic Plast Reconstr Surg 9: 182–190
Borodic GE, Ferrante R, Pearce LB, Smith K (1994) Histologic assessment of dose-related diffusion and muscle fiber response after therapeutic botulinum A toxin injections. Mov Disord 9: 31–39
Brewer GJ, Yuzbasiyan-Gurkan V (1992) Wilson disease. Medicine 71: 139–164
Brin MF, Fahn S, Moskowitz C, Friedman A, Shale HH, Greene PE, Blitzer A, List T, Lange D, Lovelace RE, et al. (1987) Localized injections of botulinum toxin for the treatment of focal dystonia and hemifacial spasm. Mov Disord 2: 237–254
Brin MF, Blitzer A, Fahn S, Gould W, Lovelace RE (1989) Adductor laryngeal dystonia (spastic dysphonia): treatment with local injections of botulinum toxin (BOTOX). Mov Disord 4: 287–296
Brin MF, Blitzer AB, Stewart C, Fahn S (1990) Botulinum toxin: now for abductor laryngeal dystonia. Neurology 40 [Suppl 1]: 381
Brin MF, Blitzer A, Stewart C, Fahn S (1992a) Treatment of spasmodic dysphonia (laryngeal dystonia) with local injections of botulinum toxin: review and technical aspects. In: Blitzer A, Brin MF, Sasaki CT, Fahn S, Harris KS (eds) Neurologic disorders of the larynx. Thieme, Stuttgart, pp 214–228
Brin MF, Stewart C, Viswanath NS et al. (1992b) Pilot study: laryngeal botulinum toxin (Botox) injections for disabling stuttering in adults. Neurology 42 [Suppl 3]: 376
Cardoso F, Jankovic J (1995) Blepharospasm. In: Tsui JKC, Calne DB (eds) Handbook of dystonia. Dekker, New York, (in press)
Carruthers A, Carruthers JDA (1994) Botulinum toxin in the treatment of glabellar frown lines and other facial wrinkles. In: Jankovic J, Hallet M (eds) Therapy with botulinum toxin. Dekker, New York, pp 577–596
Chan J, Brin MF, Fahn S (1991) Idiopathic cervical dystonia: clinical characteristics. Mov Disord 6: 119–126
Clarke CE (1992) Therapeutic potential of botulinum toxin in neurological disorders. Q J Med 299: 197–205
Clarke JR, Spalton DJ (1988) Treatment of senile entropion with botulinum toxin. Br J Ophthalmol 72: 361–362

Cohen DA, Savino PJ, Stern MB, Hurtig HI (1986) Botulinum injection therapy for blepharospasm: a review and report of 75 patients. Clin Neuropharmacol 9: 415–429

Cohen LG, Hallet M, Geller BD, Hochberg F (1989) Treatment of focal dystonias of the hand with botulinum toxin injections. J Neurol Neurosurg Psychiatry 52: 355–363

Comella CL, Buchman AS, Tanner CM, Brown-Toms NC, Goetz CG (1992a) Botulinum toxin injections for spasmodic torticollis: increased magnitude of benefit with electromyographic assistance. Neurology 42: 878–882

Comella CL, Tanner CM, DeFoor-Hill L, Smith C (1992b) Dysphagia after botulinum toxin injections for spasmodic torticollis: clinical and radiologic findings. Neurology 42: 1307–1307

Critchley EMR, Hayes PJ, Isaacs PET (1989) Outbreak of botulism in north west England and Wales. Lancet 2: 849–853

Das TK, Park DM (1989) Effect of treatment with botulinum toxin on spasticity. Postgrad Med J 65: 208–210

DasGupta BR (1989) The structure of botulinum neurotoxin. In: Simpson LL (ed) Botulinum neurotoxin and tetanus toxin. Academic San Diego pp 53–66

DasGupta BR (1994) Structures of botulinum neurotoxin, its functional domains, and perspectives on the crystalline type A toxin. In: Jankovic J, Hallett M (eds) Therapy with botulinum toxin. Dekker, New York, 15–40

Davis D, Jabbari B (1993) Significant improvement of stiff-person syndrome after paraspinal injection of botulinum toxin A. Mov Disord 8: 371–373

Dedo HH (1976) Recurrent laryngeal nerve section for spastic dysphonia. Ann Otol Rhinol Laryngol 85: 451–459

Deuschl G, Heinen F, Kleedorfer B, Wagner M, Lucking CH, Poewe UV et al (1992) Clinical and polymyographic investigation of spasmodic torticollis. J Neurol 239: 9–15

Dezfulian M, Bartlett JG (1984) Detection of Clostridium botulinum type A toxin toxin by enzyme-linked immunosorbent assay with antibodies produced in immunologically tolerant animals. J Clin Microbiol 19: 645–648

Dressler D, Schonle PW (1991) Hyperkinesias after hypoglossofacial nerve anastomosis—treatment with botulinum toxin. Eur Neurol 31: 44

Dubinsky RM, Gray CS, Vetere-Overfield B, Koller WC (1991) Electromyographic guidance of botulinum toxin treatment in cervical dystonia. Clin Neuropharmacol 14: 14: 262–267

Duchen LW (1970) Changes in motor innervation and cholinesterase localization induced by botulinum toxin in skeletal muscle of the mouse: differences between fast and slow muscles. J Neurol Neurosurg Psychiatry 33: 40–54

Duchen LW (1972) Motor nerve growth induced by botulinum toxin as a regenerative phenomenon. Proc R Soc Med 65: 196–197

Dunn WJ, Arnold AC, O'Connor PS (1986) Botulinum toxin for the treatment of dysthyroid ocular myopathy. Ophthalmology 93: 470–475

Dutton JJ, Buckley EG (1988) Long-term results and complications of botulinum A toxin in the treatment of blepharospasm. Ophthalmology 95: 1529–1534

Dykstra DD (1994) Effects of botulinum A toxin on detrusor-sphincter dyssinergia in spinal cord injury patients. In: Jankovic J, Hallet M (eds) Therapy with botulinum toxin. Dekker, New York, pp 535–542

Dykstra DD, Sidi AA (1990) Treatment of detrusor-sphincter dyssinergia with botulinum A toxin: a double-blind study. Arch Phys Med Rehabil 71: 24–26

Dykstra DD, Sidi A, Scott AB, Pagel JM, Goldfish GD (1988) Effects of botulinum A toxin on detrusor-sphincter dyssenergia in spinal cord injury patients. J Urol 139: 912–922

Elston JS (1986) Botulinum toxin treatment of hemifacial spasm. J Neurol Neurosurg Psychiatry 49: 827–829

Elston J (1992) The management of blepharospasm and hemifacial spasm. J Neurol 239: 5–8

Elston JS (1994) Botulinum toxin for blepharospasm. In: Jankovic J, Hallet M (eds) Therapy with botulinum toxin. Dekker, New York pp 191–198

Erenberg G (1992) Treatment of Tourette syndrome with neuroleptic drugs. Adv Neurol 58: 241–243

Fitzsimons R, Lee J, Elston J (1989) The role of botulinum toxin in the management of sixth nerve palsy. Eye 3: 391–400

Ford CN, Bless DM, Lowery JD (1990) Indirect laryngoscopic approach for injection of botulinum toxin in spasmodic dysphonia. Otolaryngol Head Neck Surg 103: 752–758

Frueh BR, Nelson CC, Kapustiak JF, Musch DC (1988) The effect of omitting botulinum toxin from the lower eyelid in blepharospasm treatment. Am J Ophthalmol 106: 46–47

Gacek RR (1987) Botulinum toxin for relief of spasmodic dysphonia. Arch Otolaryngol Head Neck Surg 113: 1240
Gandras F, Elston J, Quinn N, Marsden CD (1988) Blepharoespasm: a review of 264 patients. J Neurol Neurosurg Psychiatry 51: 767–772
Garner CG, Straube A, Gasser T et al. (1990) Early-single-fibre-EMG findings in patients with therapeutic use of botulinum toxin distant to the site of treatment. Mov Disord 5: 76
Garner CG, Straube A, Witt TN, Gasser T, Oertel WH (1993) Time course of distant effects of local injections of botulinum toxin. Mov Disord 8: 33–37
Gelb DJ, Lowenstein DH, Aminoff MF (1989) Controlled trial of botulinum toxin injections in the treatment of spasmodic torticollis. Neurology 39: 80–84
Gill DM (1982) Bacterial toxins: a table of lethal amounts. Microbiol Rev 46: 86–94
Glanzman KL, Gelb DJ, Drury MB, Bromberg MB, Truong DD (1990) Brachial plexopathy after botulinum toxin injections. Neurology 40: 1143
Gonnering RS (1988) Negative antibody response to long-term treatment of facial spasms with botulinum toxin. Am J Ophthalmol 105: 313–315
Green J, Spear H, Brinson RR (1983) Human botulism (type F)—a rare type. Am J Med 75: 893–895
Greene PE, Fahn S (1993) Use of botulinum toxin type F to treat torticollis patients with immunity to botulinum toxin type A. Mov Disord 8: 479–483
Greene P, Fahn S, Diamond B (1994) Development of resistance to botulinum toxin type A in patients with tortivollis. Mov Disord 9: 213–217
Greene P, Shale H, Fahn S (1988) Analysis of open-labels trials in torsion dystonia using high dosages of anticholinergics and other drugs. Mov Disord 3: 46–60
Greene P, Kang U, Fahn S, Brin M, Moskowitz C, Flaster E (1990) Double-blind, placebo-controlled trial of botulinum toxin injections for the treatment of spasmodic torticollis. Neurology 40: 1213–1218
Grossman RG, Hamilton WJ (1993) Surgery for movement disorders. In: Jankovic J, Tolosa E (eds) Parkinson's disease and movement disorders—2. Williams and Wilkins, Baltimore, pp 531–548
Guyton AC, McDonald MA (1947) Physiology of botulinus toxin. Arch Neurol Psychiatry 57: 578–592
Hallan RI, Williams NS, Melling J, Waldron DJ et al (1988) Treatment of anismus in intractable constipation with botulinum A toxin. Lancet 2: 714–717
Hambleton P (1992) Clostridium botulinum toxins: a general review of involvement in disease, structure, mode of action and preparation for clinical use. J Neurol 239: 16–20
Hambleton P, Cohen HE, Palmer BJ, Melling J (1992) Antitoxins and botulinum toxin treatment. BMJ 304: 959–960
Hatheway CH, Snyder JD, Seals JE, Edell TA, Lewis GE Jr et al (1984) Antitoxin levels in botulism patients treated with trivalent equine botulism antitoxin to toxin types A, B, and E. J Infect Dis 150: 407–412
Helveston EM, Pogrebiniak AE (1988) Treatment of acquired nystagmus with botulinum A toxin. Am J Opthalmol 106: 584–586
Herrero BA, Ecklung AE, Streett CS, Ford DF, King JK (1967) Experimental botulism in monkeys—a clinical pathological study. Exp Mol Pathol 6: 84–95
Huttner WB (1993) Snappy exocytotoxins. Nature 365: 104–105
Jahanshahi M, Marion M-H, Marsden CD (1990) Natural history of adult-onset idiopathic torticollis. Arch Neurol 47: 548–552
Janetta PJ, Abbasy M, Maroon JC, Ramos FM, Albin MS (1977) Etiology and definitive microsurgical treatment of hemifacial spasm. J Neurosurg 47: 321–328
Jankovic J (1988) Etiology and differential diagnosis of blepharospasm and oromandibular dystonia. Adv Neurol 49: 103–116
Jankovic J (1994a) Treatment of tics with botulinum toxin. In: Jankovic J, Hallet M (eds) Therapy with botulinum toxin. Dekker, New York, pp 503–510
Jankovic J (1994b) Botulinum toxin in the treatment of dystonic tics. Mov Disord 9: 347–349
Jankovic J, Brin MF (1991) Therapeutic uses of botulinum toxin. N Engl J Med 324: 1186–1194
Jankovic J, Fahn S (1993) Dystonic syndromes. In: Jankovic J, Tolosa E (eds) Parkinson's disease and movement disorders—2. Williams and Wilkins, Baltimore, pp 337–374
Jankovic J, Hallett M (eds) (1994) Therapy with botulinum toxin. Dekker, New York
Jankovic J, Orman J (1987) Botulinum A toxin for cranial-cervical dystonia: a double-blind, placebo-controlled study. Neurology 37: 616–623
Jankovic J, Schwartz K (1990) Botulinum toxin injections for cervical dystonia. Neurology 40: 277–280
Jankovic J, Schwartz K (1991a) Clinical correlates of response to botulinum toxin injections. Arch Neurol 48: 1253–1256

Jankovic J, Schwartz K (1991b) Botulinum toxin treatment of tremors. Neurology 41: 1185–1188
Jankovic J, Schwartz K (1993a) Longitudinal follow-up of botulinum toxin injections for treatment of blepharospasm and cervical dystonia. Neurology 43: 834–836
Jankovic J, Schwartz K (1993b) The use of botulinum toxin in the treatment of hand dystonias. J Hand Surg 18A: 883–887
Jankovic J, Shale H (1989) Dystonia in musicians. Semin Neurol 9: 131–135
Jankovic J, Friedman D, Pirozzolo FJ, McCrary JA (1990a) Progressive supranuclear palsy: motor, neurobehavioral, and neuro-opthalmic findings. Adv Neurol 53: 293–304
Jankovic J, Schwartz K, Donovan D (1990b) Botulinum toxin treatment of cranial-cervical dystonia, spasmodic dysphonia, other focal dystonias and hemifacial spasms. J Neurol Neurosurg Psychiatry 53: 633–639
Jankovic J, Leder S, Warner D, Schwartz K (1991) Cervical dystonia: clinical findings and associated movement disorders. Neurology 41: 1088–1091
Jayne D, Lees AJ, Stern GM (1984) Remission in spasmodic torticollis. J Neurol Neurosurg Psychiatry 47: 1236–1237
Jordan DR, Anderson RL, Thiese SM (1989) Intractable orbicularis myokymia: treatment alternatives. Ophthalmic Surg 20: 280–283
Kao I, Drachman DB, Price DL (1976) Botulinum toxin: mechanism of presynaptic blockade. Science 193: 1256–1258
Karp BI, Cole RA, Cohen LG et al (1994) Long-term botulinum toxin treatment of focal hand dystonia. Neurology 44: 70–76
Katz B, Rosenberg JH (1987) Botulinum therapy for apraxia of eyelid opening. Am J Ophthalmol 103: 718–719
Koay CE, Alun-Jones T (1989) Pharyngeal paralysis due to botulinum toxin. J Laryngol Otol 103: 698–699
Kraft SP, Lang AE (1988) Botulinum toxin injections in the treatment of blepharospasm, hemifacial spasm, and eyelid fasciculations. Can J Neurol Sci 15: 276–280
Lange DJ, Brin MF, Warner CL, Fahn S, Lovelace RE (1987) Distant effects of local injection of botulinum toxin. Muscle Nerve 10: 552–555
Lees AJ, Turjanski N, Rivest J, Whurr R, Lorch M, Brookes G (1992) Treatment of cervical dystonia, hand spasms, and laryngeal dystonia with botulinum toxin. J Neurol 239: 1–4
Lorentz IT, Subramanian SS, Yiannikas C (1991) Treatment of idiopathic spasmodic torticollis with botulinum toxin A: a double-blind study on twenty-three patients. Mov Disord 6: 145–150
Ludlow CL (1990) Treatment of speech and voice disorders with botulinum toxin. JAMA 264: 2671–2675
Ludlow CL, Naunton RF, Sedory SE, Schulz GM, Hallett M (1988) Effects of botulinum toxin injections on speech in adductor spasmodic dysphonia. Neurology 38: 1220–1225
Ludlow CL, Sedory SE, Fujita M, Naunton RF (1989) Treatment of voice tremor with botulinum toxin injection. Neurology 39 [Suppl 1]: 353
Ludlow CL, Naunton RF, Jujita M, Sedory SE (1990) Spasmodic dysphonia: botulinum injection after recurrent nerve surgery. Otolaryngol Head Neck Surg 102: 122–131
Ludlow CL, Hallett M, Rhew K, Cole R, Shimizu T, Sakaguchi G, Bagley JA, Schulz GM, Yim SG, Koda S (1992) Therapeutic use of type F botulinum toxin. N Engl J Med 326: 349–350
Magoon EH (1984) Botulinum toxin chemo-denervation for strabismus in infants and children. J Pediatr Ophthalmol Strabismus 21: 110–113
Magoon EH (1989) Chemodenervation of strabismic children: a 2- to 5-year follow-up study compared with shorter follow-up. Ophthalmology 96: 931–934
Magoon EH, Dakoske C (1985) Botulinum toxin injection for vertical strabismus. Am Orthop J 35: 48–52
Marsden CD, Sheehy MP (1990) Writer's cramp. Trends Neurosci 13: 148–153
Mauriello JA (1985) Blepharospasm, Meige syndrome, and hemifacial spasm: treatment with botulinum toxin. Neurology 35: 1499–1500
Mauriello JA Jr, Coniaris H, Haupt EJ (1987) Use of botulinum toxin in the treatment of one hundred patients with facial dyskinesias. Ophthalmology 94: 976–979
Melling J, Hambleton P, Shone CC (1988) Clostridium botulinum toxins: nature and preparation for clinical use. Eye 2: 16–23
Miller RH, Woodson GE, Jankovic J (1987) Botulinum toxin injection of the vocal fold for spasmodic dysphonia. A preliminary report. Arch Otolaryngol Head Neck Surg 113: 603–605
National Institutes of Health Consensus Development Conference (1991) Statement on Clinical Use of Botulinum Toxin. Arch Neurol 48: 1294–1298
Newman NJ, Lambert SR (1992) Botulinum toxin treatment of supranuclear ocular motility disorders. Neurology 42: 1391–1393

Olney RK, Aminoff MJ, Gelb DJ, Lowenstein DH (1988) Neuromuscular effects distant from the site of botulinum neurotoxin injection. Neurology 38: 1780–1783
Ozelius L, Kramer PL, Moskowitz CB, Kwiatkowski DJ, Brin MF, Bressman SB, Schuback DE, Falk CT, Risch M, de Leom D, et al. (1989) Human gene for torsion dystonia located on chromosome 9q 32–34. Neuron 2: 1427–1434
Ozelius LJ, Kramer PL, de Leon D, Risch N, Bressman SB, Schuback DE, Brin MF, Kwiatkowski DJ, Burke RE, Gusella JF et al. (1992) Strong allelic association between the torsion dystonia gene (DYT1) and loci in chromosome 9q34 in Ashkenazi Jews. Am J Hum Genet 50: 619–628
Pamphlett R (1989) Early terminal and nodal sprouting of motor axons after botulinum toxin. J Neurol Sci 92: 181–192
Pasricha PJ, Kaloo AN (1994) Botulinum toxin in the treatment of gastrointestinal disorders. In: Jankovic J, Hallet M (eds) Therapy with botulinum toxin. Dekker, New York, pp 543–556
Pestronk A, Drachman DB (1978) Motor nerve sprouting and acetylcholine receptors. Science 199: 1223–1225
Pestronk A, Drachman DB, Griffin JW (1976) Effect of botulinum toxin on trophic regulation of acetylcholine receptors. Nature 264: 787–789
Poewe W, Schelosky L, Kleedorfer B, Heimen F, Wagner M, Deuschl G et al. (1992) Treatment of spasmodic torticollis with local injections of botulinum toxin. One-year follow-up in 37 patients. J Neurol 239: 21–25
Poewe W, Wissel J (1994) Experience with botulinum toxin in cervical dystonia. In: Jankovic J, Hallet M (eds) Therapy with botulinum toxin. Dekker, New York, pp 267–278
Polo KB, Jabberi B (1994) Effectiveness of botulinum toxin type A against painful limb myoclonus of spinal cord origin. Mov Disord 9: 233–235
Pool KD, Freeman FJ, Finitzo T, Hayashi MM, Chapman SB, Devous MD, Close LG Jr, Kondrasue GV, Memdelsohn D, Schaefer SB et al. (1991) Heterogeneity in spasmodic dysphonia. Neurologic and voice findings. Arch Neurol 48: 305–309
Putterman AM (1990) Botulinum toxin injections in the treatment of seventh nerve misdirection. Am J Ophthalmol 110: 205
Quinn N, Hallett M (1989) Dose standardization of botulinum toxin. Lancet 1: 964
Report of the Therapeutics and Technology Assessment Subcommittee of the American Academy of Neurology (1990) Assessment: the clinical usefulness of botulinum toxin-A in treating neurologic disorders. Neurology 40: 1332–1336
Rhoad RCM, Stern PJ (1993) Writer's cramp—a focal dystonia: etiology, diagnosis, and treatment. J Hand Surg 18A: 542–544
Rivest J, Lees AJ, Marsden CD (1991) Writer's cramp: treatment with botulinum toxin injection. Mov Disord 6: 55–59
Rondot P, Marchand MP, Dellatolas G (1991) Spasmodic torticollis—review of 220 patients. Can J Neurol Sci 18: 143–151
Rosenbaum F, Jankovic J (1988) Focal task-specific tremor and dystonia: categorization of occupational movement disorders. Neurology 38: 522–527
Rosenbaum AL, Kushner BJ, Kirschen D (1989) Vertical rectus muscle transposition and botulinum toxin (Oculinum) to medial rectus for abducens palsy. Arch Opthalmol 107: 820–823
Sachdev P (1992) Golfers' cramp: clinical characteristics and evidence against it being an anxiety disorder. Mov Disord 7: 326–332
Sakaguchi G (1983) Clostridium botulinum toxins. Pharmacol Ther 19: 165–194
Sanders DB, Massey EW, Buckley EG (1986) Botulinum toxin for blepharospasm: single-fiber EMG studies. Neurology 36: 545–547
Schantz EJ, Johnson EA (1990) Dose standardization of botulinum toxin. Lancet 335: 421
Schantz EJ, Johnson EA (1992) Properties and use of botulinum toxin and other microbial neurotoxins in medicine. Microb Rev 56: 80–99
Schiavo G, Benfenati F, Poulain B et al (1992) Tetanus and botulinum-B neurotoxins block neurotransmitter release by proteolytic of synaptobrevin. Nature 359: 3577–3583
Schiavo G, Shone CC, Rossetto O, Alexandre FCG, Montecucco C (1993a) Botulinum neurotoxin serotype F is a zinc endopeptidase specific for VAMP/synaptobrevin. J Biol Chem 268: 11516–11519
Schiavo G, Rossetto O, Catsicas S, Polverino de Laureto P, DasGupta BR, Benfenati F, Montecucco C (1993b) Identification of the nerve-terminal targets of botulinum neurotoxins serotypes A, D and E. J Biol Chem 268: 23784–23787
Schiavo G, Santucci A, DasGupta BR, Metha PP, Jontes J, Benfenati F, Wilson MC, Montecucco C (1993c) Botulinum neurotoxins serotypes A and E cleave SNAP-25 at distinct COOH-terminal peptide bonds. FEBS Lett 335, 99–103

Schiavo G, Malizio C, Trimble WS, Polverino de Laureto P, Milan G, Sugiyama H, Johnson EA, Montecucco C (1994). Botulinum G neurotoxin cleaves VAMP/synaptobrevin at a single Ala/Ala peptide bond. J Biol Chem 269 (in press)
Scott AB (1980) Botulinum toxin injection into extraocular muscles as an alternative to strabismus surgery. Ophthalmology 87: 1044–1049
Scott AB (1981) Botulinum toxin injection of eye muscles to correct strabismus. Trans Am Opthalmol Soc 79: 734–770
Scott AB, Kraft SP (1985) Botulinum toxin in injection in the management of lateral rectus paresis. Opthalmology 92: 676–683
Scott AB, Suzuki D (1988) Systemic toxicity of botulinum toxin by intramuscular injection in the monkey. Mov Disord 3: 333–335
Scott AB, Rosenbaum A, Collins CC (1973) Pharmacologic weakening of extraocular muscles. Invest Ophthalmol 12: 924–927
Scott AB, Kennedy RA, Stubbs HA (1985) Botulinum A toxin injection as a treatment for blepharospasm. Arch Opthalmol 103: 347–350
Scott AB, Magoon EH, McNeer KW, Stager DR (1990) Botulinum treatment of strabismus in children. Trans Am Opthalmol Soc 87: 174–180
Sellin LC (1981) The action of botulinum toxin at the neuromuscular junction. Med Biol 59: 10–11
Shone CC, Hambleton P, Melling J (1987) A 50 kDa fragment from the NH_2 terminus of the heavy subunit of Clostridium botulinum type A neurotoxin forms channels in lipid vesicles. Eur J Biochem 167: 175–180
Simpson LL (1981) The origin, structure, and pharmacologic activity of botulinum toxin. Pharmacol Rev 33: 155–188
Simpson LL (1989) Peripheral actions of the botulinum toxins. In: Simpson LL (ed) Botulinum neurotoxin and tetanus toxin. Academic. San Diego pp 153–178
Sketelj J, Örne-Finderle N, Sket D, Dettbarn WD, Brzin M (1993) Comparison between the effects of botulinum toxin-induced paralysis and denervation on molecular forms of acetylcholinesterase in muscles. J Neurochem 61: 501–508
Snow BJ, Tsui JKC, Bhatt MH, Varelas M, Hashimoto SA, Calne DB (1990) Treatment of spasticity with botulinum toxin: a double-blind study. Ann Neurol 28: 512–515
Söllner T, Whiteheart SW, Brunner M et al (1993) SNAP receptors implicated in vesicle targeting and fusion. Nature 362: 318–324
Stacy M, Cardoso F, Jankovic J (1993) Tardive stereotypy and other movement disorders in tardive dyskinesias. Neurology 43: 937–941
Stanley EF, Drachman DB (1983) Botulinum toxin blocks quantal but not non-quantal release of ACh at the neuromuscular junction. Brain Res 261: 172–175
Talbot JF, Gregor Z, Bird AC (1982) The surgical management of essential blepharoespasm. In: Marsden CD, Fahn S (eds) Movement disorders. Butterworth, London, pp 322–329
Tolosa E, Marti MJ, Kulisevsky J (1988) Botulinum toxin injection therapy for hemifacial spasm. Adv Neurol 49: 479–491
Trosch RM, Pullman SL (1994) Botulinum Toxin A Injections for the treatment of hand tremors. Mov Disord 9: 601–609
Tse CK, Dolly JO, Hambleton P, Wray D, Melling J (1982) Preparation and characterization of homogeneous neurotoxin type A from Clostridium botulinum. Eur J Biochem 122;: 493–500
Tsui JKC, O'Brien C (1994) Clinical trials for spasticity. In: Jankovic J, Hallet M (eds) Therapy with botulinum toxin. Dekker, New York, 523–534
Tsui JK, Eisen A, Stoessl AJ, Calne S, Calne DB (1986) Double-blind study of botulinum toxin in spasmodic torticollis. Lancet 2: 245–247
Tsui JK, Eastmann M, Mak E, Calne S, Calne DB (1987a) Botulinum toxin in writer's cramp: a pilot study. Ann Neurol 22: 147
Tsui JKC, Fross RD, Calne S, Calne DB (1987b) Local treatment of spasmodic torticollis with botulinum toxin. Can J Neurol Sci 14: 533–535
Tsui JK, Wong NLM, Wong E, Calne DB (1988) Production of circulating antibodies to botulinum-A toxin in patients receiving repeated injections for dystonia. Ann Neurol 23: 181
Tsui JK, Snow B, Bratt M et al. (1990) New applications of botulinum toxin in the lower limbs. Neurology 40 (Suppl 1): 382
Tsui JKC, Bhatt M, Calne S, Calne DB (1993) Botulinum toxin in the treatment of writer's cramp: a double-blind study. Neurology 43: 183–185
van Ermengem E (1897) Über einen neuen anaeroben Bacillus und seine Beziehungen zum Botulismus. Z Hyg Infektionskrankh 26: 1–56 [English translation, Rev Infect Dis 1: 701–719]

Waddy HM, Fletcher NA, Harding AE, Marsden CD (1991) A genetic study of idiopathic focal dystonias. Ann Neurol 29: 320–324
Whurr R, Lorch M, Fontana H, Brookes G, Lees A, Marsden CD (1993) The use of botulinum toxin in the treatment of adductor spasmodic dysphonia. J Neurol Neurosurg Psychiatry 56: 526–530
Wiegand H, Wellhoner HH (1977) The action of botulinum A neurotoxin on the inhibition by antidromic stimulation of the lumbar monosynaptic reflex. Naunyn Schmied Arch Pharmacol 298: 235–238
Wiegand H, Erdmann G, Wellhoner HH (1976) ^{125}I-Labelled botulinum A neurotoxin: pharmacokinetics in cats after neuromuscular injection. Naunyn Schmiedebergs Arch Pharmacol 292: 161–165
Yamasaki S, Binz T, Hayashi T, Szabo E, Yamasaki N, Eklund M, Jahn R, Niemann H (1994) Botulinum neurotoxin type G proteolyses the Ala^{81} Ala^{82} bond of rat synaptobrevin 2. Biochem Biophys Res Commun 200: 829–835
Zuber M, Sebald M, Bathien N, de Recondo J, Rondot P (1993) Botulinum antibodies in dystonic patients treated with type A botulinum toxin: frequency and significance. Neurology 43: 1715–1718
Zwirner P, Murry T, Swenson M, Woodson GE (1991) Acoustic changes in spasmodic dysphonia after botulinum toxin injection. J of Voice 5: 1: 78–84

Growth of Clostridia and Preparation of Their Neurotoxins

C.C. SHONE and H.S. TRANTER

1 Introduction

Various strains of *Clostridium botulinum* and *C.tetani* elaborate a family of eight structurally similar protein neurotoxins. The seven botulinum neurotoxins (types A–G) are distinguished on the basis of their reactions with specific

---

Centre for Applied Microbiology and Research, Porton Down, Salisbury, Wiltshire, SP4 0JG, UK

antibodies and are the causative agents of the disease botulism. These toxins act on the peripheral nervous system where they block neuromuscular transmission and induce widespread flaccid paralysis (SHONE 1986; NIEMANN 1991). Tetanus toxin, on the other hand, acts primarily on the central nervous system to disrupt the control mechanisms of polysynapses resulting in the characteristic muscular spasms and convultions of tetanospasm. Although symptomatically very different, the mechanisms of action of botulinum and tetanus toxins are very similar; both are presynaptically acting neurotoxins which inhibit the calcium-mediated secretion of various neurotransmitter substances (BIGALKE et al. 1981).

The mechanism by which these toxins block the transmitter release process has recently become clearer. Tetanus toxin and most of the botulinum neurotoxins have been characterised as highly specific zinc endopeptidases which act intracellularly on various protein components of a putative vesicle-docking apparatus (SÖLLNER et al. 1993). Tetanus toxin and botulinum type B, D and F neurotoxins act to cleave vesicle-associated membrane protein (synaptobrevin; SCHIAVO et al. 1992, 1993a). Botulinum types A and E toxins cleave protein SNAP-25 (BLASI et al. 1993, SCHIAVO et al. 1993b) and botulinum type C toxin cleaves syntaxin, a plasma membrane protein. Interestingly, the zinc endopeptidase specificities of these toxins all differ to a greater or lesser extent.

Applications of the tetanus and botulinum neurotoxins in both research and clinical medicine have grown rapidly in the past 10 years. The botulinum neurotoxins are now widely used clinically to treat a range of ophthalmic disorders and muscular spasms. Toxoids of tetanus and botulinum toxins have a long history of use as vaccines against their respective neurological diseases and, now that the mechanism of action is better understood, the use of these toxins as tools to study the exocytotosis process will surely increase. Many of these applications depend on large supplies of purified neurotoxins. The present report describes protocols for the growth and purification of tetanus and several of the botulinum neurotoxins.

2 Safety Precautions

Botulinum and tetanus neurotoxins are extremely potent neuroparalytic agents with a human lethal dose in the order of a few micrograms. A pentavalent vaccine is available to botulinum neurotoxin types A–E, and research staff involved in the growth of the bacteria and purification of the toxins should be immunised. This vaccine presently does not cover botulinum toxins types F and G, and particular care should be exercised when handling these organisms or their toxins. Manipulations of the bacterial cultures and solutions of toxins should ideally be performed within a class III microbiological safety cabinet. Centrifugation steps should be carried out only in sealed centrifuge tubes contained within a sealed rotor. As far as possible, non-breakable plastic ware should be used in place of glassware.

3 Assays for Clostridial Toxins

At present the only method of assay for the routine measurement of the total biological activity of the clostridial neurotoxins is the acute toxicity test performed in mice (HATHEWAY 1988). Methods for measuring the zinc endopeptidase activities of the neurotoxins using isolated synaptic vesicles (SCHIAVO and MONTECUCCO 1994) and synthetic peptides (SHONE et al. 1993) are now becoming available. When it is not essential to measure the biological activity of the neurotoxin, enzyme-linked immunosorbant assays provide a convenient method for quantifying levels of toxin in culture and column chromatography fractions. The antibody reagents used for such assays, however, whether polyclonal or monoclonal, must be raised against pure neurotoxin preparations in order to eliminate non-specific reactions (NOTERMANS et al. 1982; SHONE et al. 1985). Sodium dodecyl sulphate polyacrylamide gel electrophoresis may also be used as a crude method of assay for neurotoxin preparations greater than 20% pure. The neurotoxin appears as a protein band of approximately 150 kDa which, depending on the toxin type, may dissociate into the individual subunits (100 and 50 kDa) in the presence of dithiothreitol.

4 Growth of Bacteria

4.1 Characteristics of Neurotoxigenic Clostridia

C. botulinum organisms are anaerobic, spore-forming and ubiquitous in their distribution. In culture the cells may be seen as straight or slightly curved, motile, rods which vary quite considerably in size from 0.5–2.0 μm in width and 1.6–22.0 μm long (CATO et al. 1982); such variation depends upon the type, strain, growth medium and physiological status of the culture. Although *C. botulinum* cells usually appear singly or in pairs, they sometimes form long chains during growth in particular growth media. The vegetative cells in young cultures stain Gram positive while older cultures tend to become Gram negative. Spores are formed subterminally and are larger in diameter than the vegetative cell causing swelling of the latter in the area of the spore. The heat resistance of these spores varies considerably between the different types of *C. botulinum* (LYNT et al. 1984).

C. botulinum is difficult to identify biochemically due to its inability to ferment many carbohydrate sources and the most reliable means of confirming the presence of this organism is still by detection of toxin in culture or extracts from unknown specimens by mouse toxicity and subsequent neutralisation tests (HATHEWAY 1988). Indeed some strains of *C. sporogenes* and *C. novyi* are indistinguishable from *C. botulinum* by culture or DNA homology but are non-toxigenic. The taxonomic classification of botulism causing organisms has been further complicated by the isolation of other clostridia that produce botulinum neurotoxins. These include *C. butyricum* (AURELI et al. 1986; MCCROSKEY et al. 1986) and *C. barati* (HALL et al. 1985) which produce botulinum neurotoxin types E and F respectively and *C. novyi* which can produce either type C or type D neurotoxins (EKLUND et al. 1974). More recently the *C. botulinum* that produced type G neurotoxin has been renamed *C. argentinense* (SUMMANEN 1993).

C. tetani is a Gram-positive, anaerobic, spore-forming bacillus which, as *C. botulinum*, may appear Gram negative in cultures older than 24 h. The cell dimensions are 0.5–1.7 μm wide by 2.1–18.1 μm long, and most strains are motile by means of peritrichous flagella (CATO et al. 1986). Spores tend to appear in older cultures in a terminal position giving the organism its characteristic "drumstick-like" appearance. The formation of tetanus spores may be promoted or inhibited by a number of factors including medium composition, temperature and pH. These spores are highly resistant to alcohol, phenol and even formaldehyde solutions; aqueous and alcoholic iodine, gluteraldehyde and hydrogen peroxide are some of the more effective sporicidal agents.

While there may be some aerotolerance when the redox potential is low, *C. tetani* can generally be regarded as a strict anaerobe and as such grows only on media such as blood or egg yolk agar in an anaerobic environment. As with *C. botulinum* types C and D, growth tends to be better on plates that have been preincubated anaerobically for 24 h prior to use. As with *C. botulinum*, the biochemical identification of *C. tetani* is unreliable. *C. tetani* does not ferment any sugars, does not digest milk or other complex proteins, does not produce lipase or lecithinase and does not reduce nitrate. It does, however, liquify gelatine albeit slowly and produces H_2S, DNase and indole. The biochemical characterisation of the neurotoxigenic clostridia is shown in Table 1.

4.2 Growth of *C. botulinum*

4.2.1 Strain Selection and Maintenance

Several different bacterial strains and cultural conditions have been used to prepare small quantities of the different botulinum neurotoxins. The successful production of large amounts of toxin for research use or pharmaceutical manufacture, however, depends on the selection of highly toxinogenic strains of *C. botulinum* since the amount of toxin produced by strains of a particular serotype can vary as much as 10- to 1000-fold. Careful screening of isolates may

Table 1. Biochemical identification of *C. botulinum* and *C. tetani*

	C. botulinum	*C. tetani*
Gelatin hydrolysis	+	+
Milk digestion	v	v
Glucose acid	+[a]	–
Sucrose acid	v	–
Starch acid	–	–
Lactose acid	–	–
Mannose acid	v	–
Xylose acid	–	–
Mannitol acid	–	–
Indole production	–	v
Lecithinase	–	–
Lipase	+[a]	–
Urease	–	–
Esculin hydrolysis	v	–

v, Variable reactions between types and strains.
[a]Except *C. argentinense* (formerly *C. botulinum* type G).

therefore be required before production of toxin can begin. This may be particularly important for strains of *C. botulinum* types D, F, and G in which the levels of toxin production are usually significantly lower than strains of type A and B. Generally levels of toxin in the region of 10^5–10^6 mouse LD_{50} /ml are sufficient to allow purification of large amounts of botulinum neurotoxins (see below).

Viable stock cultures of all types of *C. botulinum* can be maintained for several years in cooked meat medium (Oxoid, Columbia, MD) at 4°C. However, under these conditions the toxin-producing ability of many strains can decrease with prolonged storage, and it is advisable to store cultures at –70°C or in liquid nitrogen. Following initial growth in cooked meat medium at 30°C (types C, D, E, and F) or 35°C (types A and B) for 24 h, cultures can be stored in liquid N_2 in 2- or 5-ml cryovials as 1.8- or 3-ml aliquots each containing 10% (v/v) glycerol.

4.2.2 Growth Media

Production of toxin by the different types of *C. botulinum* has been achieved in a variety of exotic media such as boiled vegetable extracts (Sugii and Sakaguchi 1977) and corn steep liquor (Lamanna et al. 1946). However, growth and high yields of toxin can be achieved in complex culture media (DasGupta 1983) containing combinations of meat hydrolysates, casein hydrolysate, yeast hydrolysate and yeast extract to which glucose (0.5%–1.0% w/v) may be added just prior to inoculation. The growth medium is supplemented with one or more reducing agents such as sodium thioglycollate (0.05%–0.1% w/v) or L-cysteine (0.1%–0.2%). One such medium, brain heart infusion broth has proved useful in the production of large amounts of *C. botulinum* type F toxin (Wadsworth et al. 1990).

Comparable yields of botulinum toxins can also be achieved in a range of simpler media based on hydrolysates of casein, e.g. NZ case plus (Sheffield Products, Norwich, NY; SHONE et al. 1992). These media do not contain meat proteins and as such are suitable for the production of pharmaceutical grade botulinum toxins. Although a number of chemically defined culture media have been used to grow *C. botulinum* the levels of toxin produced in these are generally 10- to 100-fold lower than those obtained in complex media (GULLMAR and MOLIN 1967; BOROFF and DASGUPTA 1971).

4.2.3 Toxin Production

Small amounts of botulinum neurotoxins may be produced by growing cultures in dialysis sacs surrounded by growth medium (GERWING et al. 1966; HATHEWAY et al. 1976). Whilst this method can facilitate the subsequent purification of the toxin by excluding many of the contaminating medium proteins, it is not suitable for large-scale production of neurotoxins. Furthermore, it presents a number of technical difficulties which prevent it from becoming used for pharmaceutical manufacture of these toxins.

To produce botulinum toxins in large quantities a number of previous workers have grown *C. botulinum* statically in large bottles or carboys (HEPPLE 1965). However, this method of growth does not allow the operator to change any cultural parameters and has a number of inherent difficulties including sampling of the culture and inadequate mixing. Furthermore, toxin production using static cultures can take up to 4–5 days, incubation (SIEGAL 1981). In contrast, microbial fermenters allow precise, continuous measurement and control of parameters such as temperature, pH, redox potential and agitation and as such are the preferred means of growing larger volumes of *C. botulinum.* Furthermore, optimisation of growth and toxin production including definition of culture conditions can be carried out on a small scale (1–2 l) before scale up to larger fermenters. Although fermenters up to approximately 50 l volume can be accomodated in specially designed microbiological safety cabinets to prevent exposure of the operator, larger production of these neurotoxins should be performed in fermenters in which high containment features have been engineered into the fermenter (HAMBLETON et al. 1991).

4.2.4 Fermentation of *C. botulinum* Types A, B, C, D and F

Fermentation of *C. botulinum* types A, B and F appears to follow a similar pattern in that exponential growth occurs in the first 12 h, reaches a maximum after about 24 h, followed by lysis of the cultures over the next 48 h. This pattern may, however, vary when using different strains and different growth media. For example, the Langeland strain of *C. botulinum* type F grows rapidly over the first 10 h in NZ case medium followed by lysis in the period 10–15 h (SHONE et al. 1992).

However, the same strain does not show any appreciable lysis when grown over 48 h in brain heart infusion broth.

During the initial growth phase of *C. botulinum* types A, B and F the pH drops rapidly to pH 5.5–5.8 following the utilisation of glucose. Thereafter the pH may remain constant or show a gentle rise probably following the release of amino groups during peptide/protein degradation. Toxin is produced during the growth phase and is generally complete between 24 and 48 h. Continued incubation after this time does not enhance the levels of toxin produced by these types. Indeed maximum levels of toxin may have been produced prior to lysis indicating that autolysis is not always required for the release of these toxin types. Lysis of early (< 24h) *C. botulinum* type A cultures can be induced by the addition of cell-free culture filtrate prepared from 1-week-old homologous cultures (BONVENTRE and KEMPE 1960). The lytic enzyme contained in these older filtrates has been purified and shown to be strain specific (TAKUMI et al. 1971).

Although production of types A and B botulinum toxins can be enhanced by a nitrogen overlay rather than by sparging during fermentation (SIEGAL 1981), a similar effect has not been observed during fermentation of type F (SHONE et al. 1992).

At the end of fermentation the toxins can be recovered by precipitation after lowering the pH of the cultures to 3.4–4.2 with 3 M H_2SO_4. The acidified culture is harvested from small fermenters by centrifugation or, in the case of larger fermenters, by tangential flow filtration using a Millipore (Bedford, MA) Pellicon system fitted with 0.45 μm Durapore membranes (Millipore). The concentrated slurry resulting from the latter operation is harvested by centrifugation. Acidified paste containing botulinum toxin can be stored in this form for several months at 4°C without any significant loss of activity.

Botulinum type C and D toxins have also been produced using this procedure (KURAZONO et al. 1985). However, the addition of RNA (0.4 mg/ml) to culture supernatants is required for efficient precipitation of toxin (IWASAKI and SAKAGUCHI 1978). The RNA can subsequently be removed by precipitation with protamine sulphate.

Since the production of toxins in *C. botulinum* types C and D is related to the presence of bacteriophages within these cultures (INOUE and IIDA 1970), an additional problem frequently encountered in the growth and preparation of these toxins is the loss of these phages and corresponding toxigenicity.

4.2.5 Fermentation of *C. botulinum* Type E

Non-proteolytic strains of *C. botulinum* types B, E and F are characterised by lack of autolysis and as a consequence do not efficiently release toxin out of the cell during fermentation. This feature has been used effectively for the production of toxin from cultures of *C. botulinum* type E.

During fermentation of *C. botulinum* type E growth is rapid and complete in less than 24 h. The exponential growth phase occurs over the first 12 h, and consequently the level of toxin produced into the culture supernatant fluid is low (< 10% of the total toxin). Continued incubation does not cause appreciable lysis, but the levels of toxin secreted into the culture fluid increases after 48 h. The bacterial cells containing trypsin-activated type E toxin can be harvested from the fermenter by centrifugation (small scale) or filtration (large scale) using the technology outlined above. Once the cells have been harvested type E toxin can be extracted with 0.2 *M* phosphate buffer, pH 6.0 (see below).

4.3 Growth of *C. tetani*

4.3.1 Strain Maintenance

The ability of strains of *C. tetani* to lose their toxigenicity after repeated subculture is not unknown (NIELSEN 1966), and care should be taken when preparing and storing stock cultures. As with *C. botulinum*, cultures may be maintained in cooked meat medium although cultures may be best stored in liquid nitrogen. Although *C. tetani* has been shown to posses lysogenic phages (PRESCOTT and ALTENBERN 1967), there is no evidence to suggest they are involved in toxin production. Indeed it now appears that the gene for toxin production is contained on a single large (75-kb) plasmid (LAIRD et al. 1980), and it may be the loss of this plasmid that results in the loss of toxigenicity during subculture.

4.3.2 Growth Media

Although *C. tetani* can be cultured in a variety of growth media, such as thioglycollate and casein hydrolysate, tetanus toxin production for vaccine purposes has been shown to occur in both semi-synthetic (MUELLER and MILLER 1945) and synthetic media (LATHAM et al. 1962). The most widely used growth medium for production purposes (LATHAM et al. 1962) consists of a tryptic digest of casein which is free of peptones and other allergenic substances thus simplifying the problem of separating tetanus toxin from other proteins in the culture filtrate.

The synthesis of tetanus toxin can be influenced by the presence of certain amino acids in the culture medium (TSUNASHIMA et al. 1964). For example, an excess of glutamate in the medium accelerates the growth of the bacteria and reduces the time necessary for autolysis. Since less toxin is produced in the presence of excess glutamate, it appears that toxin production occurs mainly during the stationary growth phase.

4.3.3 Toxin Production

As with the production of botulinum neurotoxins, tetanus can be produced in large quantities in fermenter culture. Under the correct conditions the quantity of tetanus toxin may account for up to 10% of the whole bacterial cell weight. Single-chain tetanus toxin may be prepared by growing the highly toxinogenic

Harvard strain of *C. tetani* and harvesting the bacterial cells just before the end of exponential growth by centrifugation or, on a larger scale, by membrane filtration using conventional tangential flow systems. The uncleaved toxin may then be prepared from the bacteria by hypertonic extraction (HELTING et al. 1979). The protease-cleaved toxin or di-chain toxin normally associated with vaccine production is prepared by growing *C. tetani* for 6–8 days until the bacteria lyse and release the toxin into the culture fluid. In either case tetanus toxin can be recovered from the cell-free culture filtrate or the bacterial cell extract by precipitation using ammonium sulphate.

The average yield from 6–8 days, fermentation of *C. tetani* is about 120–200 μg/ml crude tetanus toxin. In the preparation of tetanus vaccine the toxin-containing culture filtrate is detoxified by the addition of formaldehyde (final concentration 0.5%) for 4 weeks at 37°C and a pH of 7.6.

5 Purification of Clostridial Neurotoxins

5.1 Botulinum Neurotoxins

5.1.1 Purification Strategies

Since the botulinum neurotoxins were first purified in 1967 (DASGUPTA and BOROFF 1967), a number of purification strategies have been adopted to obtain homogeneous toxin preparations. Toxin purification using combinations of ion-exchange and gel filtration chromatography can be used (DASGUPTA and SATHYAMOORTHY 1984), but such procedures are slow and are difficult to perform on a large scale. More recent purification strategies take advantage of the fact that the botulinum toxins occur as complexes which, depending on the toxin type, contain at least one non-toxic protein component in addition to the neurotoxin moiety. At pH values below 6.5 these complexes appear to be stable. When the pH is raised above 7.5, however, the complexes dissociate, liberating the free neurotoxin. This property of these protein complexes in particular make them particularly useful for purifying the neurotoxin.

One such method takes advantage of the presence of a haemagglutinin component in several of the botulinum toxin complexes, notably types A and B. Using an immobilised haemagglutinin inhibitor (ρ-aminophenyl-β, D-thiogalacto-pyranoside) an affinity chromatography-based method has been developed for the purification of these neurotoxin types (MOBERG and SUGIYAMA 1978; EVANS et al. 1986; TSE et al. 1981). Neurotoxin in its complex form is first adsorbed onto the affinity medium at low pH. The neurotoxin is then eluted using a combination of high pH and ionic strength. Although an effective strategy, this method suffers from a number of drawbacks. Firstly, it cannot be used with the haemagglutinin-negative botulinum toxin types such as type E. Secondly, in the case of type A and

B, a portion of the toxin complex fails to bind to the affinity matrix, perhaps due to the haemagglutinin component becoming detached from the complex, and this reduces yields.

A useful and more generally applicable purification strategy was developed by SCHMIDT and SEGAL (1986) and was first used to purify type E neurotoxin. This method takes advantage of the different ionic properties of the neurotoxin in its complexed and free form and involves three ion exchange chromatography steps. Crude toxin is first chromatographed by ion exchange at low pH (5.5–6.0), and fractions containing the toxin complex are collected. The components of the complex are then separated by a second ion exchange chromatography step at a higher pH (8–8.5). During this chromatography step neurotoxin in its free form generally elutes from the column first followed by the non-toxic components of the complex. Finally, fractions containing the free form of the neurotoxin are re-chromatographed by ion exchange at the lower pH. Since the neurotoxin is the only component in the mixture to have changed its ionic characteristics, it is readily separated from the contaminating proteins. The latter method has been successfully applied to the purification of botulinum types A, B, C, E and F neurotoxins and can easily be adapted to production of toxins on a large scale.

The purification of botulinum neurotoxins types D (MURAYAMA et al. 1984) and G (NUKINA et al. 1988) are particularly problematic due to the low levels of toxins produced by these strains of *C. botulinum*.

5.1.2 Purification Procedure of the Botulinum Type A Neurotoxin Using Ion Exchange Chromatography

This section describes in detail the purification procedure based on ion exchange chromatography outlined above using botulinum type A neurotoxin as an example. After fermentation of *C. botulinum* type A, the toxin present in the supernatant fluid is most conveniently concentrated by the addition of acid to pH 3.5 and collected by centrifugation as detailed above. Neurotoxin in this precipitated form appears stable for several months and may be stored at 4°C prior to downstream processing. The starting material for the purification stage is assumed to be the acid-precipitated toxin obtained from a total of 200 l culture.

Step 1. Extraction of the toxin. The acid precipitate of the toxin is homogenised in a total of 1400 ml 0.2 *M* sodium phosphate buffer pH 6.0. Homogenisation is best performed in a stomacher type homogeniser (Seward Laboratories, London, UK) which minimises frothing of the solution. The homogenised crude toxin is then transferred to a 5-l plastic beaker and the pH adjusted to pH 6.0 by the slow addition of 2 *M* NaOH to a rapidly stirred solution. After pH adjustment the solution is stirred for a further 1 h at 22°C before being centrifuged at 25000 *g* for 1 h. The supernatant fluid is removed and stored at 4°C. The pellets are collected and re-extracted with a further 1400 ml 0.2 *M* sodium phosphate buffer pH 6.0 as described above but without the pH adjustment step. The homogenised solution may be incubated at 4°C overnight prior to centrifugation if convenient.

Step 2. Ribonuclease treatment and ammonium sulphate precipitation. The combined extracts from stage 1 are treated with protease-free ribonuclease A (Sigma, St. Louis, MO) at a concentration of 0.1 mg/ml extract at 37°C for 2 h and then solid ammonium sulphate added to 60% saturation whilst the solution is rapidly stirred at room temperature. After stirring for 30 min the precipitated toxin is collected by centrifugation at 25000 *g* for 30 min. The above extraction and precipitation steps are applicable without modification to the purification of botulinum type B and F toxins where the toxins are recovered from cultures by acid precipitation. For botulinum neurotoxins types which retain the neurotoxin within the bacterial cells, such as type E toxin, only minor modifications to the extraction procedure are involved. For these toxins whole cells are first washed with ice-cold 0.05 *M* sodium acetate buffer pH 5.0, and then the toxin is extracted from the cells by incubation with 0.2 *M* sodium phosphate pH 6.0 buffer for 3 h at 37°C and then overnight at 4°C.

Step 3. Chromatography on Sepharose Q at pH 5.8. The ammonium sulphate precipitated toxin is re-suspended in 800 ml of 50 m*M* bis-Tris-HCl pH 5.8 buffer and dialysed against 2 × 10 l of the same buffer for 48 h at 4°C. After dialysis the toxin solution is centrifuged at 25000 *g* for 1 h and the supernatant fluid applied (50 ml/min) to a column (10 × 10 cm) of Sepharose Q equilibrated in 0.05*M* bis-Tris-HCl pH 5.8. After washing the column with a further 1100 ml bis-Tris-HCl buffer the toxin is eluted with 0.05 *M* bis-Tris-HCl pH 5.8 buffer containing 0.15 *M* NaCl. The type of ion exchanger (either anion or cation) used for the first chromatography step depends on the neurotoxin type and strain and should be selected for its ability to bind the neurotoxin complex at pH 5.5–6.0. The above chromatography step has been successfully applied to the purification of botulinum type B, and F toxins (SHONE et al. 1992, 1993) without modification. In the case of botulinum type E toxin, a cation exchanger is employed for the first column step (SCHMIDT and SEGAL 1986).

Step 4. Chromatography on Sepharose Q at pH 8.5. Toxin eluted from the pH 5.8 Sepharose Q column is dialysed against 2 × 10 l of 0.05 *M* triethanolamine pH 8.5 buffer containing 0.15 *M* NaCl at 4°C and then chromatographed (20 ml/min) on a column (5 × 9 cm) of Sepharose Q equilibrated in the triethanolamine buffer. The toxin is eluted in the wash-through fraction. During this high pH ion exchange step the components of the neurotoxin complex are separated and the neurotoxin component of the toxin complex elutes from the anion exchange column before the non-toxic complex components (DASGUPTA and BOROFF 1967; KITAMURA et al. 1969). In step 4, for type A neurotoxin, conditions were chosen which retained the non-toxic complex components whilst allowing the type A neurotoxin to pass through the column. It is essential during this chromatography step that conditions are chosen which adequately separate the components of the neurotoxin complex. The separation of the toxin complex components for botulinum type B (SHONE et al. 1993), type E (SCHMIDT and SEGAL 1986) and type F (SHONE et al. 1992) require slight modifications to those described here for type A neurotoxin.

Step 5. Chromatography on Sepharose S at pH 5.5. Toxin recovered from step 4 is dialysed against 2 × 10 l 0.05 *M* succinate buffer pH 5.5 at 4°C and then applied (5 ml/min) to a column (2.6 × 6 cm) of Sepharose S equilibrated in the succinate buffer. The column is washed with a further 100 ml buffer and then the toxin eluted using a linear NaCl gradient (0–0.4 M NaCl; 300 ml) in 0.05 *M* succinate pH 5.5 buffer. The botulinum type A neurotoxin is recovered as the major protein peak during the salt gradient. During this chromatography step the botulinum type A neurotoxin is virtually the only protein to bind to the cation exchange column since potential contaminating proteins with a net positive charge at pH 5.5 are removed during the first low pH chromatography step. To maximise yields during this step it is recommended that a column size no larger than necessary be employed. For botulinum types A and E neurotoxin about 1.5 ml gel per 10 ml starting culture may be used a guide. Botulinum type B toxin appears to bind with a lower capacity to Sepharose S than type A toxin and a column size of approximately 5 ml gel per 10 l starting culture is required (Shone et al. 1993).

Step 6. Dialysis against a suitable buffer for storage. The purified neurotoxin is not particularly soluble at pH 5.5, and threadlike precipitation may begin to form after a few hours. The toxin should therefore be dialysed immediately against a suitable storage buffer such as 0.05 *M* HEPES pH 7.4 buffer containing 0.15 *M* NaCl. Stored frozen at –80°C in this bufer, the neurotoxin is stable for several years. Toxin may also be stored frozen in other buffers such as 0.1 *M* Tris-HCl pH 8.0 at temperatures below –40°C. Sodium phosphate buffers should be avoided if the toxin is to be stored frozen. The above procedure normally yields 400–500 mg botulinum type A neurotoxin with a specific toxicity of 1–2 × 10^8 mouse LD_{50}/mg and has been used successfullywith *C. botulinum* type A strains NCTC 7272, ATCC 3502 and the Hall strain. Purity is usually greater than 95% when assessed by sodium dodecyl sulphate polyacrylamide gel electrophoresis. If neurotoxin of higher purity is required, the toxin may be further purified by an additional ion exchange chromatography step on columns of Mono Q (HR 10/10; Pharmacia). Portions of neurotoxin are applied to the column in 0.05 *M* Tris-HCl pH 8.0 buffer and then eluted with a linear gradient of NaCl (0–0.3 *M*, 150 ml) in the Tris-HCl buffer. This procedure may be carried out successfully on a smaller scale than that described above but to avoid low yields it is not recommended that purification be undertaken from cultures with a total toxicity of less than 2 × 10^{10} mouse LD_{50}. Conditions and column sizes for the purification of botulinum type A toxin from 30 l of culture are summarised in Table 2.

5.2 Subunits of the Botulinum Neurotoxins

5.2.1 Subunits of Botulinum Type A Neurotoxin

The individual subunits of the botulinum neurotoxins may be purified by the ion exchange method developed by Kozaki et al. (1981) for botulinum type A neurotoxin.

Table 2. Small-scale purification of botulinum type A neurotoxin

Stage	Vol (ml)	LD_{50}/ml	Total LD_{50}	Protein (mg/ml)	Yield (%)
Re-suspended acid precipitate	480	7×10^7	3.3×10^{10}	n.d.	100
Combined buffer extract	640	4×10^7	2.6×10^{10}	7.1	78
Re-suspended ammonium sulphate ppt.	420	4×10^7	1.7×10^{10}	8.0	51
Sepharose Q column pH 5.8 (5 × 5 cm)	300	4×10^7	1.2×10^{10}	4.4	36
Sepharose Q column pH 8.5 (2.6 × 9 cm)	100	9×10^7	9×10^9	2.6	27
Sepharose S column pH 5.5 (1.6 × 2.5 cm)	25	3×10^8	7.5×10^9	1.6	23

The acid-precipitated toxin obtained from a 30-l culture of *C. botulinum* ATCC strain 3502 was extracted twice with 400 ml volumes of 0.2 *M* sodium phosphate buffer pH 6.0 and processed as described in text. Chromatography column elution volumes and flow rates were reduced in proportion to the smaller column sizes indicated.

Step 1. Botulinum type A neurotoxin (5 mg total) is dialysed against borate/phosphate buffer pH 8.5 (29 m*M* $Na_2B_4O_7$; 42 m*M* NaH_2PO_4 titrated to pH 8.5 with NaOH) and then applied (0.5 ml/min) to a column (1 × 5 cm) of QAE-Sephadex (Pharmacia) equilibrated in the borate/phosphate pH 8.5 buffer. The column is then washed with a further 10 ml buffer followed by 10 ml borate/phosphate buffer containing 10 m*M* dithiothreitol. After the latter buffer has been allowed to run almost completely onto the column, 3 ml borate/phosphate pH 8.5 buffer containing 100 m*M* dithiothreitol and 2 *M* urea are carefully applied to the column and 2.5 ml of this buffer allowed to run onto the gel. The column is then sealed and incubated at 4°C overnight.

Step 2. The light subunit of the neurotoxin is eluted with 20 ml borate/phosphate pH 8.5 buffer containing 2 *M* urea and 10 m*M* dithiothreitol collecting 1.5 ml fractions. The oxidized form of dithiothreitol absorbs strongly at 280 nm, and it is not possible to locate the fractions containing the light-chain by UV measurements at this stage. Chemical protein assays, however, may be employed. Alternatively the first six fractions may be dialysed separately against a desired buffer and then the protein content assessed by UV measurement after dialysis. It is essential to keep both column and eluted fractions at 4°C or below at this stage to avoid precipitation of the light subunit which has a low solubility.

Step 3. The heavy subunit of the neurotoxin is eluted with 20 ml borate/phosphate pH 8.5 buffer containing 2 *M* urea, 10 m*M* dithiothreitol and 0.2 *M* NaCl. The presence of heavy chain in eluted fractions (1 ml) may be detected by UV measurement and then dialysed against a desired buffer. The separated heavy subunit of botulinum type A neurotoxin has a low solubility in low ionic strength buffers, and it is not recommended that the heavy chain be transferred to buffers containing less than 0.15 *M* NaCl or the equivalent. For both the heavy band light subunits of the neurotoxin prolonged storage in the buffers containing urea should be avoided as some chemical modification of the protein may occur.

5.2.2 Subunits of Botulinum Types B, C, E and F Neurotoxin

The above procedure describing the purification of the subunits of botulinum type A toxin can be applied directly to type F neurotoxin without modification. Purification of the subunits of botulinum types B and E neurotoxin, however, requires an additional step since both these toxin types contain a proportion of the single chain form of the neurotoxin which must be converted to the di-chain form. Trypsin treatment of these neurotoxins is the most convenient method for producing the di-chain form of the toxins. The incubation conditions depend on both the neurotoxin concentration and activity of the trypsin preparation, and conditions should be optimised by experiment. As a starting point, neurotoxin (1 mg/ml) in 0.1 *M* sodium phosphate pH 6.0 buffer can be treated with 20 μg/ml trypsin (tosyl-phenylethyl-chloro-methyl ketone treated trypsin, Sigma) for 30 min at 37°C and the proteolysis stopped by the addition of a 10 *M* excess of soybean trypsin inhibitor. The separated light subunits of the neurotoxin are particularly sensitive to proteases, and it is advisable at this stage to remove traces of trypsin by ion exchange chromatography on a Mono Q column (Pharmacia) before proceeding with the subunit separation protocol. Botulinum type B neurotoxin is sensitive to trypsin, and some degradation of the heavy subunit may be observed if the trypsin treatment is allowed to proceed too long.

A similar ion exchange method for the purification of the subunits of botulinum type C neurotoxin has been described by Syuto and Kubo (1981).

5.3 Tetanus Neurotoxin

5.3.1 Purification Strategy

Tetanus toxin differs from the botulinum toxins in that it is not produced as a protein complex in association with non-toxin proteins but is liberated from the bacterium as the free neurotoxin. Fortunately, it has been possible to define growth media for *C. tetani* which contain no protein constituents, thus greatly simplifying the purification of tetanus neurotoxin. Purification of tetanus neurotoxin from such media can be effected by a combination of ion exchange and gel filtration chromatography (Schiavo and Montecucco 1994).

5.3.2 Di-Chain Form of Tetanus Toxin

Cultures of *C. tetani* grown for 6–8 days, as described above, release the di-chain form of tetanus toxin into the culture medium. Culture supernatant fluids obtained by centrifugation or filtration therefore provide a convenient starting material from which to purify the neurotoxin.

Step 1. Ammonium sulphate precipitation. To a stirred culture supernatant fluid at 4°C, solid ammonium sulphate is added slowly to 43% saturation (250 g/l). After stirring for a further 30 min the resulting precipitate is recovered by centrifugation 16 000 *g* for 10 min and dissolved in 10 m*M* sodium phosphate buffer pH 7.4. The ammonium sulphate precipitation step is then repeated on this solution, raising the final concentration to 46%. The resulting pellet is dissolved in 10 m*M* sodium phosphate buffer pH 7.4 and dialysed thoroughly against the same buffer.

Step 2. Chromatography on DE-52 cellulose. The dialysed toxin solution is centrifuged at 25000 *g* for 30 min and then applied to a column of DE-52 (Whatman, 15 mg protein per milliliter of gel) equilibrated in 10 m*M* sodium phosphate buffer pH 7.4. The tetanus toxin is eluted with a linear gradient of 10–100 m*M* sodium phosphate. Fractions (4 ml) are then analysed by sodium dodecyl sulphate gel electrophoresis, and those containing the 150-kDa toxin band are pooled and precipitated with ammonium sulphate added to 60% saturation at 4°C.

Step 3. The resulting precipitate from step 2 is recovered by centrifugation at 27 000 *g* for 15 min and then dissolved in a minimum volume of 10 m*M* sodium phosphate buffer pH 7.4 and applied at 25ml/h to an ACA-34 gel filtration column (3 × 130 cm) equilibrated in the phosphate buffer. The first peak to elute from the column contains tetanus toxin approximately 95% pure; the second peak contains the neurotoxin light subunit, approximately 70% pure.

Step 4. Tetanus neurotoxin recovered from step 3 may be concentrated by ammonium sulphate precipitation (60% saturation at pH 7) and stored at 4°C as a suspension. Alternatively the toxin may be stored frozen at –80°C as described above for the botulinum neurotoxins. Further purification of tetanus neurotoxin may be effected by metal chelating chromatography (Schiavo and Montecucco 1994) or by high-pressure anion exchange chromatography on Mono Q (Pharmacia) as described above for botulinum type A neurotoxin. A small percentage of the single-chain form of tetanus toxin may be present in the neurotoxin preparation, and this may be cleaved into the di-chain form by limited treatment with trypsin as described above.

5.3.3 Single-Chain Form of Tetanus Neurotoxin

The single-chain form of tetanus neurotoxin may be obtained by extracting the toxin from cultures before the end of the exponential phase of growth. Bacterial cells are first washed twice with 10 m*M* sodium phosphate buffer pH 7.5 containing 0.145 *M* NaCl and then extracted overnight with 0.1 *M* sodium citrate buffer pH 7.5 containing 1 *M* NaCl, 2 m*M* benzamidine and 1 m*M* diisopropylfluorophosphate at 4°C with stirring. The cells are then removed by centrifugation at 10 000 *g* for 30 min, and the tetanus toxin in the resulting supernatant fluid is precipitated by the addition of ammonium sulphate to 43% saturation at 4°C. The precipitated toxin is washed with 40% saturated ammonium sulphate in a buffer containing 2 m*M* sodium citrate, 100 m*M* sodium potassium phosphate, 1 m*M* EDTA, 1 m*M* sodium azide and 1 m*M* benzamidine pH 7.5 and is then dissolved in the same buffer but without the ammonium sulphate. After dialysis against 10 m*M* sodium phosphate pH 6.8 buffer the single-chain tetanus toxin may be purified as described above for the di-chain form (Schiavo and Montecucco 1994).

6 Concluding Remarks

The procedures outlined above provide the basis for preparing highly purified clostridial neurotoxins. While such procedures work consistently for a given bacterial strain in a given laboratory, we have demonstrated in our own laboratory that different growth conditions and variations within bacterial strains can lead to differences in the chromatographic behaviour of these toxins. This is particularly true for the botulinum toxins where the neurotoxin moiety is a component of a protein complex consisting of several different protein components. In addition, the distribution of the neurotoxin between the cells and the medium fluid may also vary with different growth conditions. It is therefore essential, when embarking on the purification of a neurotoxin for the first time, to validate each step in the procedure before continuing with the purification process.

References

Aureli P, Fenicia L, Pasolini B, Gianfranceschi M, McCroskey LM, Hatheway CL (1986) Two cases of type E infant botulism caused by neurotoxigenic Clostridium butyricum in Italy. J Infect Dis 154: 207–211

Blasi J, Chapman ER, Link E, Binz T, Yamasaki S, De Camilli P, Sudhof TC, Niemann H, Jahn R (1993) Botulinum neurotoxin A selectively cleaves the synaptic protein SNAP-25. Nature 365: 160–163

Bigalke H, Heller I, Bizzini B, Habermann E (1981) Tetanus toxin and botulinum toxin inhibit the release and uptake of various transmitter, as studied with particulate fractions from rat brain and spinal chord. Naunyn Schmiedebergs Arch Pharmacol 316: 244–251

Bonventre PF, Kempe LL (1960) Physiology of toxin production by Clostridium botulinum types A and B.I. Growth, autolysis and toxin production. J Bacteriol 79: 18–23

Boroff DA, DasGupta BR (1971) Botulinum toxin. In: microbial toxins. Kadis S, Montie TC, Ajl SJ (eds) vol IIA p1-68 Academic, New York

Cato EP, Hash DE, Holdeman LV, Moore WEC (1982) Electrophoretic study of Clostridium species. J Clin Microbiol 15: 688–702

Cato EP, George WL, Finegold SM (1986) Genus Clostridium. In: Bergey's manual of systematic bacteriology vol. 2 Sneath PHA, Nair NS, Sharpe ME, Holt JG (eds) Williams and Wilkins, Baltimore, pp 1141–1200

DasGupta BR (1983) Microbial food toxicants: Clostridium botulinum toxins. CRC handbook of foodborne diseases of biological origin. Rechcigl M (ed) CRC Boca Raton, pp 25–56

DasGupta BR, Boroff DA (1967) Chromatographic isolation of haemagglutinin-free neurotoxin from crystalline toxin of Clostridium botulinum type A toxin. Biochem Biophys Acta 147: 603–609

DasGupta BR, Sathyamoorthy V (1984) Purification and amino acid composition of type A botulinum neurotoxin. Toxicon 3: 415–424
Eklund MW, Poysky FT, Meyers JA, Pelroy GA (1974) Interspecies conversion of Clostridium botulinum type C to Clostridium novyi type A by bacteriophage. Science 186: 456–458
Evans DM, Williams RS, Shone CC, Hambleton P, Melling J, Dolly JO (1986) Botulinum type B neurotoxin. Its purification, iodination and binding to rat brain synaptosomes. Eur J Biochem 154: 409–416
Gerwing J, Dolman CE, Kason DV, Tremaine JH (1966) Purification and characterisation of C. botulinum type B toxin. J Bacteriol 91: 484–487
Gullmar B, Molin N (1967) Effect of nutrients on physiological properties of Clostridium botulinum type E. J Bacteriol 94: 1924–1929
Hall JD, McCroskey LM, Pincomb BJ, Hatheway CL (1985) Isolation of an organism resembling Clostridium baratii which produces type F botulinal toxin from an infant with botulism. J Clin Microbiol 21: 654–655
Hambleton P, Griffiths JB, Cameron R, Melling J (1991) A high containment polymodal pilot-plant fermenter- design concepts. J Chem Tech Biotechnol 50: 167–180
Hatheway CL (1976) Toxoid of C. botulinum type F: purification and immunogenicity studies. Appl Environ Microbiol 31: 234–242
Hatheway CL (1988) Botulism. In: Barlows A, Hansler WJ Jr, Ohashi M, Turano A (eds) Laboratory diagnosis of infectious diseases: principles and practice vol 1, bacterial, mycotic and parasitic diseases. Springer, Berlin Heidelberg New York, pp 111–133
Helting TB, Parschat S, Engelhardt H (1979) Structure of tetanus toxin. J Biol Chem 254: 10728–10733
Hepple JR (1965) Large-scale cultivation of Clostridia. J Appl Bacteriol 28: 52–55
Inoue K, Iida H (1970) Conversion to toxigenicity in C. botulinum type C. Jpn J Microbiol 14: 87–89
Iwasaki M, Sakaguchi G (1978) Acid precipitation of C. botulinum type C and D toxins from whole culture by addition of RNA as a precipitation aid. Infect Immun 19: 749–751
Kitamura M, Sakaguchi S, Sakaguchi G (1969) Significance of the 12S toxin of C. botulinum type E. J Bacteriol 98: 1173–1178
Kozaki S, Togashi S, Sakaguchi G (1981) Separation of Clostridium botulinum type A derivative toxin into two fragments. Jpn J Med Sci Biol 34: 61–68
Kuarzono H, Shimozowa K, Hosokawa M, Sakaguchi G (1985) Procedure for the large-scale production and purification of C. botulinum C1 toxin for the preparation of toxoid. FEM Microbiol Lett 30: 47–51
Laird WJ, Aaronson W, Silver RP, Habig WH, Hardegree MC (1980) Plasmid-associated toxigenicity in Clostridium tetani. J Infect Dis 142: 623
Lamanna C, Eklund HW, McElroy OE (1946) Botulinum toxin (type A); including a study of shaking with chloroform as a step in the isolation procedure. J Bacteriol 52: 1–13
Latham WC, Bent DF, Levine L (1962) Tetanus toxin production in the absence of protein. Appl Microbiol 10: 146–152
Lynt RK, Solomon HM, Kautter DA (1984) Heat resistance of Clostridium botulinum type G in phosphate buffer. J Food Prot 47: 463–466
McCroskey LM, Hatheway CL, Fenicia L, Pasolini B, Aureli P (1986) Characterisation of an organism that produces type E botulinal toxin but which resembles Clostridium butyricum from the faeces of an infant with type E botulism. J Clin Microbiol 23: 201–202
Moberg LJ, Sugiyama H (1978) Affinity chromatography purification of type A botulinum neurotoxin from crystalline toxic complex. Appl Environ Microbiol 35: 828–831
Mueller JH, Miller PA (1945) Production of tetanal toxin. J Immunol 50: 377–384
Murayama S, Syuto B, Oguma K, Iida H, Kubo S (1984) Comparison of C. botulinum toxins type D and C1 in molecular property, antigenicity and binding to rat-brain synaptosomes. Eur J Biochem 142: 487–492
Nielsen KE (1966) Growth and toxin production by Clostridium tetani on repeated transfer in culture media. Acta Pathol. Microbiol Scand 77: 542–554
Neimann H (1991) Molecular biology of Clostridial neurotoxins. In: Alouf JE, Freer JH (eds) A sourcebook of bacterial protein Toxins. Academic, New York, pp. 303–348
Notermans S, Hagenaars AM, Kozaki S (1982) The enzyme-linked immunosorbant assay for the detection and determination of C. botulinum toxins A, B and E. Meth Enzymol 84: 223–228
Nukina M, Mochida Y, Sakaguchi S, Sakaguchi G (1988) Purification of botulinum type G progenitor toxin. Zentralbl Bakterial Hyg A 268: 270–272
Prescott LM, Altenbern RA (1967) Detection of bacteriophages from two strains of Clostridium tetani. J Virol 1: 1085–1086

Schiavo G, Montecucco C (1994) Tetanus and botulinum neurotoxins: isolation and assay. Methods Enzymol (in press)

Schiavo G, Benefenati F, Poulani B, Rosetto O, Polverino de Laureto P, DasGupta BR, Montecucco C (1992) Tetanus and botulinum-B neurotoxins block transmitter release by proteolytic clevage of synaptobrevin. Nature 359: 832–835

Schiavo G, Shone CC, Rossetto O, Alexander FCG, Montecucco C (1993a) Botulinum neurotoxin serotype F is a zinc endopeptidase specific for VAMP/synaptobrevin. J Biol Chem 268: 11516–11519

Schiavo G, Santucci A, DasGupta BR, Mehta PP, Jontes J, Benefeati F, Wilson MC, Montecucco C (1993b) Botulinum neurotoxins serotypes A and E cleave SNAP-25 at distinct COOH-terminal peptide bonds. FEBS Lett 335: 99–103

Schmidt JJ, Segal LS (1986) Purification of type E botulinum neurotoxin by high performance ion exchange chromatography. Anal Biochem 156: 213–219

Shone C, Quinn CP, Wait R, Hallis B, Fooks SG, Hambleton P (1993) Proteolytic cleavage of synthetic fragments of vesicle-associated membrane protein isoform-2 by botulinum type B neurotoxin. Eur J Biochem 217: 965–971

Shone CC (1986) Clostridium botulinum neurotoxins, their structures and modes of action. In: Watson D (ed) Natural toxicants in foods. Horwood Chichester, pp 11–57

Shone CC, Wilton-Smith P, Appleton N, Hambleton P,Modi N, Gatley S, Melling J (1985) Monoclonal antibody-based immunoassay for C. botulinum type A toxin is comparable to the mouse bioassay. Appl Environ Microbiol 50: 63–67

Shone CC, Tranter HS, Alexander FCG (1992) Purification and radiolabelling of Clostridium botulinum type F neurotoxin. Methods Neurosci 8: 165–179

Siegal LS (1981) Fermentation kinetics of botulinum toxin production (types A, B and E). In: Lewis GE (ed) Biomedical aspects of botulism. Academic, New York, pp 121–128

Söllner T, Whiteheart SW, Brunner M, Erdjument-Bromage H, Geromanus S, Tempst P, Rothman JE (1993) SNAP receptors implicated in vesicle fusion and targeting. Nature 362: 318–324

Sugii S, Sakaguchi G (1977) Botulogenic properties of vegetables with special reference to the molecular size of the toxin in them. J Food Safety 1: 53–65

Summanen P (1993) Recent taxonomic changes for anaerobic gram-positive and selected gram-negative organisms. Clin Infect Dis 16 [Suppl 4]: S168–174

Syuto B, Kubo S (1981) Separation and characterisation of heavy and light chains from Clostridium botulinum type C and their reconstitution. J Biol Chem 256: 3712–3717

Takumi K, Kawata T, Hisatsune K (1971) Autolytic enzyme system of C. botulinum, vol II mode of action of autolytic enzymes in C. botulinum type A. Jpn J Microbiol 15: 131–141

Tse T-K, Dolly JO, Hambleton P, Wray D, Melling J (1981) Preparation and characterisation of homogenious preparation of toxin type A from Clostridium botulinum. Eur J Biochem 122: 493–500

Tsunashima I, Sato K, Shoji K, Yoneda M, Amono T (1964) Excess supplementation of certain amino acids to medium and its inhibitory effect on toxin production by Clostridium tetani Biken J 7: 161–163

Wadsworth JDF, Desai M, King HJ, Tranter HS, Hambleton P, Melling J, Dolly JO, Shone CC (1990) Botulinum type F neurotoxin, its purification and binding to rat brain synaptosomes. Biochem J 268: 123–128

Molecular Genetics of Clostridial Neurotoxins

N.P. Minton

1 Introduction

Tetanus and botulinum neurotoxins (TeNT and BoNT, respectively) are produced exclusively by bacterial species belonging to the genus *Clostridium*. Their extreme potency to man and his animals has ensured their extensive analysis by the scientific community. In recent years these studies have been strengthened by the application of recombinant procedures. As a result the genes encoding TeNT and all seven BoNT serotypes have been cloned and their entire nucleotide sequences determined. This in turn has led to significant advances in our understanding of the relationships between clostridial neurotoxins and their

Department of Molecular Microbiology, Research Division, Centre for Applied Microbiology and Research, Porton Down, Salisbury, SP4 OJG, UK

mode of action. The availability of these genes has also provided the starting point from which the molecular basis of bacterial toxin production may be unravelled.

2 Characterisation of Neurotoxin Genes

The properties of those bacterial isolates which produce clostridial neurotoxin are discussed by Popoff (this volume). However, because the characteristics of neurotoxin genes depend very much on their source, the salient features of producer organisms are worth reiterating. In essence, *C. botulinum* may be divided into four distinct groups (see Table 1). This subdivision, initially made on the basis of shared metabolic traits and DNA homologies, is fully supported by phylogenic analysis of 16S RNA sequences (Hutson et al. 1993a). Each of the four groups encompasses isolates which resemble their toxinogenic counterparts in all aspects but the ability to produce botulinum toxin. In addition to the four *C. botulinum* groups, clostridial strains otherwise indistinguishable from *C. butyricum* and *C. barati* have been isolated from cases of infant botulism which produce BoNT/E and BoNT/F, respectively (Aureli et al. 1986; Hall et al. 1985; McCroskey et al. 1986, 1991). These six groups, together with *C. tetani*, comprise all the currently known neurotoxic clostridia (Table 1). The group to which an isolate belongs influences the nucleotide sequence of the cloned gene, the nature of the ancillary genes involved in progenitor toxin formation, and the type of genetic element on which the determinants of botulism are found.

Table 1. Characteristics of the neurotoxigenic clostridia (from McCroskey et al. 1986; Hall et al. 1985; Cato et al. 1986)

Species	Group	Toxin type	Proteolysis	Glucose fermentation	Lipase	Optimum growth temperature (°C)	Related clostridia
C. botulinum	I	A,B,F	a^+	+	+	30–37	*C.sporogenes*
C. botulinum	II	B,E,F	a^-	+	+	25–30	"E-like" organisms
C. botulinum	III	C,D	$a^{+/-}$	+	+	30–37	*C. novyi*
C. botulinum	IV	G	a^+	–	–	30–37	*C. subterminale/ C. hastiforme*
C. butyricum	V	E	b	–	–	30–37	*C. butyricum*
C. barati	VI	F	a^-	+	–	30–45	*C. barati*
C. tetani		TeNT	c	–	–	37	*C.tetanomorphum*

Proteolysis: a, refers to digestion of milk (strains of group I may also produce two distinct botulinum serotypes, e.g. A_B, A_F, B_A and B_F); b, *C. butyricum* partially digests milk protein giving a "stormy" result, c, *C. tetani* is obligately proteolytic but cannot degrade complex proteins such as those in milk. :"E-like" organisms refer to various clostridia which have not been given a species name but share characteristics with *C. botulinum* group II bacteria.

2.1 Cloning and Nucleotide Sequence Analysis of the Structural Genes

The first clostridial neurotoxin gene to be characterised was that encoding TeNT. It was independently cloned and sequenced by two separate research groups (Eisel et al. 1986; Fairweather and Lyness 1986). However, despite isolating the gene from two different strains of *C. tetani*, the sequences obtained were identical. This reflects the longstanding observation that the toxin produced by different strains of *C. tetani* is homogeneous in nature. A complete nucleotide sequence of a BoNT gene (*bot*) was not available until several years later, when the sequence of the BoNT/A gene of *C. botulinum* NCTC 2916 was determined (Thompson et al. 1990). Shortly thereafter the sequence of a second BoNT/A gene was published (Binz et al. 1990a), and in the few short years since this date representative sequences of genes encoding all the remaining six serotypes have been derived. Those strains whose neurotoxin genes have been sequenced are presented in Table 2. This list does not include those strains from which partial gene sequences have been derived.

Regardless of source all the determined neurotoxin nucleotide sequences exhibit a codon usage characteristic of clostridial genes (Young et al. 1989a). Thus, codons ending in A or T are generally preferred. This codon bias reflects the low G+C content of the genes (24.51%–27.84%). All initiate with an AUG translational start codon and are preceded by sequence motifs typical of procaryotic ribosome binding sites. With one exception they all terminate with the ochre codon UAA. The exception, the *botF* gene of *C. barati* ATCC43756, terminates with an umber codon (UGA) not generally used by clostridia (Young et al. 1989b). Sequence motifs with the potential to act as transcriptional termination sequences have only been noted 3' to the *botA* genes of the *C. botulinum* strains 62A and NCTC 2916, and downstream of the TeNT gene. In most other cases, however, insufficient 3' non-coding sequence has been reported for any subjective judgments to be made as to their presence or absence.

2.2 Relationships Amongst Neurotoxins

Translation of the sequenced genes has allowed the primary amino acid sequences of all clostridial neurotoxins to be deduced. Not altogether unexpectedly, the various BoNT serotypes exhibit a considerable degree of amino acid sequence heterogeneity. Comparative alignment (Fig. 1) shows that they, along with TeNT, are composed of highly conserved amino acid domains interspersed with amino acid tracts exhibiting little overall similarity (Fig. 1). Among the various L chains (average size 439) 63 amino acids are absolutely conserved. Throughout the H chains (average size 843) 97 amino acids are identical. By reference to BoNT/A of Fig. 1, the most notable areas of conservation include: the two cysteine residues (position 430 and 454) involved in the disulphide bond formation between the L and H chains, the histidine-rich motif between positions 216 and 234 associated

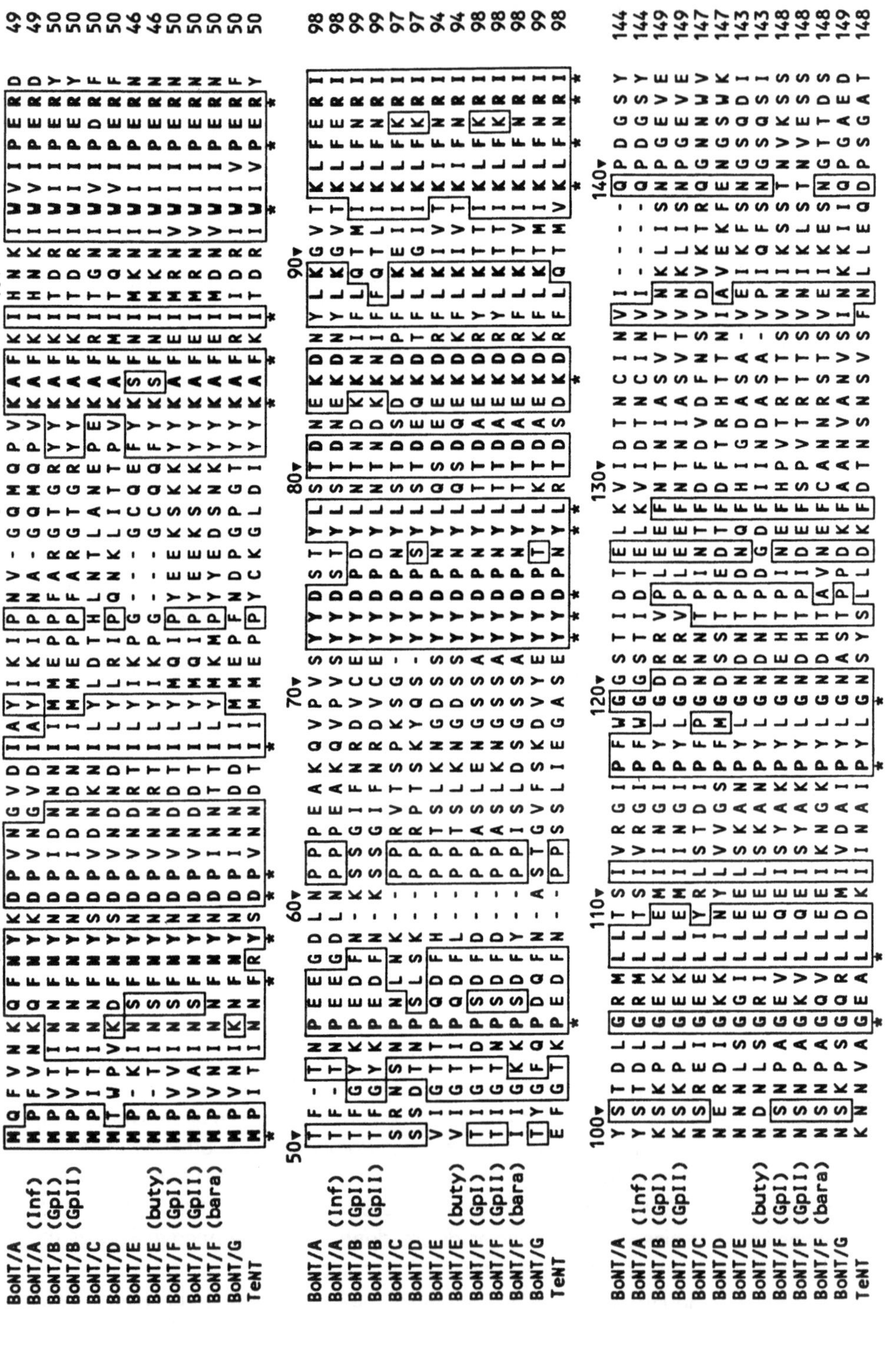

BoNT/A
BoNT/A (Inf)
BoNT/B (GpI)
BoNT/B (GpII)
BoNT/C
BoNT/D
BoNT/E
BoNT/E (buty)
BoNT/F (GpI)
BoNT/F (GpII)
BoNT/F (bara)
BoNT/G
TeNT

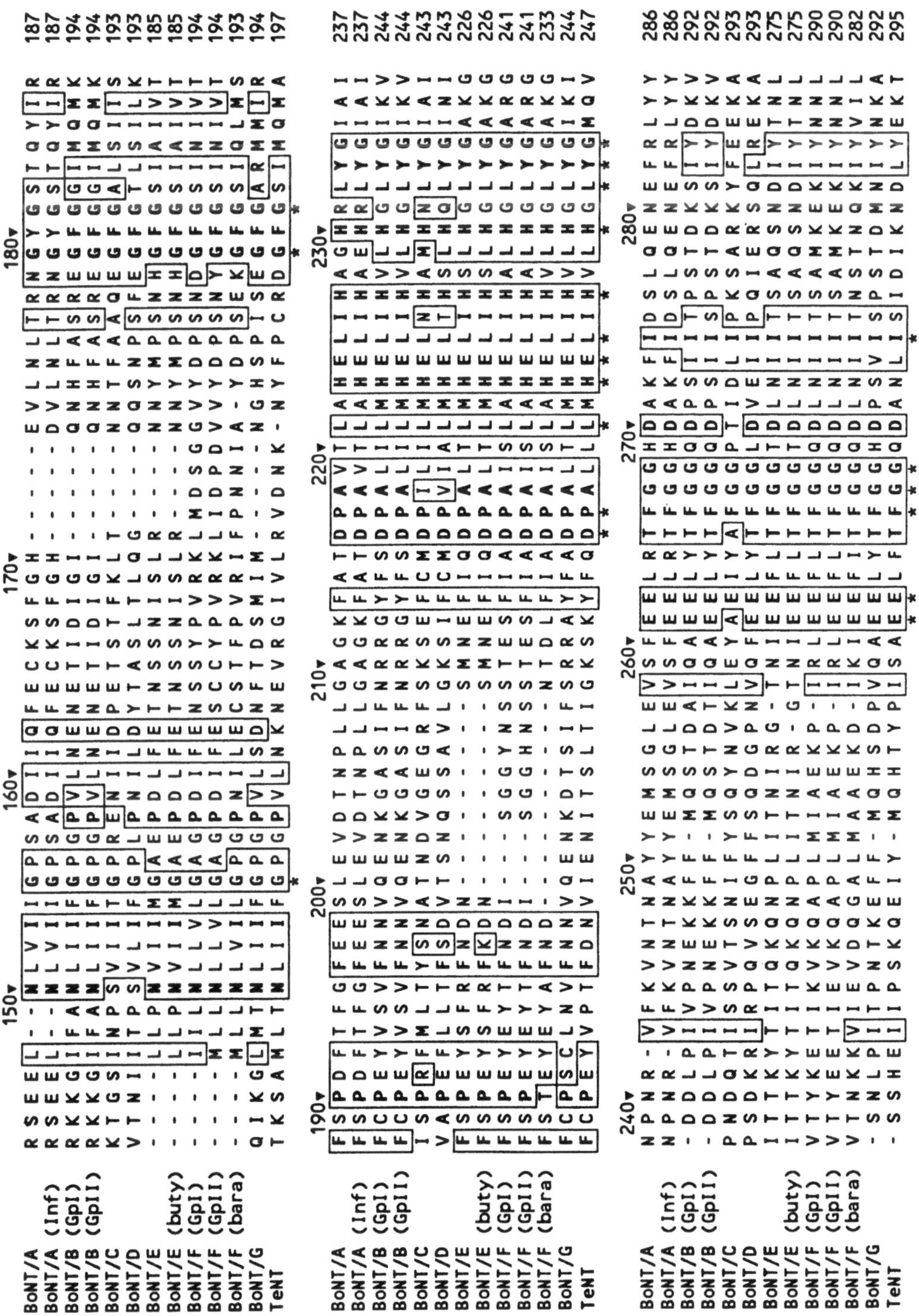

BoNT/A	YNKFKDIASTLNKAKSIVGTT--ASLQYMKNVFKEKYLLSEDTSGKFSVD	334
BoNT/A (Inf)	YNKFKDVASTLNKAKSIIGTT--ASLQYMKNVFKEKYLLSEDTSGKFSVD	334
BoNT/B (GpI)	LQNFRGIVDRLNKVLVCISDPNIN-ININIYKNKFKDKYKFVEDSEGKYSID	341
BoNT/B (GpII)	LQNFRGIVDRLNKVLVCISDPNIN-ININIYKNKFKDKYKFVEDSEGKYSID	341
BoNT/C	LDYYRSIAKRLNSITTANPSSFNKYIGEYKQKLIRKYRFVVESSGEVTVN	343
BoNT/D	LGHYKDIAKRLNNINKTIPSSWISNIDKYKKIFSEKYNFDKDNTGNFVVN	343
BoNT/E	LADYKKIASKLSKVQ--VSNPLLNP---YKDVFEAKYGLDKDASGIYSVN	320
BoNT/E (buty)	LADYKKIASKLSKVQ--VSNPLLNP---YKDVFEAKYGLDKDASGIYSVN	320
BoNT/F (GpI)	LANYEKIATRLSRVNSA---PPEYDINEYKDYFQWKYGLDKNADGSYTVN	337
BoNT/F (GpII)	LANYEKIATRLSEVNSA---PPEYDINEYKDYFQWKYGLDKNADGSYTVN	337
BoNT/F (bara)	LSNYTAIASRLSQVNRN---NSALNTTYYKNFFQWKYGLDQDSNGNYTVN	329
BoNT/G	LQNFQDIANRLN-IVSSAQGSGID-ISLYKQIYKNKYDFVEDPNGKYSVD	340
TeNT	LNDYKAIANKLSQVTSC-NDPNID-IDSYKQIYQQKYQFDKDSNGQYIVN	343

BoNT/A	KLKFDKLYKMLTEIYTEDNFVKFFKVLNRKTYLNFDK-AVFKINIVPKVN	383
BoNT/A (Inf)	KLKFDKLYKMLTEIYTEDNFVNFFKVINRKTYLNFDK-AVFRINIVPDEN	383
BoNT/B (GpI)	VESFDKLYKSLMFGFTETNIAENYKIKTRASYFSDSLPPVKIKNLLDNEI	391
BoNT/B (GpII)	VESFNKLYKSLMLGFTEINIAENYKIKTRASYFSDSLPPVKIKNLLDNEI	391
BoNT/C	RNKFVELYNELTQIFTEFNYAKIYNVQNRKIYLSNVYTPVTA-NILDDNV	392
BoNT/D	IDKFNSLYSDLTNVMSEVVYSSQYNVKNRTHYFSRHYLPVFA-NILDDNI	392
BoNT/E	INKFNDIFKKL-YSFTEFDLATKFQVKCRQTYIGQYK-YFKLSNLLNDSI	368
BoNT/E (buty)	INKFNDIFKKL-YSFTEFDLATKFQVKCRQTYIGQYK-YFKLSNLLNDSI	368
BoNT/F (GpI)	ENKFNEIYKKL-YSFTEIDLANKFKVKCRNTYFIKYG-FLKVPNLLDDDI	385
BoNT/F (GpII)	ENKFNEIYKKL-YSFTESDLANKFKVKCRNTYFIKYE-FLKVPNLLDDDI	385
BoNT/F (bara)	ISKFNAIYKKL-FSFTECDLAQKFQVKNRSNYLFHFK-PFRLLDLLDDNI	377
BoNT/G	KDKFDKLYKALMFGFTETNLAGEYGIKTRYSYFSEYLPPIKTEKLLDNTI	390
TeNT	EDKFQILYNSIMYGFTEIELGKKFNIKTRLSYFSMNHDPVKIPNLLDDTI	393

BoNT/A	YTIYDGFNLRNTNLAANFNGQNTEINNMNFTKLKNFTGLFEFYKLL-CVR	432
BoNT/A (Inf)	YTIKDGFNLKGANLSTNFNGQNTEINSRNFTRLKNFTGLFEFYKLL-CVR	432
BoNT/B (GpI)	YTIEEGFNISDKDMEKEYRGQNKAINKQAYEEISK--EHLAVYKIQMCKS	439
BoNT/B (GpII)	YTIEEGFNISDKNMGKEYRGQNKAINKQAYEEISK--EHLAVYKIQMCKS	439
BoNT/C	YDIQNGFNIPKSNLNVLFMGQNLSRNPALRKVNPENMLYLFTKF---CHK	439
BoNT/D	YTIRDGFNLTNKGFNIENSGQNIERNPALQKLSSESVVDLFTKV---CLR	439
BoNT/E	YNISEGYNIN--NLKVNFRGQNANLNPRIITPITGRG--LVKKIIRFCKN	414
BoNT/E (buty)	YNISEGYNIN--NLKVNFRGQNANLNPRIITPITGRG--LVKKIIRFCKN	414
BoNT/F (GpI)	YTVSEGFNIG--NLAVNNRGQNIKLNPKIIDSIPDKG--LVEKIVKFCKS	431
BoNT/F (GpII)	YTVSEGFNIG--NLAVNNRGQSIKLNPKIIDSIPDKG--LVEKIVKFCKS	431
BoNT/F (bara)	YSISEGFNIG--SLRVNNNGQNINLNSRIVGPIPDNG--LVERFVGLCKS	423
BoNT/G	YTQNEGFNIASKNLKTEFNGQNKAVNKEAYEEISL--EHLVIYRIAMCKP	438
TeNT	YNDTEGFNIESKDLKSEYKGQNMRVNTNAFRNVDGSG--LVSKLIGLCKK	441

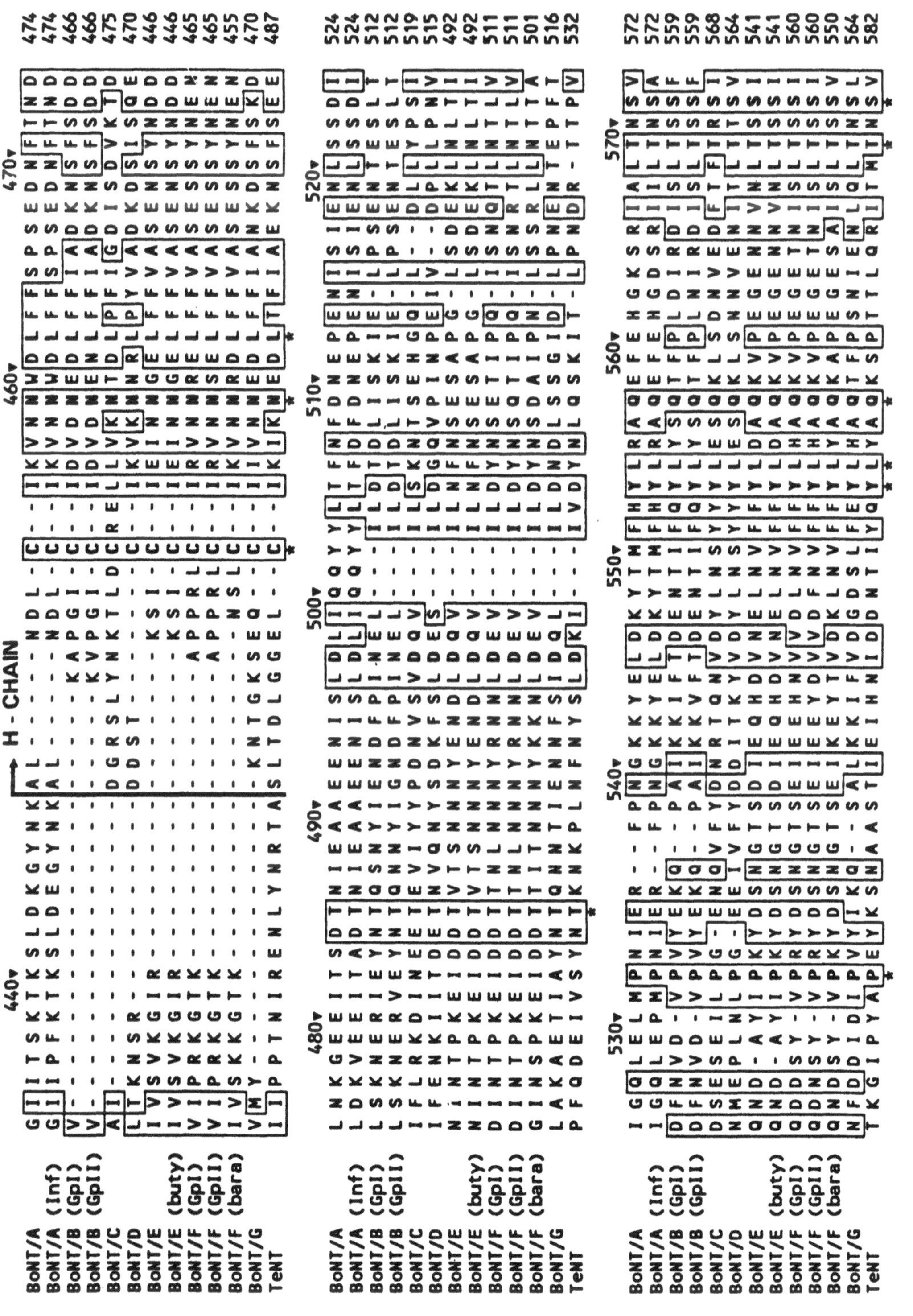
H - CHAIN
BoNT/A
BoNT/A (Inf)
BoNT/B (GpI)
BoNT/B (GpII)
BoNT/C
BoNT/D
BoNT/E
BoNT/E (buty)
BoNT/F (GpI)
BoNT/F (GpII)
BoNT/F (bara)
BoNT/G
TeNT

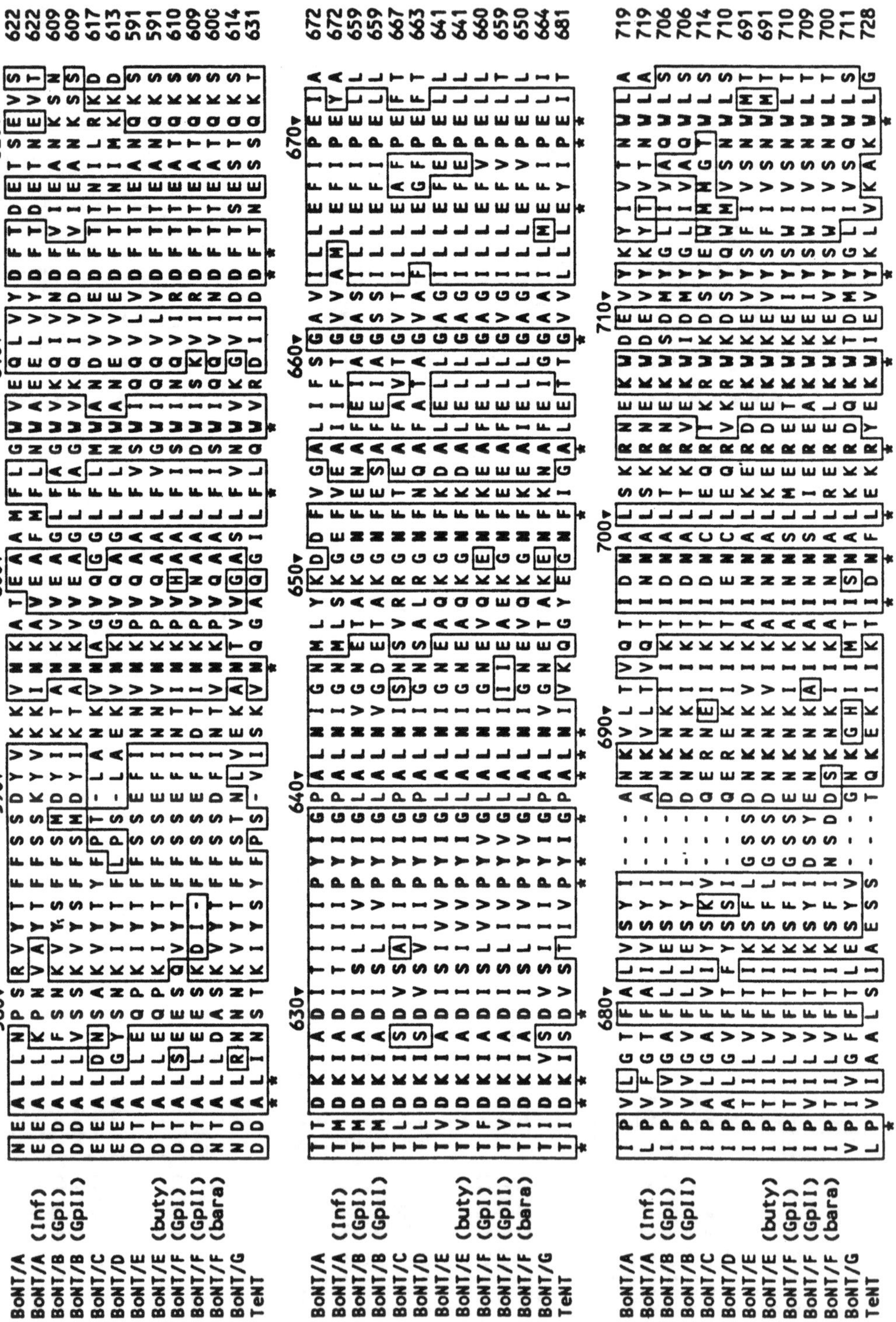

```
                720▾                730▾                740▾                750▾                760▾
BoNT/A          K V N T Q I D L I R K K M K E A L E N Q A E A T K A I I N Y Q Y N Q Y T E E E K N N I - - N F N I D   767
BoNT/A (Inf)    K V N T Q I D L I R E K M K K A L E N Q A E A T K A I I N Y Q Y N Q Y T E E E K N N I - - N F N I D   767
BoNT/B (GpI)    T V N T Q F Y T I K E G M Y K A L N Y Q A Q A L E E I I K Y R Y N I Y S E K E K S N I - - N I D F N   754
BoNT/B (GpII)   T V N T Q F Y T I K E G M Y K A L N Y Q A Q A L E E I I K Y K Y N I Y S E E E K S N I - - N I N F N   754
BoNT/C          R I I T Q F N N I S Y Q M Y D S L N Y Q A G A I K A K I D L E Y K K Y S G S D K E N I K S - - Q V E   762
BoNT/D          R I T T Q F N H I N Y Q M Y D S L S Y Q A D A I K A K I D L E Y K K Y S G S D K E N I K S - - Q V E   758
BoNT/E          K I N T Q F N K R K E Q M Y Q A L Q N Q V N A I K T I I E S K Y N S Y T L E E K N E L T N K Y D I K   741
BoNT/E (buty)   K I N T Q F N K R K E Q M Y Q A L Q N Q V N A L K A I I E S K Y N S Y T L E E K N E L T N K Y D I E   741
BoNT/F (GpI)    R I N T Q F N K R K E Q M Y Q A L Q N Q V D A I K T V I E Y K Y N N Y T S D E R N R L E S E Y N I N   760
BoNT/F (GpII)   R I N T Q F N K R K E Q M Y Q A L Q N Q V D A I K T A I E Y K Y N N Y T S D E K N R L E S E Y N I N   759
BoNT/F (bara)   R I N T Q F N K R K E Q M Y Q A L Q N Q V D G I K K I I E Y K Y N N Y T L D E K N R L R A E Y N I Y   750
BoNT/G          T V N T Q F Y T I K E R M Y N A L N N Q S Q A I E K I I E D Q Y N R Y S E E D K M N I - - N I D F N   759
TeNT            T V N T Q F Q K R S Y Q M Y R S L E Y Q V D A I K K I I D Y E Y K I Y S G P D K E Q I A D - - E I N   776
                      * *               *             *               *       *     *

                    770▾                780▾                790▾                800▾                810▾
BoNT/A          D L S S K L N E S I N K A M I N I N K F L N Q C S V S Y L M N S M I P Y G V K R L E D F D A S L K D   817
BoNT/A (Inf)    D L S S K L N E S I N S A M I N I N K F L D Q C S V S Y L M N S M I P Y A V K R L K D F D A S V R D   817
BoNT/B (GpI)    D I N S K L N E G I N Q A I D N I N N F I N G C S V S Y L M K K M I P L A V E K L L D F D N T L K K   804
BoNT/B (GpII)   D I N S K L N D G I N Q A M D N I N D F I N E C S V S Y L M K K M I P L A V K K L L D F D N T L K K   804
BoNT/C          N L K N S L D V K I S E A M N N I N K F I R E C S V T Y L F K N M L P K V I D E L N E F D R N T K A   812
BoNT/D          N L K N S L D V K I S E A M N N I N K F I R E C S V T Y L F K N M L P K V I D E L N K F D L R T K T   808
BoNT/E          Q I E N E L N Q K V S I A M N N I D R F L T E S S I S Y L M K L I N E V K I N K L R E Y D E N V K T   791
BoNT/E (buty)   Q I E N E L N Q K V S I A M N N I D R F L T E S S I S Y L M K L I N E V K I N K L R E Y D E N V K T   791
BoNT/F (GpI)    N I R E E L N K K V S L A M E N I E R F I T E S S I F Y L M K L I N E A K V S K L R E Y D E G V K E   810
BoNT/F (GpII)   N I E E E L N K K V S L A M K N I E R F M T E S S I S Y L M K L I N E A K V G K L K K Y D N H V K S   809
BoNT/F (bara)   S I K E E L N K K V S L A M Q N I D R F L T E S S I S Y L M K L I N E A K I N K L S E Y D K R V N Q   800
BoNT/G          D I D F K L N Q S I N L A I N N I D D F I N Q C S I S Y L M N R M I P L A V K K L K D F D D N L K R   809
TeNT            N L K N K L E E K A N K A M I N I N I F M R E S S R S F L V N Q M I N E A K K Q L L E F D T Q S K N   826
                          *             *     * *     *         *       *                       *       *

                    820▾                830▾                840▾                850▾                860▾
BoNT/A          A L L K Y I Y D N R G T - L I G - Q V D R L K D K V N N T L S T D I P F Q L S K Y V D N Q R L L S T   865
BoNT/A (Inf)    V L L K Y I Y D N R G T - L V L - Q V D R L K D E V N N T L S A D I P F Q L S K Y V D N K K L L S T   865
BoNT/B (GpI)    N L L N Y I D E N K L - Y L I G - S A E Y E K S K V N K Y L K T I M P F D L S I Y T N D T I L I E M   852
BoNT/B (GpII)   N L L N Y I D E N K L - Y L I G - S V E D E K S K V D K Y L K T I I P F D L S T Y S N I E I L I K I   852
BoNT/C          K L I N L I - D S H N I I L V G - E V D K L K A K V N N S F Q N T I P F N I F S Y T N N S L L K D I   860
BoNT/D          E L I N L I - D S H N I I L V G - E V D R L K A K V N E S F E N T M P F N I F S Y T N N S L L K D I   856
BoNT/E          Y L L N Y I I Q H G S - I L G E - S Q Q E L N S M V T D T L N N S I P F K L S S Y T D D K I L I S Y   839
BoNT/E (buty)   Y L L D Y I I K H G S - I L G E - S Q Q E L N S M V I D T L H N S I P F K L S S Y T D D K I L I S Y   839
BoNT/F (GpI)    Y L L D Y I S E H R S - I L G N - S V Q E L N D L V T S T L N N S I P F E L S S Y T N D K I L I L Y   858
BoNT/F (GpII)   D L L N Y I L D H R S - I L G E - Q T N E L S D L V T S T L N S S I P F E L S S Y T N D K I L I I Y   857
BoNT/F (bara)   Y L L N Y I L E N S S - T L G T S S V P E L N N L V S N T L N N S I P F E L S E Y T N D K I L I H I   849
BoNT/G          D L L E Y I D T N E L - Y L L D - E V N I L K S K V N R H L K D S I P F D L S L Y T K D T I L I Q V   857
TeNT            I L M Q Y I K A N S K F I G I T - E L K K L E S K I N K V F S T P I P F S Y S K N L D - - C W V D N   873
                  *       *                                                         * *
```

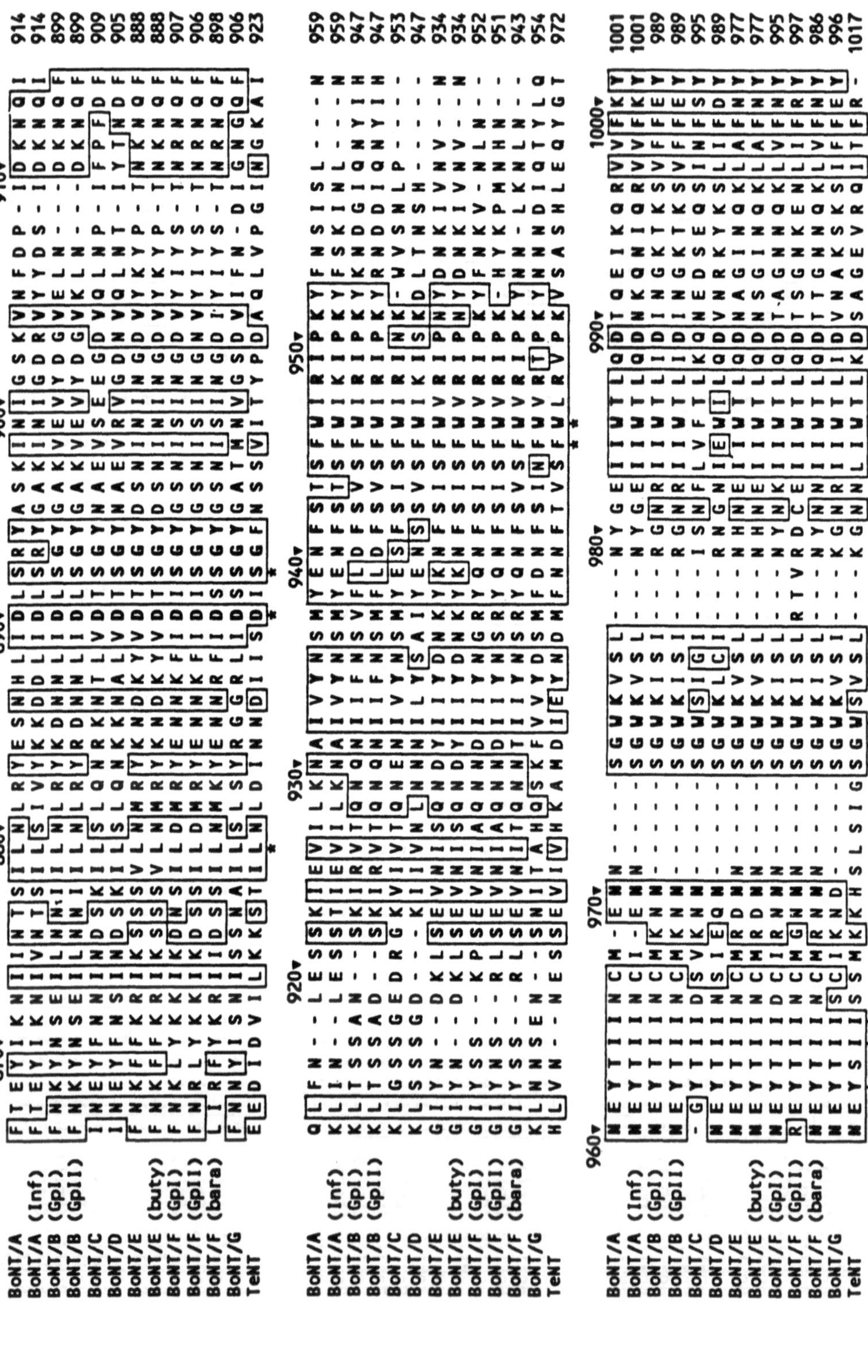
BoNT/A
BoNT/A (Inf)
BoNT/B (GpI)
BoNT/B (GpII)
BoNT/C
BoNT/D
BoNT/E
BoNT/E (buty)
BoNT/F (GpI)
BoNT/F (GpII)
BoNT/F (bara)
BoNT/G
TeNT

```
                              1010▼               1020▼               1030▼               1040▼               1050▼
BoNT/A           S Q M I N I S D Y I - N R W I F V T I T N N R L N N S K I Y I N G R L I D Q K P I S N L G N I H A S  1050
BoNT/A (Inf)     S Q M V N I S D Y I - N R W I F V T I T N N R L T K S K I Y I N G R L I D Q K P I S N L G N I H A S  1050
BoNT/B (GpI)     N I R E D I S E Y I - N R W F F V T I T N N - L N N A K I Y I N G K L E S N T D I K D I R E V I A N  1037
BoNT/B (GpII)    N I R E D I S E Y I - N R W F F V T I T N N - L D N A K I Y I N G T L E S N M D I K D I G E V I V N  1037
BoNT/C           D I S N N A P G Y - - N K W F F V T V T N N M M G N M K I Y I N G K L I D T I K V K E L T G I N F S  1043
BoNT/D           S E S L S H T G Y T - N K W F F V T I T N N I M G Y M K L Y I N G E L K Q S Q K I E D L D E V K L D  1038
BoNT/E           G N A N G I S D Y I - N K W I F V T I T N D R L G D S K L Y I N G N L I D Q K S I L N L G N I H V S  1026
BoNT/E (buty)    G N A N G I S D Y I - N K W I F V T I T N D R L G D S K L Y I N G N L I D K K S I L N L G N I H V S  1026
BoNT/F (GpI)     T Q M I S I S D Y I - N K W I F V T I T N N R L G N S R I Y I N G N L I D E K S I S N L G D I H V S  1044
BoNT/F (GpII)    E E L N R I S N Y I - N K W I F V T I T N N R L G N S R I Y I N G N L I V E K S I S N L G D I H V S  1046
BoNT/F (bara)    T Q M I D I S D Y I - N K W T F V T I T N N R L G H S K L Y I N G N L T D Q K S I L N L G N I H V D  1035
BoNT/G           S I K D N I S D Y I - N K W F S I T I T N D R L G N A N I Y I N G S L K K S E K I N N L D R I N S S  1045
TeNT             D L P D K F N A Y L A N K W V F I T I T N D R L S S A N L Y I N G V L M G S A E I T G L G A I R E D  1071
                                 *     *   *       *   * *                 * * * *   *

                                                    1060▼               1070▼               1080▼               1090▼
BoNT/A           N N I M F K L D G - - - - - - - - C R D T H R Y I W I K Y F N L F D K E L N E K E I K D L Y D N Q S  1092
BoNT/A (Inf)     N K I M F K L D G - - - - - - - - C R D P R R Y I M I K Y F N L F D K E L N E K E I K D L Y D S Q S  1092
BoNT/B (GpI)     G E I I F K L D G D I D R - - - - - - - - T Q F I W M K Y F S I F N T E L S Q S N I E E R Y K I Q S  1079
BoNT/B (GpII)    G E I T F K L D G D V D R - - - - - - - - T Q F I W M K Y F S I F N T Q L N Q S N I K E I Y K I Q S  1079
BoNT/C           K T I T F E I N K I P D T G L I T S D S D N I N M W I R D F Y I F A K E L D G K D I N I L F N S L Q  1093
BoNT/D           K T I V F G I D E N I D - - - - - - - - E N Q M L W I R D F N I F S K E L S N E D I N I V Y E G Q I  1080
BoNT/E           D N I L F K I V N - - - - - - - - C S Y T - R Y I G I R Y F N I F D K E L D E T E I Q T L Y S N E P  1067
BoNT/E (buty)    D N I L F K I V N - - - - - - - - C S Y T - R Y I G I R Y F N I F D K E L D E T E I Q T L Y N N E P  1067
BoNT/F (GpI)     D N I L F K I V G - - - - - - - - C N D T - R Y V G I R Y F K V F D T E L G K T E I E T L Y S D E P  1085
BoNT/F (GpII)    D N I L F K I V G - - - - - - - - C D D E - T Y V G I R Y F K V F N T E L D K T E I E T L Y S N E P  1087
BoNT/F (bara)    D N I L F K I V G - - - - - - - - C N D T - R Y V G I R Y F K I F N M E L D K T E I E T L Y H S E P  1076
BoNT/G           N D I D F K L I N - - - - - - - - C T D T T K F V W I K D F N I F G R E L N A T E V S S L Y W I Q S  1087
TeNT             N N I T L K L D R - - - - - - - - C N N N N Q Y V S I D K F R I F C K A L N P K E I E K L Y T S Y L  1109
                     *                                                     *     *       *

                            1100▼               1110▼               1120▼               1130▼               1140▼
BoNT/A           N S G I L K D F W G D Y L Q Y D K P Y Y M L N L Y D P N K Y V D V N N V G I R G Y M Y L K G P R G S  1142
BoNT/A (Inf)     N S G I L K D F W G N Y L Q Y D K P Y Y M L N L F D P N K Y V D V N N I G I R G Y M Y L K G P R G S  1142
BoNT/B (GpI)     Y S E Y L K D F W G N P L M Y N K E Y Y M F N A G N K N S Y I K L K K D S P V G E I - - - L T R S K  1146
BoNT/B (GpII)    Y S E Y L K D F W G N P L M Y N K E Y Y M F N A G N K N S Y I K L V K D S S V G E I - - - L I R S K  1146
BoNT/C           Y T N V V K D Y W G N D L R Y N K E Y Y M V N I - - - - D Y L N R - - - - - - - Y M Y - - - A N S R  1129
BoNT/D           L R N V I K D Y W G N P L K F D T E Y Y I I N D - - - - N Y I D R - - - - - - - Y I A - - - P E S N  1116
BoNT/E           N T N I L K D F W G N Y L L Y D K E Y Y L L N V L K P N N F I D R R K D S - - - T L S I N N I R S T  1114
BoNT/E (buty)    N A N I L K D F W G N Y L L Y D K E Y Y L L N V L K P N N F I N R R T D S - - - T L S I N N I R S T  1114
BoNT/F (GpI)     D P S I L K D F W G N Y L L Y N K R Y Y L L N L L R T D K S I T Q N S N - - - - F L N I N Q Q R G V  1131
BoNT/F (GpII)    D P S I L K N Y W G N Y L L Y N K K Y Y L F N L L R K D K Y I T L N S G - - - - I L N I N Q Q R G V  1133
BoNT/F (bara)    D S T I L K D F W G N Y L L Y N K K Y Y L L N L L K P N M S V T K N S D - - - - I L N I N R Q R G I  1122
BoNT/G           S T N T L K D F W G N P L R Y D T Q Y Y L F N Q G M Q N I Y I K Y F S K A S M G E T A - - - P R T N  1134
TeNT             S I T F L R D F W G N P L R Y D T E Y Y L I P V A S S S K D V Q L K N - - I T D Y M Y L T N A P S Y  1157
                                 * *     *           * *
```

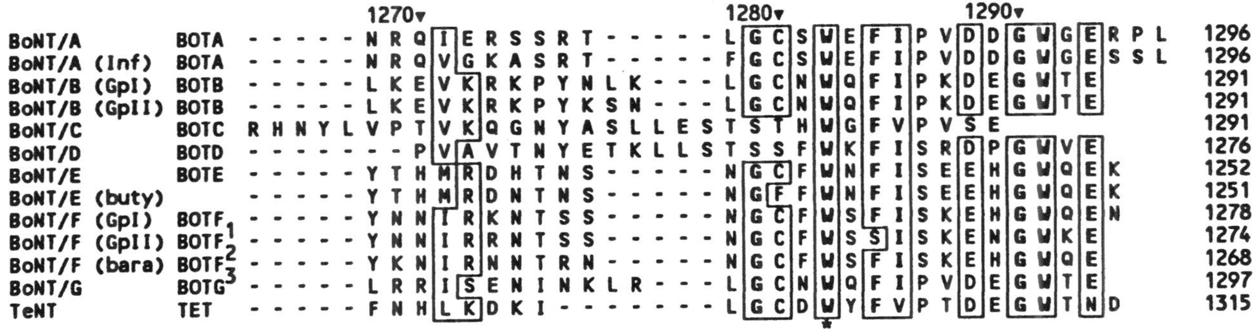

Fig. 1. Comparative alignment of clostridial neurotoxins. The illustrated alignment was derived essentially using the computer programme CLUSTAL (HIGGINS and SHARP 1988) and has been gapped to maximise homology. *Boxes*, highly conserved regions, including areas in which conservative replacements have occurred, in addition to sequence identity; *boldface*, amino acids conserved in at least 11 of 13 toxins; *asterisks*, amino acids conserved amongst all toxins. *Numbering above alignment*, corresponding to BoNT/A. By reference to Table 1, the sequenced toxins were derived from the following strains: BoNT/A, *C. botulinum* NCTC 2916; BoNT/A (*Inf*), *C. botulinum* Kyoto-F; BoNT/B (GpI) *C.botulinum* Danish; BoNT/B (*GpII*), *C. botulinum* Eklund; BoNT/C, *C. botulinum* C-St/468C; BoNT/E *C. botulinum* NCTC 11219; BoNT/E (*buty*), *C. butyricum ATCC* 43181/43755; BoN/F (*GpI*), *C. botulinum* Langeland; BoNT/F (GpII), *C. botulinum* ATCC 23387; BoNT/F (*bara*), *C. barati* ATCC 43756; BoNT/G, *C. botulinum* 89G/ NCFB 3012; TeNT, *C. tetani* CN3911

with metalloprotease activity (see Benfenati and Valtorta, this volume), and a highly conserved PYI/VXALN motif (635–642) found adjacent to regions identified as possessing membrane spanning potential (see Shone and Tranter, this volume). The most notable tract of sequence divergence amongst toxins is localised to the COOH terminus of their respective H chains (amino acid 1124 onwards of BoNT/A). This appears consistent with the notion that this domain is involved in neurotoxin binding, and that different toxins target different acceptors on neural cell surfaces. Given the diversity, the apparent conservation of the amino acid sequence at the extreme COOH terminus is intriguing.

As can be seen from Table 2, in the majority of cases more than one representative of a particular neurotoxin gene has been sequenced. In many instances little sequence divergence is evident. Thus, the sequenced pairs of TeNT, *C. butyricum botE,* and the *C. botulinum botG* genes are identical, while the two *botC* genes and the *botA* genes of *C. botulinum* NCTC 2916 and 62A diverge by only one and two nucleotides, respectively. In other instances, quite substantial differences are apparent. However, with one exception (the sequence divergence seen between the *botA* gene of *C. botulinum* Kyoto-F and NCTC 2916), these larger degrees of sequence divergence occur between genes which, although encoding the same BoNT serotype, are carried by a different member of the six neurotoxic clostridial groups. This is most apparent amongst the three sequenced *botF* genes, where the encoded neurotoxins of the group II and group VI strains differ from the BoNT/F of the group I strain Langeland by 151 and 341 amino acids, respectively (Table 2). Similarly, BoNT/B produced by the group I strain Danish and the group II strain Eklund diverge by 93 amino acids. The lowest level of divergence is seen between the BoNT/E of the group II *C. botulinum* strains and that of group V *C. butyricum* strains, for example, only 34 amino acids between BoNT/E of ATCC 11219 and ATCC 43181.

Comparative alignments of the respective light (L) and heavy (H) chain components of all the characterised neurotoxins demonstrates that the amino acid sequence identity amongst the different serotypes generally does not exceed 40%. Notable exceptions are the level of identity seen between the L and H chains of BoNT/C1 and BoNT/D (47% and 56%, respectively), the L chain of BoNT/B and those of BoNT/G and TeNT (61% and 50%, respectively), and the L and H chains of BoNT/E and all type F BoNTs (57% and 68% identity to the BoNT/F of strain Langeland). A distance matrix analysis illustrating the genealogical relationships amongst L and H chains is presented in Fig. 2. From these it is apparent that whereas the L chain of TeNT is relatively closely related to many BoNTs, that of BoNT/A represents the most divergent member of the tree. A complete reversal in relative relatedness is seen between H chains. In contrast, BoNT/C1 and BoNT/D remain distant to the other neurotoxins regardless of which di-chain component is compared.

The presence of BoNT genes within various the genotypically distinct clostridial groupings indicates that lateral transfer of the genes has occurred. Assuming similar rates of nucleotide change, the degree of divergence between genes provides some indication of the relative timing of such a event in

Table 2. Characterised clostridial neurotoxin genes

Toxin	*Clostridum* species	Strain	Group[a]	Percentage G+C	Ribosome binding site	Stop codon	Amino acids	Differences[b]		Reference
								NT	AA	
BoNT/A	*C. botulinum*	NCTC 2916	I	26.2	TTATAAGAGGTGTTAAAT ATG	TAA	1296	–	–	THOMPSON et al. (1990)
BoNT/A	*C. botulinum*	62A	I	26.3	TTATAAGAGGTGTTAAAT ATG	TAA	1296	2	2	BINZ et al. (1990a)
BoNT/A	*C. botulinum*	Kyoto-F	I	26.8	TTATAAGAGGTGTTAAAT ATG	TAA	1296	213	132	WILLEMS et al. (1993)
BoNT/B	*C. botulinum*	Danish	I	25.5	ATAAGGAGGAGAATATTT ATG	TAA	1291	–	–	WHELAN et al. (1992a)
BoNT/B	*C. botulinum*	Eklund	II	25.5	ATAAGGAGGAGAATATTT ATG	TAA	1291	153	93	HUTSON et al. (1993b)
BoNT/C1	*C. botulinum*	Phage 468C	III	26.2	TTAGGAGATGTTAGTATT ATG	TAA	1291	–	–	HAUSER et al. (1990)
BoNT/C1	*C. botulinum*	Phage C-St	III	26.2	TTAGGAGATGTTAGTATT ATG	TAA	1291	1	0	KIMURA et al. (1990)
BoNT/D	*C. botulinum*	Phage DVD/-3	III	25.6	TTAGGAGATGTTAGTATT ATG	TAA	1291	–	–	BINZ et al. (1990b)
BoNT/E	*C. botulinum*	NCTC 11219	II	25.0	AATTAGGAGATGCTGTAT ATG	TAA	1252	–	–	WHELAN et al. (1992b)
BoNT/E	*C. botulinum*	Beluga	II	25.1	AATTAGGAGATGCTGTAT ATG	TAA	1251	13	8	POULET et al. (1992)
BoNT/E	*C. butyricum*	ATCC 43181[c]	V	24.8	AATTAGGAGATGCTGTAT ATG	TAA	1251	58	34	POULET et al. (1992)
BoNT/F	*C. botulinum*	Langeland	I	26.5	AAGTATAGGGGGGATTTT ATG	TAA	1278	–	–	This laboratory
BoNT/F	*C. botulinum*	ATCC 23387	II	26.9	AGGTATAGGGGGGATTTT ATG	TGA	1274	359	151	EAST et al. (1992)
BoNT/F	*C. barati*	ATCC 43756	VI	24.5	TATAGGAGGTTGATATAT ATG	TAA	1268	700	341	THOMPSON et al. (1993)
BoNT/G	*C. botulinum*	89G	IV	27.1	AATTAGGAGGTATATATT ATG	TAA	1297	–	–	This laboratory
BoNT/G	*C. botulinum*	NCFB 3012	IV	27.1	(not determined)	TAA	1297	0	0	CAMPBELL et al. (1993a)
TeNT	*C. tetani*	Plasmid pE88	–	27.8	AATTAGGAGATGATACGT ATG	TAA	1315	–	–	EISEL et al. (1986)
TeNT	*C. tetani*	CN3911	–	27.8	AATTAGGAGATGATACGT ATG	TAA	1315	0	0	FAIRWEATHER and LYNESS et al. (1986)

[a] Neurotoxic clostridial groups are listed in Table 1.
[b] Where more than one sequence has been determined, differences—in nucleotides (*nt*) and amino acids (*aa*)—are expressed relative to the first gene listed.
[c] The nucleotide sequence of the *botE* gene of *C. butyricum* ATCC 43755 is identical to that of strain ATCC 43756. The nucleotide sequences of *C. botulinum* strains Langeland and 89G were determined in this laboratory (S.M. Whelan, N.J. Bodsworth, M.J. Elmore and N.P.M., unpublished data).

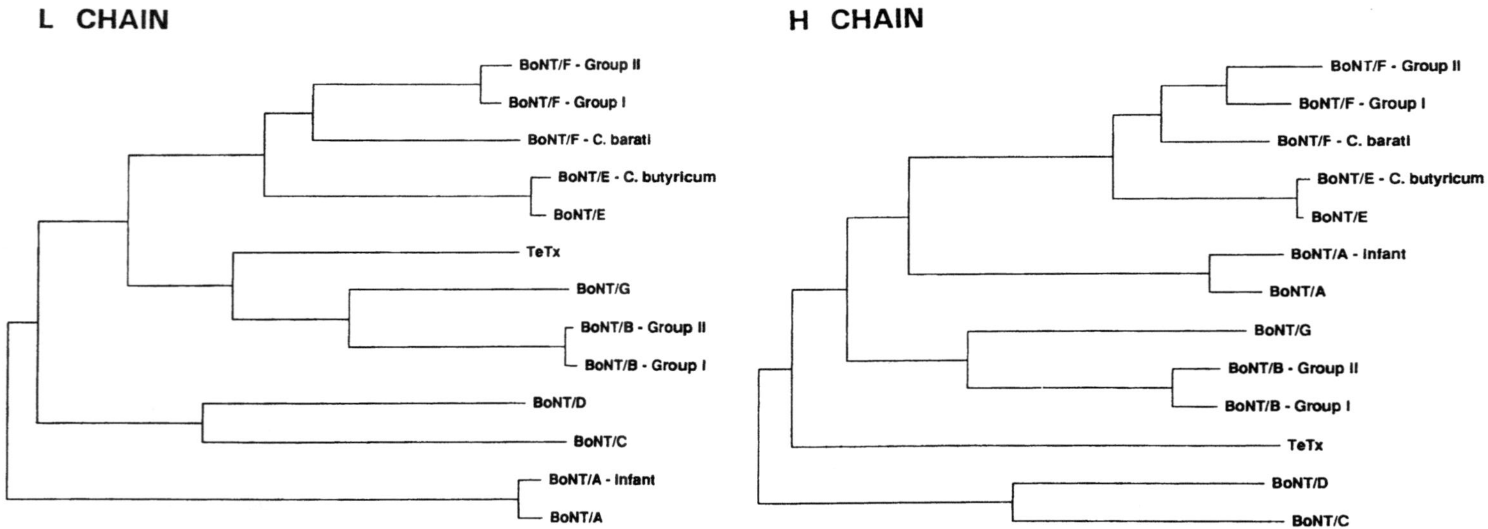

Fig. 2. Distance matrix trees depicting the relationships of light (*L*) and heavy (*H*) neurotoxin chains. The toxins compared are those illustrated in Fig. 1. Analysis was undertaken using the programme of VAN DE PEER and DE WACHTER (1993)

evolutionary history. Thus, the high level of amino acid sequence identity (97%) between the type E neurotoxins of *C. botulinum* and *C. butyricum* and between BoNT/B of groups I and II *C. botulinum* genotypes (93%) indicates that gene transfer took place relatively recently. BoNT/F of groups I and II *C. botulinum* genotypes share 87.3% identity between themselves and, respectively, 74% and 69.4% identity to BoNT/F of *C. barati*. Transfer of BoNT/F genes between group I and group II *C. botulinum* genotypes and between *C. botulinum* and *C. barati* therefore apparently took place at progressively earlier evolutionary dates.

2.3 Derivation of DNA Probes

Amongst the many driving forces for elucidating the nucleotide sequences of the various neurotoxin genes has been the desire to design DNA probes. These may be used for detection of neurotoxic clostridia in clinical, environmental and food samples. This is particularly pertinent to botulinogenic clostridia, where the only method of sufficient sensitivity for the detection of toxin is the mouse bioassay. The availability of a DNA test would considerably speed up diagnosis and could in certain instances circumvent the need for the use of undesirable animal lethality tests. As the sequences of the various BoNT genes have become available, a number of reports have appeared describing the design of pairs of oligonucleotide primers which can be employed in polymerase chain reaction (PCR) to amplify specifically *bot* gene specific DNA fragments.

Of these the most comprehensive studies to date have been those of SZABO et al. (1993) and CAMPBELL et al. (1993b). In the study of SZABO et al. (1993) five pairs of oligonucleotide primers were described which enabled the PCR-mediated generation of DNA fragments from clostridial strains carrying genes encoding BoNT/A to BoNT/E. The sensitivity of the test was demonstrated by the generation of a fragment from as little as 10 fg DNA (three clostridial cells). Similar probes were used by CAMPBELL et al. (1993b) for the detection of BoNT genes of *C. botulinum* type B, E and F strains and the BoNT/F gene of *C. barati*. They additionally designed a mixture of degenerate primers which could be employed to amplify a 1.1-kb fragment from the H chain of *C. botulinum* types A, B, E, F and G. In the case of type A strains a smaller than expected DNA fragment was generated. This was apparently due to significant mismatch between one set of the degenerate primers and the intended target in BoNT/A genes.

In our own laboratory we have also designed pairs of oligonucleotide primers for the detection of each toxin type (S.M. Whelan, M.J. Elmore and N.P.M., unpublished data). Uniquely, these primers amplify a DNA fragment which is of a size specific to the encoded toxin type and can be used as a primer cocktail. Using such a primer mixture, we have been able to show in a single PCR reaction that certain group I *C. botulinum* strains do indeed carry two serotypically distinct BoNT genes. In one instance a type A strain has been shown to carry a *botB* gene in addition to *botA*. As there is no evidence for the presence of BoNT/B, the *botB* gene is presumed to be silent.

3 Genetic Localisation of the Genes

The ease with which any particular toxin gene undergoes lateral transfer is undoubtedly linked to its mobility. This in turn depends on the extent of any association with a transferable genetic element. There is much evidence to suggest that clostridial neurotoxins are located on extrachromosomal elements, being encoded by either bacteriophage or plasmids. For TeNT, BoNT/C1 and BoNT/D the evidence is irrefutable. With the exception of BoNT/G the case for the other BoNTs remains largely circumstantial.

3.1 Neurotoxin Genes Encoded by Extrachromosomal Bacteriophages

Many different strains of *C. botulinum* have been shown to carry bacteriophages (Inoue and Iida 1968; Eklund et al. 1969; Dolman and Chang 1972). Only in the case of BoNT/C1 and BoNT/D, however, has a direct relationship between bacteriophage and toxinogenicity been established. This relationship was demonstrated initially in two independent studies (Inoue and Iida 1970; Eklund et al. 1971) when it was shown that when *C. botulinum* type C or type D is cured of bacteriophage following exposure to UV light or acridine orange, there is a concomitant loss of the ability to produce BoNT. Loss of toxicity was permanent, but non-toxinogenic derivatives could be re-infected with phage from the culture supernatants of their progenitors, whereupon they were converted back to toxinogenicity. As many type C and D strains can carry more than one bacteriophage, it was necessary to purify each phage type individually. Subsequent re-infection experiments allowed the identification of the two phages (TOX^+) responsible for toxicity in either type C or type D strains. Other phages, termed TOX^-, did not affect toxicity.

The non-toxigenic derivatives of type C and type D *C. botulinum* strains were found to be indistinguishable from one another, and it soon became apparent that certain derivatives of each type become sensitive to infection by the TOX^+ phage of the other serotype (Eklund and Poysky 1974). Thus a *C. botulinum* type C strain could be cured of its TOX^+ phage, and the resultant non-toxigenic derivative converted to a toxinogenic *C. botulinum* type D strain by infection with a type D TOX^+ phage. It became further apparent that the non-toxinogenic derivatives were also indistinguishable from strains of *C. novyi* type A which had been cured of their own TOX^+ phage (Eklund et al. 1974), responsible for the production of α-toxin (Eklund et al. 1976). It transpired that a single non-toxinogenic *C. botulinum* type C strain could be converted either back to *C. botulinum* type C, to *C.botulinum* type D or to *C. novyi* type A, depending on the TOX^+ phage with which the cells had been infected. These experiments indicated that the genes encoding both BoNT/C1 and BoNT/D and α-toxin are encoded by separate bacteriophages, and that it is the presence of one or another of these phages within the cell which determines whether a toxin is produced and thereby the species/strain designation. A phage location for the toxin gene of BoNT/C1 was established directly through the experiments of FUJII et al. (1988). In this study an oligonucleotide probe based on available NH_2 terminal amino acid sequence data was shown to hybridize specifically to phage DNA purified from toxinogenic *C. botulinum* strains. The phage location of the neurotoxin genes was definitively demonstrated when both the *botC* and *botD* genes were cloned from their respective bacteriophage genomes and their nucleotide sequences determined (BINZ et al. 1990b; HAUSER et al. 1990).

The relative ease with which TOX^+ phages are lost by group III clostridial cells suggests that the prophage is not stably integrated into the host chromosome. Rather, the phage exists in an extrachromosomal state, somewhere between virulence and lysogeny, in so-called "pseudolysogeny". This assumption is directly supported by the observation, following cloning of the *botC* and *botD* genes, that gene-specific probes hybridise only with phage DNA and not host chromosome (BINZ et al. 1990b; HAUSER et al. 1990). In contrast to the situation in group III strains, *C. botulinum* strains of groups I and II maintain their phages in the true lysogenic state. Phage-sensitive bacteria are therefore seldom isolated. Despite these difficulties cultures of group I type A and type F strains have been cured of multiple phages. There were, however, no apparent consequences with regard to neurotoxin production (EKLUND et al. 1989). Similarly, phage-sensitive derivatives of *C. botulinum* group II strains have been isolated, but they also continue to produce toxin (EKLUND et al. 1989). Despite these findings the involvement of bacteriophages in the toxigenesis of *C. botulinum* groups I and II cannot be discounted.

3.2 Plasmid-Encoded Neurotoxin Genes

Failure to identify a phage responsible for the toxigenic phenotype of neurotoxic clostridia has focussed attention on the possible involvement of plasmids. LAIRD et al. (1980) examined 21 toxigenic *C. tetani* strains of diverse origin and found the loss of a single 75-kb plasmid (pCL1) to be correlated with loss of toxicity. However, it was unclear whether this plasmid encoded the toxin structural gene or genes whose products are concerned with regulatory control of a chromosomally encoded toxin gene. Subsequently, however, an oligonucleotide probe based on the NH_2-terminal amino acid sequence of TeNT was used in hybridisation studies to demonstrate that the 75-kb plasmid of three toxigenic strains does indeed encode the TeNT structural gene (FINN et al. 1984). Furthermore, loss of toxicity in at least one instance was shown not to be due to the loss of the entire plasmid but to the spontaneous deletion of 22 kb of DNA from pCL1. The deleted plasmid still hybridised to the TeNT gene probe, indicating that at least the 5'-end of the toxin gene remained.

Cryptic plasmids have been isolated from many group I and II *C. botulinum* strains (SCOTT and DUNCAN 1978; STROM et al. 1984). Six *C. botulinum* type A strains were cured of their toxigenicity by culturing in media containing sodium deoxycholate. Toxigenic and non-toxigenic organisms were then screened for plasmids, but toxigenicity was not correlated with any specific plasmid (WEICKERT et al. 1986). STROM et al. (1984), however, noted that all six of the group IV *C. botulinum* type G strains examined harboured a single 81-MDa plasmid. Restriction enzyme analysis showed that all strains carried an identical plasmid. In a subsequent study (EKLUND et al. 1988) it proved possible to induce loss of this plasmid by consecutive daily transfers of growing cells in complex medium at 44°C. The resultant plasmid-free strains ceased to produce BoNT/G. Loss of toxigenicity was a permanent characteristic and could not be reversed by incubation of non-toxigenic strains with culture filtrates from toxigenic strains. These results suggest that either the structural *botG* gene or some positive regulatory element is plasmid encoded. Hybridisation of *botG*-specific probe to the plasmid in question should confirm this conclusion.

3.3 Neurotoxin Genes of Clostridia Other than *C. botulinum*

With the exception of the cases discussed above, the location of BoNT genes in all other *C. botulinum* strains remains unclear. Recent discoveries that clostridial strains classified as other than *C. botulinum* can produce either BoNT/E (*C. butyricum*) or BoNT/F (*C. barati*) has been touted as evidence for the extra-chromosomal location of the encoding genes in these strains. Efforts focussed on demonstrating such a location of *botE* in strains of *C. butyricum* have, however, resulted in conflicting evidence.

HAUSER et al. (1992) presented data in support of *butE* being located on a large plasmid in two toxinogenic strains of *C. butyricum*, ATCC 43181 and ATCC 43755. Essentially, two *botE*-specific oligonucleotide primers were used to PCR-amplify a 1.2-kb fragment encoding the L chain from either "plasmid" or "chromosomal" DNA prepared from the two strains. The subjective observation that a greater quantity of product was obtained when the PCR template was "plasmid" DNA than when "chromosomal" DNA was used is taken as indicating that the gene is plasmid encoded. More substantive data are required before the *botE* genes of these strains can be confidently assigned to an extrachromosomal plasmid. In particular it will need to be demonstrated that a *botE*-specific probe does indeed hybridise to a more fully defined plasmid species.

In a more recent study ZHOU et al. (1993) presented data indicative of a chromosomal location for *botE* in the toxinogenic *C. butyricum* strains ATCC 5839 and 5521. Thus, in Southern blots of electrophoresed total DNA a *botE*-specific probe hybridised to the smear of DNA attributed to chromosomal DNA and not to discrete bands corresponding to plasmid DNA. Hybridisation did not occur to the equivalent DNA of a nontoxinogenic *C. butyricum* strain, ATCC 18398. Both toxinogenic and non-toxinogenic strains were subsequently shown to harboured prophage, released from the cells following mitomycin C induction. Although of a similar genomic size (34 kb), these phage were deemed unrelated on the basis of dissimilar restriction fragment profiles for *Hin*dIII and *Eco*RI, and following the demonstration that radiolabelled ATCC 5839 phage DNA hybridises to the chromosomal of its host but not to that of ATCC 18398.

Intriguingly, both a *botE*-specific probe and radiolabelled phage DNA hybridised to similarly sized *Nde*I and *Hin*dIII fragments of the ATCC 5839 chromosome. Although such a finding could well be coincidence, it has been suggested (ZHOU et al. 1993) that *botE* and the phage genome are in some way linked on the host chromosome. Indeed, in subsequent experiments it was shown that the ability to produce BoNT/E can be transferred from *C. butyricum* ATCC 5839 to a non-toxinogenic E-like *C. botulinum* strain by a phage-mediated process. Transfer would occur, however, only when the recipient was incubated in filter-sterilised culture filtrate obtained by growing a helper strain (a non-toxinogenic *C. butyricum* strain ATCC 19398) in broth containing phage induced by mitomycin C from the donor strain. In such experiments approximately one in 10^4 recipient cells became toxinogenic, and transfer was never observed without the participation of a helper strain. It was concluded that the phage particles produced by ATCC 5839 are defective, becoming infective through the participation of the helper strain.

These experiments appear to indicate that the toxin gene is encoded by a lysogenic phage present within the host chromosome. However, while purified phage DNA was shown to act as a template for the PCR-mediated amplification of a *botE*-specific DNA fragment, it did not hybridise to a radiolabelled *botE* DNA probe. This major contradiction is not fully explored by ZHOU et al. (1993), other than to make the suggestion that the toxin gene is present at a concentration below the detection limit of the dot blot assay employed. The lack of sensitivity is apparently not due to insufficient material, as the quantity of phage DNA isolated in their experiments was adequate for restriction enzyme analysis. It therefore implies that only a very minor proportion of the phage genomes probed contain the *botE* gene. This could be the result of the presence of more than one type of phage, with the *botE*-encoding phage being in the minority. Alternatively, the identified prophage could have an attachment site in the *C. butyricum* chromosome next to the *botE* gene. The DNA isolated by ZHOU et al. (1993) could therefore have consisted of wild-type phage genomes and derivatives carrying host chromosome from adjacent to the prophage attachment site. Such derivatives would arise following imperfect excision, in much the same way as in other specialised transducing phages such as phage λ. The frequency with which such an event occurs would then dictate the abundance of *botE* in phage DNA preparations, for example, one in 10^{-6} in the case of phage λ. As the gene could not be detected in phage DNA preparations by standard DNA/DNA hybridisation techniques, such imperfect excisions would need to be relatively rare. They must, however, occur at a sufficient frequency to give the observed rates of toxin transfer.

3.4 "Chromosomally Encoded" Neurotoxin Genes

In the absence of evidence to suggest that either a phage or plasmid is involved it has been common practice to assume a chromosomal location for the neurotoxin genes of groups I and II *C. botulinum* strains. The fact that their genes were cloned from chromosomal DNA preparations is frequently quoted as supportive of this conclusion. Such assumptions are clearly unwise, as any extrachromosomal element is co-isolated along with chromosome DNA and is therefore able to be cloned. Such co-purification presumably happened during the cloning procedure adopted by FAIRWEATHER et al. (1986) when cloning the TeNT gene of *C. tetani* strain CN3911. In a recent study, however, we have obtained data which provide some clues as to the probable location of BoNT genes in Group I *C. botulinum* strains.

In addition to sequencing upstream of the *botA* gene of *C. botulinum* NCTC 2916, we have also turned our attention to the region of the chromosome immediately 3'. An open reading frame (ORF) has been identified (WHELAN et al. 1994) approximately 1.0 kb 3' to the *botA* stop codon (Fig. 3) encoding a polypeptide which exhibits significant homology with the *Charalopsis* group of lysozymes. Such proteins include the chromosomally located autolysins of *Clostridium acetobutylicum* and *Streptomyces globisporus* (CROUX et al. 1991; LICHENSTEIN et al. 1990) and the bacteriophage-encoded lysozymes of *Lactobacillus bulgaricus* and *Streptococcus pneumoniae* (BOIZET et al. 1990; GARCIA et al. 1988). The identity of the encoded protein was confirmed by placing the gene under the inducible control of the *lpp* promoter carried by the expression vector pINIII A3. Isopropyl thiogalactoside mediated induction of the *lpp* promoter of the resultant recombinant plasmids caused the *Escherichia coli* cells to lyse. Furthermore, cell extracts prepared from such clones were shown to be active against clostridial cell walls. Activity was stimulated by conditions of high ionic strength (0.2–0.5 *M* NaCl) and occurred over a broad pH range (4–7).

Oligonucleotide primers, based on two conserved amino acids motifs, were used in PCR to show that *lycA* was also present in five other group I *C. botulinum* strains which produce BoNT/A and BoNT/F. In three of the five cases it was established that *lyc* is located in an equivalent position relative to the BoNT gene. These same primers were used to amplify a 450-bp internal fragment from the *lyc* gene of the type F strain Langeland. Only seven nucleotide differences were noted, indicating that *lycA* is highly conserved between strain Langeland and NCTC 2916. The use of the same primers in a PCR using template DNA from members of groups II, III and IV *C. botulinum* strains failed to generate a DNA fragment. Similarly, a DNA fragment encompassing the majority of the gene did not hybridise to chromosomal preparations of the *C. botulinum* type B strains Danish (group I), ATCC 43756 (group II), *C. botulinum* NCTC 11219 (type E), *C. botulinum* 89G (type G), or purified DNA from phage C-St (type C) or DVD/-3 (type D).

Given that the release of toxin from the cell is intimately linked with autolysis (BONVENTRE and KEMPE 1960), the proximity of these genes may be more than just coincidental. Furthermore, the occurrence of similar genes on bacteriophage

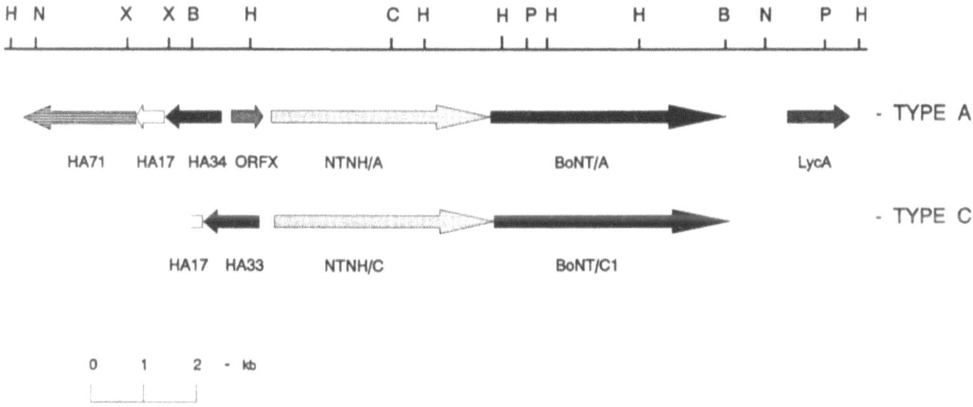

Fig. 3. Arrangement of genes with the vicinity of the BoNT/A gene of the type A *C. botulinum* strain NCTC 2916 compared to that of the type C *C. botulinum* phage C-St. The identity of the illustrated genes are discussed in the text. Type C data are taken from KIMURA et al. (1990) and TSUZUKI et al. (1990, 1992); type a sequence was determined in this laboratory (S.M. Whelan, T.O. Davis and N.P.M., unpublished data). Sites: *B, Bgl* II; *C, Cla* I; *H, Hind* III; *N, Nde* I; *P, Pvu* II; X, *Xba* I

genomes raises the possibility that *lycA*, and the adjacent toxin genes, forms part of a prophage genome present in the chromosome of certain group I *C. botulinum* strains.

3.5 Involvement of Transposons

The fact that BoNT genes are localised to different genetic elements suggests that toxin genes may themselves be mobile through the agency of transposable elements. There is no direct evidence to support this view. It is, however, noticeable that while sequences of certain BoNT structural genes are highly conserved, complete sequence divergence can occur at defined points downstream of the translational stop codon. The most dramatic example of this is the abrupt divergence which occurs 100 bp downstream of the stop codons of the *botA* genes of *C. botulinum* strains NCTC 2916 and 62A. These two genes may therefore have transposed to two different genome environments. Conversely, the divergence seen between the *botA* gene of NCTC 2916 and *botF* of *C. botulinum* strain Langeland continues until some 600 bp past their stop codons before becoming essentially 99% homologous (S.M. Whelan and N.P Minton, unpublished data). This level of homology continues for at least a further 1.2 kb and encompasses the *lcyA* gene described above. In this instance, the two BoNT genes have apparently "transposed" to the same region of the genome of these *C. botulinum* group I strains.

A similar conclusion has been reached for the genes encoding exoenzyme C3, which are carried by the botulinogenic type C and type D bacteriophages. An investigation undertaken by HAUSER et al. (1993) led them to suggest that the C3 gene is localised to a 21.5-kb mobile element which in common with the site-specific transposon family of Tn*554* is flanked by a 6 bp core sequence (in this case 5'-AAGGAG-3'). This suggestion is based largely on the fact that a mutant type C phage (CN) which no longer carries the C3 gene has lost the 21.5-kb fragment and retains only a single copy of the core motif at the point of deletion. Furthermore, while sequenced ends of this 21.5-kb fragment (along with the encompassed C3 genes) from a type C (C-St) and a type D (D-1873) phage are essentially identical, the nucleotide sequences of the C-St and D-1873 genomes flanking the 21.5-kb fragment are unrelated. However, although sequence divergence between the two phages corresponds exactly to the core motif at one end of the 21.5-kb fragment, homology continues for a further 61 nucleotides past the core sequence at the opposite end before diverging. The role of the 6-bp sequence in transposition is therefore questionable.

4 Characterisation of Botulinum Toxin Complex

In contrast to TeNT, BoNTs are found in bacterial cultures and contaminated foodstuffs complexed with other, non-toxic proteins. These complexes are sometimes referred to as progenitor toxins. Three forms are recognised: M toxin (300 kDa), L toxin (500 kDa) and LL toxin (900 kDa). The smaller M toxin is composed of a BoNT (150 kDa) in association with a similarly sized non-toxic protein (NTNH, 150 kDa). The larger L and LL progenitor toxins additionally contain an undefined number of proteins with haemagglutinin (HA) activity. The form of progenitor toxin found varies between the different toxinogenic types, and more than one form may be produced by a single strain. All three forms have been found in type A *C. botulinum* cultures, while only the L and M forms are produced by type B, C and D *C. botulinum* strains. *C. botulinum* type G strains produce the L toxin. The botulinum toxin of type E and F strains is composed exclusively of M progenitor toxin.

Progenitor toxins exhibit greater stability to high temperature and pH extremes than purified BoNT. Botulism, unlike tetanus, is caused principally by the ingestion of food in which contaminating clostridial cells have produced BoNT. It has therefore been argued (OHISHI et al. 1977; OHISHI and SAKAGUCHI 1980) that the major role of the non-toxic proteins of the botulinum complex is to aid the neurotoxin in withstanding the acidic and protease-rich environment of the stomach. Indeed, it is generally observed that the larger the progenitor the toxin complex the greater is its oral toxicity (SUGII and SAKAGUCHI 1977). To unravel the exact consistency of the complex, researchers have begun to clone and characterise the encoding genes.

4.1 Cloning of the Gene Encoding the HA33 Complex Component

Early studies by OGUMA et al. (1976) indicated that the ability to produce HA proteins is phage transmitted, and that transfer occurs concomitantly with that of toxinogenicity. A plasmid gene library of the *C. botulinum* type C phage C-St was therefore constructed in *E. coli* and screened with anti-HA sera (TSUZUKI et al. 1990). A clone was subsequently identified which directed the synthesis of 33-kDa protein and a 65-kDa protein, of which only the former reacted with the anti-HA sera and possessed HA activity. The larger protein corresponded to the NH_2 terminal portion of BoNT/C1. Deletion studies located the HA33 gene some 4.3 kb upstream of the BoNT/C1 gene (Fig. 3). Nucleotide sequencing subsequently showed that it is transcribed from the opposite DNA strand to the BoNT/C1 gene. To confirm the identity of the cloned gene the 33-kDa component of the HA preparation obtained from *C. botulinum* was subjected to NH_2 terminal amino acid sequencing. The sequence obtained was identical to that of the deduced NH_2 terminus of the polypeptide encoded by the sequenced HA33 gene.

In a parallel study SOMERS and DASGUPTA (1991) undertook a more detailed investigation of the physical properties of the non-toxic proteins of type A, B and E progenitor toxins. These were purified and their component parts separated electrophoretically. Under reduced conditions the HA component of type A and type B progenitor toxins was found to migrate as five electrophoretic bands, equating to polypeptide subunits of 17, 21.5, 35, 57 and 92 kDa in size. NH_2 terminal amino acid sequencing of the four smaller polypeptides demonstrated that each subunit is unique, but the corresponding subunits from types A and B are highly homologous. Indeed, the determined sequences of the two 21.5-kDa polypeptides were identical, whilst those of the 57-kDa proteins shared 80% identity. The 35-kDa subunits of both types A and B were found to share 84% sequence identity with the deduced NH_2 terminal sequence of the type C HA33 protein (TSUZUKI et al. 1990). They therefore appear to represent the equivalent toxin complex components.

The equivalence of the type HA33 and the 35-kDa protein of type A has been confirmed by a recent study in this laboratory, during which chromosomal DNA proximal to the *botA* gene of *C. botulinum* NCTC2916 was cloned and characterised by nucleotide sequencing. The data obtained indicated than an ORF is present some 4.55 kb upstream of *botA* of an equivalent size to the HA33 gene (Fig. 3). The encoded polypeptide was estimated to have a molecular weight of 33.669 kDa and has therefore been designated HA34. It possesses a NH_2 terminal sequence identical to that of the 35-kDa protein purified by SOMERS and DASGUPTA (1991) and furthermore exhibits 39.6% similarity with the deduced amino acid sequence of the type C HA33 protein (TSUZUKI et al. 1990). As with the type C HA33 gene, the HA34 is transcribed from the opposite DNA strand to the BoNT/A gene (Fig. 3).

4.2 Cloning of Genes Encoding NTNH

Subsequent nucleotide sequence analysis of the intervening DNA between *botC* and the gene encoding HA33 revealed the presence of a substantial ORF (TSUZUKI et al. 1992). The encoded protein was calculated to have a molecular weight of 138.758 kDa. This equates almost exactly to the size of the large NTNH/C component of the BoNT/C1 progenitor toxin. Indeed, the NH_2 terminal amino acid sequence of this protein was determined and was shown to be identical to the predicted NH_2 terminal sequence of the polypeptide encoded by the identified ORF. Thus, the gene encoding the NTNH/C resides immediately upstream of the BoNT/C1 structural gene (Fig. 3). The same research group subsequently demonstrated that an identical arrangement of genes occurs in the type E *C. botulinum* strain Mashike (FUJII et al. 1993a) and the toxinogenic *C.butyricum* strain BL6340 (FUJII et al. 1993b). Indeed, translation of the region upstream of all the BoNT genes listed in Table 1 demonstrates that the 3' end of an ORF is present, encoding polypeptides homologous to the COOH terminus of NTNH/C and NTNH/E. Not unexpectedly, a similar gene is not evident upstream of the TeNT gene sequence.

It is noticeable that NTNH proteins are more highly conserved between the various neurotoxin-producing clostridia than the respective neurotoxins themselves. Thus, the overall sequence identity shared by the NTNH/C and NTNH/E of *C. botulinum* and *C. butyricum* is approximately 56%. This compares with the 33% similarity between BoNT/C1 and BoNT/E (WHELAN et al. 1992b). We have recently completed the sequence of the gene encoding the NTNH/A of *C. botulinum* NCTC 2916 (Fig. 3), and its encoded protein displays 66% and 68% similarity with NTNH/C and NTNH/E, respectively. The equivalent homologies between BoNTs are 33% and 39%. A further intriguing feature noted following the derivation of the NTNH/C and NTNH/E was that they share amino acid sequence homology with BoNTs. The most significant regions of homology are confined to the extreme NH_2 terminal regions of both proteins. These same sequence motifs are conserved in NTNH/A (Fig. 4). The functional significance of their homology remains unclear. It may indicate that both proteins interact with a common component. Such an interaction may be required for the formation of the toxin complex. Alternatively, if the toxins are actively secreted rather than released by autolysis, they may interact at these conserved regions with a common component of the secretory system, such as a protein chaperone.

4.3 Cloning of Genes Encoding Additional Haemagglutinin Components

SOMERS and DASGUPTA (1991) noted that the HA33 gene sequenced by of TSUZUKI et al. (1990) was followed some 60 bp downstream of its stop codon by an incomplete ORF. Its encoded polypeptide shared 84% and 73% identity with the

Fig. 4. NH_2 terminal amino acid sequence homology between the BoNT and NTNH components of the botulinum toxin complex

determined NH_2 terminal sequences of the 17-kDa HA protein purified from the type A and B progenitor toxins. This means that the genes encoding at least three of the nontoxic proteins of the type C progenitor toxin are located upstream of the BoNT/C1 structural gene (Fig. 3). Data obtained in this laboratory demonstrate that the same three genes are located upstream of the *botA* gene of *C. botulinum* strain NCTC2916 (Fig. 3). Thus, the gene encoding HA34 is followed immediately by an ORF encoding a protein with a calculated molecular weight of 17.035 kDa and which shares 100% identity with the NH_2 terminal sequence of the 17-kDa protein purified by SOMERS and DASGUPTA (1991). Aside from the homologies discussed above, neither HA17 nor HA34 displays significant similarity to any polypeptide currently held in the protein database of the European Molecular Biology Laboratory.

In the case of NCTC 2916, at least, nucleotide sequencing of the genome adjacent to the HA17 gene has identified a further ORF involved in progenitor toxin formation. Encoded by the same DNA strand as the HA34 and HA17 genes, this ORF initiates 14 bp downstream of the HA17 stop codon. It encodes a polypeptide with an estimated molecular weight of 71.258 kDa. Positions 16–25 of the deduced amino acid sequence correspond exactly to the determined NH_2 terminal sequence of the 21.5-kDa protein purified from type A progenitor toxin by SOMERS and DASGUPTA (1991), while residues 204–213 match exactly the NH_2 terminal sequence of the 57-kDa component of the type A complex. This suggests that the 21.5 and 57-kDa proteins identified by SOMERS and DASGUPTA (1991) are initially produced as a single protein which subsequently becomes proteolytically cleaved. Intriguingly, the smaller of the two proteins shows some homology with the enterotoxin of *C. perfringens*.

5 Gene Expression

5.1 Transcriptional Signals

Primer extension studies using *C. tetani* mRNA have demonstrated that transcription of the TeNT gene initiates at nucleotide position −127, relative to the translation start codon (NIEMANN et al. 1988). As the gene is followed at its 3' end by a sequence motif typical of ρ-independent transcriptional terminators, TeNT is presumed to be translated from a monocistronic message. A similar conclusion was reached for the BoNT/A gene of *C. botulinum* strain 62A, following the demonstration that a mRNA transcript initiated at position −118 relative to the translational start codon (BINZ et al. 1990a). It has since become apparent that all botulinum genes are preceded immediately by the gene encoding the NTNH component of the toxin complex. The distance between the stop codon of NTNH and the start codon BoNT can be as little as 10–14 nucleotides in the case of type C, D and F strains, 24/25 nucleotides in strains producing type B and E toxin, and as great as 84 nucleotides in the case of BoNT/G. There is little or no room within this gap for the presence of a promoter element. Indeed, as the distance between NTNH/A and BoNT/A genes is 43 nucleotides in all three strains listed in Table 2, the promoter identified by BINZ et al. (1990a) lies within the COOH terminal coding region of the NTNH gene of strain 62A.

The close proximity of the NTNH/E and BoNT/E genes of the *C. botulinum* type E strain Mashike led FUJII et al. (1993a) to suggest that the two genes are transcribed by a polycistronic mRNA species initiated from a promoter located 5' to the start of the NTNH/E gene. We have analysed the mRNA of *C. botulinum* NCTC 2916, and our preliminary evidence suggests that two mRNA transcripts specifying the BoNT/A gene are produced; a monocistronic message initiating immediately 5' to *botA*, and a polycistronic message initiating 5' to the NTNH/A gene. These same experiments also indicate that an approximately 4.0 kb mRNA transcript is present in NCTC 2916 cells which encodes HA33, HA17 and HA71 (Fig. 3). The fact that these genes form an operon is consistent with their close association, being separated by 62 and 14 bp, respectively.

5.2 Regulatory Factors

It is generally recognised that tightly managed regulatory systems have evolved to control precisely the circumstances under which virulence/toxin genes are expressed. Indeed, it can be argued that since constitutive expression of a virulence gene is only rarely observed, it is probably almost never advantageous. As more and more pathogens are analysed, it is apparent that complex schemes involving regulatory cascades and acting predominantly at the level of transcription control virulence gene expression. In many cases so-called two-component systems play a major role. In this regard it is interesting to note that studies have shown that both BoNT and TeNT production are significantly affected by the availability of exogenous nitrogen (BOWERS and WILLIAMS 1963; PATTERSON et al. 1989; SCHANTZ and JOHNSON 1992), an environmental signal frequently sensed and acted upon by two-component signal transduction systems. Recently, a gene encoding the response regulator of such a system has been identified in *C. perfringens* which regulates production of perfringolysin O, collagenase and HA (SHIMIZU et al. 1994).

Although genes encoding two-component systems have yet to be identified in neurotoxic clostridia, an ORF has been identified in the vicinity of *botA* which may play a role in regulating botulinum toxin production. It lies between the NTNH/BoNT operon and the HA33/HA17/HA71 operon of *C. botulinum* NCTC 2916 (ORFX in Fig. 3). The encoded protein (molecular weight 21.654 kDa) shares amino acid sequence similarity with the plasmid encoded UviA protein of the *C.perfringens* plasmid pIP404 (GARNIER and COLE 1988) and the MsmR protein of *Streptococcus mutants* (RUSSELL et al. 1992). The former has an unspecified role in regulating the production of bacteriocin while the latter is presumed to contribute to the transcriptional regulation of a streptococcal operon involved in sugar transport. The conclusion that MsmR is a DNA binding protein is based on its amino acid sequence similarity with known transcriptional activators, such as

MelR and AraC of *E. coli* and VirF of *Yersinia*. All these proteins contain helix-turn-helix motifs presumed to bind to DNA operator sites. ORFX also contains an amino acid sequence with a high probability of being a helix-turn-helix. Significantly, the highest level of sequence similarity between the ORFX polypeptide and MsmR is between their respective helix-turn-helix motifs.

5.3 Heterologous Expression

The expression of genes derived from neurotoxic clostridial in *E. coli* occurs at variable levels but is, generally speaking, relatively inefficient. In most instances expression has formed part of the strategy employed in detecting the isolation of the desired recombinant clone in *E. coli*. Detection of recombinant protein has been reliant on the use of appropriate antibody in Western blots, for example NTNH/E and the L chain of BoNT/E (Fujii et al. 1991, 1993a) and HA33 of *C. botulinum* type C (Tsuzuki et al. 1990). In certain instances, expression of the cloned clostridial DNA has been attributed to recognition of sequence motifs 5' to the structural genes by *E. coli* RNA polymerase (RNP). In the case of the TeNT genes these motifs correspond directly to those sequences identified as the natural promoter utilised by the clostridial RNP (Eisel et al. 1986). In other instances (Fujii et al. 1993a) expression in *E. coli* has been attributed to the presence of fortuitous sequences. The high A+T content of clostridial DNA means that such sequences are likely to occur fairly frequently.

In the case of TeNT concerted efforts have been made to obtain over-expression of gene subfragments as a means of improving the formulation of tetanus vaccine. In the early studies regions coding for various portions of the H chain were fused to the *E. coli trpE* gene and the resultant genetic fusion placed downstream of the *trp* promoter (Fairweather et al. 1986). Expression levels were disappointingly low, and the protein produced insoluble. Subsequent studies demonstrated that a fragment encoding only the H_c fragment could be expressed such that the recombinant fragment represents between 3% and 4% of the total cell soluble protein (Makoff et al. 1989a). One of the reasons for the relatively low expression of the clostridial-derived DNA fragments is the extreme codon bias caused by the high A+T content of the DNA. As a consequence codons which function as modulators in *E. coli* occur frequently. This problem has been circumvented by either re-synthesising the gene to incorporate codons employed by highly expressed *E. coli* genes (Makoff et al. 1989b) or by using an expression host with a comparably low percentage G+C DNA content, such as *Lactococcus lactis* (Wells et al. 1993). These two strategies have resulted in levels of TeNT H_c fragment being attained which are equivalent to 14% and 22% of total cell protein, respectively. Similar strategies are now being employed to overexpress gene fragments of BoNTs, primarily to assist in structure/function studies but also for vaccination purposes.

Acknowledgments. I thank Sarah M. Whelan, Nicola J. Bodsworth, Michael Elmore and Tom Davis for their unpublished data, and Nicola Minion for typing this manuscript. I also wish to acknowledge the financial support of the PHLS, the U.K. Ministry of Defence (grant no. 2238/002), the U.S. Army Medical Research and Development Command (grant no. DAMD17-90-Z-0033) and the U.K. Medical Research Council (grant no. G9207831CB).

References

Aureli P, Fenicia L, Pasolini B, Gianfranceschi M, McCroskey LM, Hatheway CL (1986) Two cases of type E infant botulism caused by neurotoxigenic Clostridium butyricum in Italy. J Infect Dis 154: 207–211

Binz T, Kurazono H, Wille M, Frevert J, Wernars K, Niemann H (1990a) The complete sequence of botulinum neurotoxin type A and comparison with other clostridial neurotoxins. J Biol Chem 265: 9153–9158

Binz T, Kurazono H, Popoff MR, Eklund MW, Sakaguchi G, Kozaki S, Krieglstein K, Henschen A, Gill DM, Niemann H (1990b) Nucleotide sequence of the gene encoding Clostridium botulinum neurotoxin type D. Nucleic Acids Res 18: 5556

Boizet B, Lahbib-Mansais Y, Dupont L, Ritzenthaler P, Mata M (1990) Cloning, expression and sequence analysis of an endolysin-encoding gene of Lactobacillus bulgaricus bacteriophage mv1. Gene 94: 61–67

Bonventre PF, Kempe LL (1960) Physiology of toxin production of Clostridium botulinum types A and B.I. Growth, autolysis and toxin production. J Bacteriol 79: 18–23

Bowers LE, Williams OB (1963) Effect of arginine on the growth and lysis of Clostridium botulinum. J Bacteriol 85: 1175–1176

Campbell KD, Collins MD, East AK (1993a) Nucleotide sequence of the genecoding for Clostridium botulinum (Clostridium argentinense) type G neurotoxin: genealogical comparison with other clostridial neurotoxins. Biochim Biophys Acta 1216: 487–491

Campbell KD, Collins MD, East AK (1993b) Gene probes for identification of botulinal the neurotoxin gene and specific identification of neurotoxin types B, E, and F. J Clin Mocrobiol 31: 2255–2262

Cato EP, George WL, Finegold SM (1986) Genus Clostridium. In: Sneath PHA, Mair NS, Sharpe ME, Holt JG (eds) Bergey's manual of systematic bacteriology. Williams and Wilkins, Baltimore, pp 1141–1200

Croux C, Garcia JL (1991) Sequence of the lyc gene encoding the autolyticlysozyme of Clostridium acetobutylicum ATCC 824: comparison with otherlytic enzymes. Gene 104: 25–31

Croux C, Canard B, Goma G, Soucaille P (1991a) Autolysis of Clostridium acetobutylicum ATCC 824. J Gen Microbiol 138: 861–689

Campbell KD, Collins MD, East AK (1993b) Gene probes for identification of botulinal the neurotoxin gene and specific identification of neurotoxin types B, E, and F. J Clin Microbiol 31: 2255–2262

Cato EP, George WL, Finegold SM (1986) Genus Clostridium. In: Sneath PHA, Mair NS, Sharpe ME, Holt JG (eds) Bergey's manual of systematic bacteriology. Williams and Wilkins, Baltimore, pp 1141–1200

Croux C, Garcia JL (1991) Sequence of the lyc gene encoding the autolytic lysozyme of Clostridium acetobutylicum ATCC 824: comparison with other lytic enzymes. Gene 104: 25–31

Croux C, Canard B, Goma G, Soucaille P (1991a) Autolysis of Clostridium acetobutylicum ATCC 824. J Dolman CE, Chang E (1972) Bacteriophages of Clostridium botulinum. Can J Microbiol 18: 67–76

East AK, Richardson PT, Allaway D, Collins MD, Roberts TA, Thomson De (1992) Sequence of the gene encoding type F neurotoxin of Clostridium botulinum. FeMS Microbiol Letts 96: 225–230

Eisel U, Jarausch W, Goretzki K, Henschen A, Engels J, Weller U, Hudel M, Habermann E, Niemann H (1986) Tetanus toxin: primary structure, expression in E coli, and homology with botulinum toxins. EMBO J 5: 2495–2502

Eklund MW, Poysky FT (1974) Interconversion of type C and D strains of Clostridium botulinum by specific bacteriophages. Appl Microbiol 27: 251–258

Eklund MW, Poysky FT, Boatman ES (1969) Becteriophages of Clostridium botulinum types A, B, E and F and nontoxigenic strains resembling type E. J Virol 3: 270–274

Eklund MW, Poysky FT, Reed SM, Smith CA (1971) Bacteriophages and toxigenicity of Clostridium botulinum type D. Nature (London) New Biol 235: 16–18

Eklund MW, Poysky FT, Meyers JA, Pelroy (1974) Interspecies conversion of Clostridium botulinum type C to Clostridium novyi by bacteriophage. Science 186: 456–458

Eklund MW, Poysky FT, Peterson ME, Meyers JA (1976) Relationship of bacteriophage to alpha toxin production in Clostridium novyi types A and B. Infect Immun 14: 798–803

EklundMW, Poysky FT, Mseitif LM, Strom MS (1988) Evidence for plasmid-mediated toxin production and bacteriocin production in Clostridium botulinum type G. Appl Environ Microbiol 54: 1405–1408

Eklund MW, Poysky FT, Habig WH (1989) Bacteriophages and plasmids in clostridium botulinum and Clostridium tetani and their relationships to production of toxins. In: Simpson LL (ed) Botulinum neurotoxin and tetanus toxin. Academic San Diego, pp 26–52

Fairweather NF, Lyness VA (1986) The complete nucleotide sequence of tetanus toxin. Nucleic Acid Res 14: 7809–7812

Fairweather NF, Lyness VA, Pickyard DJ, Allen G, Thomson RO (1986) Cloning, nucleotide sequencing and expression of tetanus toxin fragment C in Eschericia coli. J Bacteriol 165: 21–27

Finn CW Jr, Silver RP, Habig WH, Hardegree MC (1984) The structural gene for tetanus neurotoxin is on a plasmid. Science 224: 881–884

Fujii N, Oguma K, Yokosawa N, Kimura K, Tsuzuki K (1988) Characterization of bacteriophage nucleic acids obtained from Clostridium botulinum types C and D. Appl Environ Microbiol 54: 69–73

Fujii N, Kimura K, Yahiki T, Indoh T, Murakami T, Tsuzuki K, Yokosawa N, Oguma K (1991) Cloning of a DNA fragment encoding the 5'-terminus of the type E toxin gene from Clostridium butyricum strain BL6340. J Gen Microbiol 137: 519–525

Fujii N, Kimura K, Murakami T, Yokosawa N, Yashiki T, Tsuzuki K, Oguma K (1993a) The complete nucleotide sequence of the gene encoding the nontoxic component of Clostridium botulinum type E progenitor toxin. J Gen Microbiol 139: 79–86

Fujii N, Kimura K, Yokosawa N, Oguma K, Yashiki T, Takeshi K, Tsuzuki K, Ohyama T, Isogai E, Isogai H (1993b) Similarity in nucleotide sequence of the gene encoding nontoxic component of botulinum toxin produced by toxigenic Clostridium butyricum strain BL6340 and Clostridium botulinum type E strain Mashike. Microbiol Immunol 37: 395–398

Garcia E, Garcia JL, Garcia P, Arraras A, Sanchez-Puelles JM, Lopez R (1988) Molecular evolution of lytic enzymes of Streptococcus pneumoniae and its bacteriophages. Proc Natl Acad Sci USA 85: 914–918

Garnier T, Cole ST (1988) Studies of UV-inducible promoters from Clostridium perfringens in vivo and in vitro. Mol Microbiol 2: 607–614

Hall JD, McCroskey LM, Pincomb BJ, Hatheway CL (1985) Isolation of an organism resembling C. barati which produces type F botulinal toxin from an infant with botulism. J Clin Microbiol 21: 654–655

Hara T, Matsuda M, Yoneda M (1977) Isolation and some properties of nontoxigenic derivaties of a strain of Clostridium tetani. Biken J 20: 105–115

Hauser D, Eklund MW, Kurazono H, Binz T, Niemann H, Gill DM, Boquet P, Popoff MR (1990) Nucleotide sequence of Clostridium botulinum C1 neurotoxin. Nucleic Acids Res 18: 4924

Hara T, Matsuda M, Yoneda M (1977) Isolation and some properties of nontoxigenic derivaties of a strain of Clostridium tetani. Biken J 20: 105–115

Hauser D, Eklund MW, Kurazono H, Binz T, Niemann H, Gill DM, Boquet P, Popoff MR (1990) Nucleotide sequence of Clostridium botulinum C1 neurotoxin. Nucleic Acids Res 18: 4924

Hauser D, Gibert M, Boquet P, Popoff MR (1992) Plasmid localization of a type E botulinal neurotoxin gene homologue in toxigenic Clostridium butyricum strains, and absence of this gene in non-toxigenic C. butyricum strains. FEMS Microbiol Letts 99: 251–256

Hauser D, Gibert M, Eklund MW, Boquet P, Popoff MR (1993) Comparative analysis of C3 and botulinal neurotoxin genes and their environment in Clostridium botulinum types C and D. J Bacteriol 175: 7260–7268

Higgins DG, Sharp PM (1988) CLUSTAL: a package for performing multiple sequence alingment on a microcomputer. Gene 73: 237–244

Hutson RA, Thompson DE, Collins MD (1993a) Genetic interrelationlships of saccharolytic Clostridium botulinum types B, E and F and related clostridia as revealed by small-subunit rRNA gene sequences. FEMS Microbiol Letts 108: 103–110

Hutson RA, Collins MD, East AK, Thompson DE (1993b) Nucleotide sequence of the gene coding for non-proteolytic Clostridium botulinum type B neurotoxin: comparison with other clostridial nuerotoxins. Curr Microbiol 28: 101–110

Inoue K, Iida H (1968) Bacteriophages of Clostridium botulinum. J Virol 2: 537–540

Inoue K, Iida H (1970) Conversion to toxigenicity in Clostridium botulinum type C. Jap J Microbiol 14: 87–89

Kimura K, Fujii N, Tsuzuki K, Murakami T, Indoh T, Yokosawa N, Takeshi K, Syuto B, Oguma K (1990) The complete nucleotide sequence of the gene coding for botulinum type C1 toxin i n the C-ST phage genome. Biochem Biophys Res Commun. 171: 1304–1311

Laird WJ, Aaronson W, Silver RP, Habig WH, Hardegree MC (1980) Plasmid-associated toxigenicity in Clostridium tetani. J Infect Dis 142: 623

Lichenstein HS, Hastings AE, Langley KE, Mendiaz EA, Rohde MF, Elmore R, Zukowski MM (1990) Cloning and nucleotide sequence of the N-acetylmuramidase MI-encoding gene from Streptomyces globisporus. Gene 88: 81–86

Makoff AJ, Ballantine SP, Smallwood AE, Fairweather NF (1989a) Expression of tetanus toxin fragment C in E. coli: its purification and potential use as a vaccine. BIO/TECHNOL 7: 1043–1046

Makoff AJ, Oxer MD, Romanos MA, Fairweather NF, Ballantine S (1989b) Expression of tetanus toxin fragment C in E. coli: high level expression by removing rare codons. Nucleiec Acids Res 17: 1019–1202

McCroskey LM, Hatheway CL, Fencia L, Pasolini B, Aureli P (1986) Characterization of an organism that produces type E botulinal toxin but which resembles Clostridium butyricum from the faeces of an infant with type E botulism. J Clin Microbiol 23: 201–202

McCroskey LM, Hatheway CL, Woodruff BA, Greenberg JA, Jurgenson P (1991) Type F botulism due to neurotoxigenic Clostridium barati from an unknown source in an adult. J Clin Microbiol 29: 2618–2620

Niemann H, Andersen-Beckh, Binz T, Eisel U, Demotz S, Mayer T, Widmann C (1988) Tetanus toxin: evaluation of the primary sequence and potential applications. Zentralbl Bakteriol [Suppl] 17: 29–38

Oguma K, Iida H, Shiozaki M (1976) Phage conversion to haemagglutinin production in Clostridium botulinum types C and D. Infect Immun 14: 567–602

Oguma K, Murayama S, Syuto B, Iida H, Kubo S (1984) Analysis of antigenicity of Clostridium botulinum type C1 and D toxins by polyclonal and monoclonal antibodies. Infect Immun 43: 584–588

Ohishi I, Sakaguchi G (1980) Oral toxicities of Clostridium botulinum type C and D toxins of different molecular size. Infect Immun 28: 303–309

Ohishi I, Sughii S, Sakaguchi G (1977) Oral toxicities of Clostridium botulinum toxins in response to molecular size. Infect Immun 16: 107–109

Patterson Curtis SI, Johnson EA (1989) Regulation of neurotoxin and protease formation in botulinum Clostridium Okra B and Hall A by arginine. Appl Environ Microbiol 55: 1544–1548

Perkins WF, Tsuji K (1962) Sporulation of Clostridium botulinum. II. Effect of arginine and its degradation products on sporulation in synthetic medium. J Bacteriol 84: 86–94

Poulet S, Hauser D, Quanz M, Niemann H, Popoff MR (1992) Sequences of the botulinal neurotoxin E derived from Clostridium botulinum type E (strain Beluga) and Clostridium butyricum (strains ATCC 43181 and ATCC 43755). Biochem Biophys Res Commun 183: 107–11

Russell RRB, Aduse-Opoku J, Sutclife IC, Tao L, Ferretti JJ (1992) A binding protein-dependent transport system in Streptococcus mutans responsible for multiple sugar metabolism. J Biol Chem 267: 4631–4637

Schantz EJ, Johnson EA (1992) Properties and use of botulinum toxin and other microbial neurotoxins in medicine. Microbiol Rev 56: 80–99

Scott VN, Duncan CL (1978) Cryptic plasmids in Clostridium botulinum and C. botulinum-like organisms. FEMS Microbiol Lett 4: 55–58

Shimizu T, Ba-Thein W, Tamaki M, Hayashi H (1994) The virR gene, a member of a class of two-component response regulators, regulates the production of perfringolysin O, collagenase, and haemagglutinin in Clostridium perfringens. J Bacteriol 176: 1616–1623

Somers E, DasGupta BR (1991) Clostridium botulinum types A, B, C1, and E produce proteins with or without hemagglutinating activity: do they share common amino acid sequences and genes. J Pro Chem 10: 415–425

Strom MS, Eklund MW, Poysky FT (1984) Plasmids in Clostridium botulinum and related Clostridium species. Appl Environ Microbiol 48: 956–963

Suen JC, Hatheway CL, Steigerwalt AG, Brenner DJ (1988) Clostridium argentinense sp. nov.: a genetically homogeneous group composed of all strains of Clostridium botulinum toxin type G and some nontoxigenic strains previously identified as Clostridium subterminale or Clostridium hastiforme. Int J Sys Bacteriol 38: 375–381

Sugii S, Sakaguchi J (1977) Botulinogenic properties of vegetables with special reference to the molecular size of the toxin in them. J Food Safety 1: 53–65

Szabo EA, Pemberton JM, Desmarchelier PM (1993) Detection of the genes encoding botulinum neurotoxin types A to E by the polymerase chain reaction. Appl Environ Microbiol 59: 3011–3020

Thompson DE, Brehm JK, Oultram JD, Swinfield TJ, Shone CC, Atkinson T, Melling J, Minton NP (1990) The complete amino acid sequence of the Clostridium botulinum type A neurotoxin, deduced by nucleotide sequence analysis of the encoding gene. Eur J Biochem 189: 73–81

Thompson DE, Hutson RA, East AK, Allaway D, Collins MD, Richardson PT (1993) Nucleotide sequence of the gene coding for Clostridium barati type F neurotoxin: comparison with other clostridial neurotoxins. FEMS Microbiol Lett 108: 175–182

Tsuzuki K, Kimura K, Fujii N, Yokosawa N, Indoh T, Murakami T, Oguma K (1990) Cloning and complete nucleotide sequence of the gene coding for the main component of hemagglutinin produced by Clostridium botulinum type C. Infect Immun 58: 3173–3177

Tsuzuki K, Kimura K, Fujii N, Yokosawa N, Oguma K (1992) The complete nucleotide sequence of the gene coding for the nontoxic-nonhemagglutinin component of Clostridium botulinum type C progenitor toxin. Biochem Biophys Res Commun 183: 1273–1279

Van de Peer Y, De Wachter R (1993) TREECON: a software package for the construction and drawing of evolutionary trees. Comput Appl Biosci 9: 177–182

Weickert MJ, Chambliss GH, Sugiyama H (1986) Production of toxin by Clostridium botuinum type A strain cured by plasmids. Appl Environ Microbiol 51: 52–56

Wells JM, Wilson PW, Norton PM, Gasson MJ, Le Page RWF (1993) Lactococcus lactis: high-level expression of tetanus toxin fragment C and protection against lethal challenge. Mol Microbiol 8: 1155–1162

Whelan SM, Elmore MJ, Bodsworth NJ, Brehm JK, Atkinson T, Minton NP (1992a) Molecular clonning of the Clostridium botulinum structural gene encoding the type B neurotoxin and determination of its entire nucleotide sequence. Appl Environ Microbiol 58: 2345–2354

Whelan SM, Elmore MJ, Bodsworth NJ, Atkinson T, Minton NP (1992b) Complete amino acid sequence of the Clostridium botulinum type E neurotoxin, deduced by nucleotide sequence analysis of the encoding gene. Eur J Biochem 204: 657–667

Whelan SM, Garcia JL, Elmore MJ, Minton NP (1994) The botulinum neurotoxin gene of the type A Clostridium botulinum strain NCTC 2916 is followed by a gene (lycA) encoding a lysozyme. Zentralbl Bakteriol [Suppl] 24: 162–163

Willems A, East AK, Lawson PA, Collins MD (1993) Sequence of the gene coding for the neurotoxin of Clostridium botulinum type A associated with infant botulism: comparison with other clostridial neurotoxins. Res Microbiol144: 547-556

Young M, Minton NP, Staudenbauer WL (1989a) Recent advances in the genetics of the clostridia. FEMS Microbiol Rev 1: 1–63

Young M, Staudenbauer WL, Minton NP (1989b) Genetics of Clostridium. In: Minton NP, Clarke DJ (eds) Clostridia. Plenum, New York, pp 63-103

Zhou Y, sugiyama H, Johnson EA (1993) Transfer of nuerotoxigenicity from Clostridium butyricum to a nontoxinogenic Clostridium botulinum type E-like strain. Appl Environ Microbiol 59: 3825–3831

Neuroexocytosis

F. Benfenati[1] and F. Valtorta[2]

Neurons are polarized cells specialized at transferring and processing information. With very few exceptions (e.g., electrical synapses), information transfer is achieved by exocytotic release of chemical messengers, the neurotransmitters, into the extracellular space upon stimulation. Two major classes of chemical messengers exist: small molecules such as amino acids or amines (the so-called classical neurotransmitters) and neuropeptides. The two types of secretory product correspond to two completely different modalities of neurosecretion. While the processing and release of neuropeptides by neurons resemble secretory events in endocrine and exocrine cells, the processing and release of classical neurotransmitters is a process exclusive of neuronal cells.

1 Structural Aspects of the Exocytotic Apparatus in Neurons

The nerve terminal is the cellular compartment specialized in storage and release of neurotransmitters and is separated from the postsynaptic neuron by a thin

[1] Department of Experimental Medicine and Biochemical Sciences, University of Rome Tor Vergata, Via O. Raimondo, 00173 Roma, and Institute of Human Physiology, University of Modena, Via Campi 287, 41100 Modena, Italy
[2] Department of Medical Pharmacology, DIBIT, CNR Center of Cytopharmacology, San Raffaele Scientific Institute, University of Milan, Via Olgettina 60, 20132 Milano, Italy

synaptic cleft. In addition to mitochondria, smooth endoplasmic reticulum, and cytoskeletal structures, nerve terminals are filled with vesicles, specific organelles which store packets (called "quanta") of neurotransmitter molecules and release them by exocytosis. Although synaptic vesicles are normally present in all regions of the nerve terminal cytoplasm, in mature nerve terminals there are discrete regions underneath the presynaptic plasma membrane in which synaptic vesicles are more concentrated and clustered via interactions among their membranes and with cytoskeletal elements.

In the areas of vesicle clustering, the presynaptic plasma membrane shows structural specializations, called "active zones," where exocytosis of classical neurotransmitters occurs (COUTEAUX and PECOT-DECHAVASSINE 1970). At the frog neuromuscular junction active zones are regularly spaced and appear to be formed by double rows of large intramembrane particles of uniform size (for review, see CECCARELLI and HURLBUT 1980) which are thought to correspond to Ca^{2+} channels (PUMPLIN et al. 1981). Opposite the active zones the postsynaptic membrane of the muscle fiber displays deep infoldings, enriched on their edges in neurotransmitter receptors, thus maximizing the efficiency of synaptic transmission.

In addition to their rich population of small, clear synaptic vesicles, neurons also contain another kind of secretory granules. Because of their size and appearance in the electron microscope these granules have been named large dense-core vesicles (LDCVs). In most synapses they are much less abundant than small synaptic vesicles and appear to be randomly distributed in nerve terminals, with no preferential accumulation at active zones.

2 Functional Features of Neuroexocytosis

Recent results obtained with complementary approaches ranging from molecular biology to biochemistry, genetics, and microbiology support the idea that neuronal exocytosis is a particular aspect of the more general processes of intracellular vesicular fusion and constitutive secretion which are shared by all eukaryotic cells (BARINAGA 1993). The same or closely related proteins which are involved in neurotransmitter release in neurons appear to play a role in the constitutive exo-endocytotic pathway as well as in the intracellular transport of vesicles from the endoplasmic reticulum to the plasma membrane through the golgi apparatus (SÖLLNER et al. 1993a). These proteins are highly conserved during evolution from yeast to mammalian neurons (BENNETT and SCHELLER 1993).

Although the basic features of the process of neurotransmitter release are shared with other kinds of exocytosis, many differences exist. Three major features distinguish neurotransmitter release from other examples of regulated secretory events: (a) the extreme rapidity of excitation-secretion coupling (exocytosis occurs in a fraction of millisecond after the invasion of the axon terminal by the action potential and the subsequent influx of Ca^{2+}) and the ability to synchronize exocytosis of many secretory vesicles; (b) a high resistance to exhaustion (fatigue) during prolonged periods of intense secretory activity; (c) the ability of varying its efficiency on the basis of the recent history of the nerve terminal and of the physicochemical environment (FESCE 1990).

These remarkable characteristics are contributed by specific structural and/or functional features of the nerve terminal. The rapidity and synchrony of exocytosis reside in the presence of preassembled complexes between the presynaptic membrane and ready-to-fuse synaptic vesicles which can undergo fusion upon Ca^{2+} entry from the adjacent Ca^{2+} channels. On the other hand, resistance to fatigue is attributable to the existence of an efficient local recycling of used synaptic vesicles and to the presence of an adequate reservoir of vesicles to be recruited upon activity.

Electrophysiological studies have indeed suggested the existence of two pools of synaptic vesicles within the nerve terminal: a reserve pool of vesicles which are presumably embedded in the cytoskeletal meshwork and a releasable pool of vesicles which are not restrained by cytoskeletal structures and which may be already docked to the presynaptic membrane. Generally the releasable pool accounts for only a small percentage of the total vesicle content of the terminal (0.5%–10% depending on the type of synapse analyzed), but its size is functionally very important to define the magnitude of the secretory response to the action potential. In fact, the number of neurotransmitter quanta released per impulse, m, equals $n \times p$, where n is the number of quanta available for release and p the average probability for a quantum to be released (DEL CASTILLO and KATZ 1954; see STEVENS 1993, for an updated discussion).

The number of quanta available for release in a nerve terminal depends on the number of release sites occupied by ready-to-fuse synaptic vesicles and on the number of vesicles released from the cytoskeletal constraint to refill the docking sites emptied upon activity (releasable pool). During repetitive stimulation, the ability of a nerve terminal to release neurotransmitter not only depends on n, but is also proportional to the size of the reserve pool and to the rate of mobilization of vesicles from the reserve to the available pool. This machinery offers a molecular explanation for the processes of synaptic plasticity involving an increased or decreased responsiveness of the nerve terminal to release neurotransmitter. The transition of synaptic vesicles between the two functional pools is thought to be caused by variations in the intraterminal concentrations of second messengers which follow the repetitive opening of voltage-dependent Ca^{2+} channels or the activation of presynaptic receptors (GREENGARD et al. 1993; KELLY 1993).

3 Molecular Mechanisms of Exocytosis: Preparatory Events

The analysis of the functional characteristics of neurotransmitter release suggests that two series of processes with different rates and regulation by intracellular messengers are involved: on the one hand, the transition of synaptic vesicles from the reserve to the releasable pool which prepares the nerve terminal for exocytosis by setting the number of release-ready vesicles; on the other hand, the obligatory events of targeting and docking of the releasable vesicles to the release sites, eventually followed by fusion with the presynaptic membrane upon stimulation and Ca^{2+} entry (Fig. 1).

In spite of the experimental evidence for the existence of two pools of synaptic vesicles, monitoring of recycling in stimulated terminals in the presence of extracellular tracers has indicated that they do not constitute physically distinct populations of vesicles, and that recycled vesicles are topologically and functionally indistinguishable from vesicles which have not undergone exocytosis and are randomly distributed within nerve terminals (CECCARELLI et al. 1972, 1973; HEUSER and REESE 1973; BETZ and BEWICK 1992). Therefore, the two pools of synaptic vesicles appear to be dynamic entities and each vesicle may become part of either pool on a stochastic basis and independently of its location within the terminal. It is therefore likely that all synaptic vesicles are potentially identical and may continuously enter and exit the reserve pool.

It is reasonable to assume, on the basis of morphological and functional data (LANDIS et al. 1988; HIROKAWA et al. 1989; BERNSTEIN and BAMBURG 1989; TRIFARO and VITALE 1993) that the vesicles in the reserve pool are those which are organized in clusters linked reversibly to the cytoskeleton. Variations in the proportion of vesicles which are part of the reserve pool might therefore reflect variations in the probability of each vesicle to be tethered to the cytoskeleton. Since the cytomatrix of the nerve terminal in the areas of axoplasm close to the release sites appears to be formed mainly by actin filaments (FIFKOVA 1985), actin and actin-binding proteins appear to be critical for the accumulation of synaptic vesicles within the nerve terminals as well as for regulating their traffic and their availability for exocytosis.

A major nerve terminal phosphoprotein, synapsin I, by interacting with both synaptic vesicles and actin in a phosphorylation-dependent manner (for review, see DE CAMILLI et al. 1990; VALTORTA et al. 1992) appears to mediate the aggregation of synaptic vesicles in clusters and their attachment to the cytoskeleton (BENFENATI et al. 1993; CECCALDI et al. 1995). Since the interaction of synapsin I both with actin and with synaptic vesicles is weakened upon synapsin I phosphorylation by Ca^{2+}/calmodulin-dependent protein kinase II (CaM kinase II), it has been hypothesized that synapsin I may reversibly regulate the availability of vesicles for exocytosis in response to changes in the intracellular levels of second messengers and protein phosphorylation/dephosphorylation processes (GREENGARD et al. 1993). Several experimental data support this hypothesis: (a) the phosphorylation of synapsin I in intact nerve terminal appears to be physiologically regulated and evoked by any stimulations known to induce Ca^{2+}-dependent neurotransmitter release (NESTLER and GREENGARD 1984); (b) dephosphorylated synapsin I injected into intact nerve terminals has an inhibitory effect on neurotransmitter release, whereas the phosphorylated form of the protein is ineffective (LLINAS et al. 1985, 1991); (c) video-enhanced microscopy has provided direct evidence that synapsin I is necessary for the

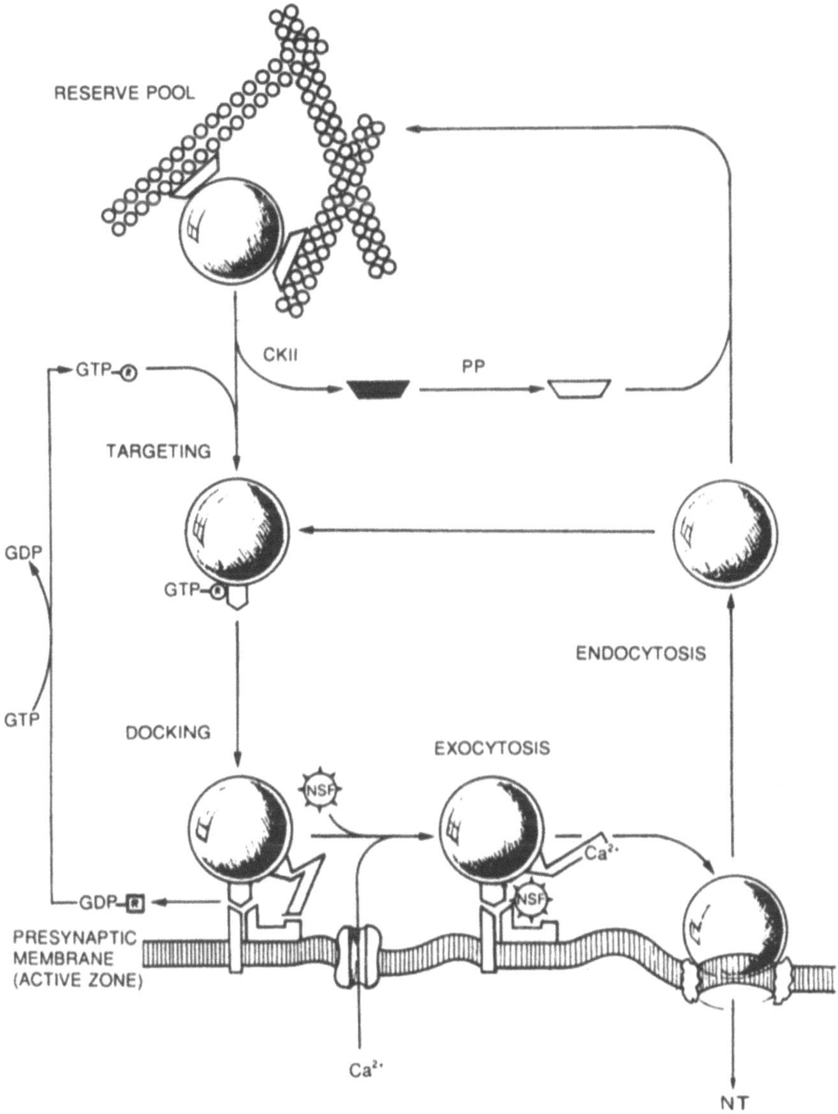

Fig. 1. Schematic representation of the main steps involved in the exo-endocytotic cycle of synaptic vesicles. Some of the nerve terminal proteins involved are shown as follows: , actin; , dephosphorylated synapsin I; , phosphorylated synapsin I; ®, Rab3A; , VAMP/synaptobrevin; , HPC-1/syntaxin; , SNAP-25; , synaptotagmin; , SNAPs/NSF; , fusion pore; CKII, CaM kinase II; PP, protein phosphatase

vesicles to interact with actin and that its phosphorylation by CaM kinase II abolishes these interactions in all-or-none manner (CECCALDI et al. 1995); (d) immunoelectron microscopy of intact nerve terminals revealed that synapsin I dissociates from, and reassociates with, the vesicle membrane during exocytosis and recycling (TORRI-TARELLI et al. 1992); (e) transgenic mice lacking either synapsin I or CaM kinase II exhibit alterations in short-term synaptic plasticity (SILVA et al. 1992; ROSAHL et al. 1993).

The direct binding of synapsin I to a synaptic vesicle-associated form of CaM kinase II, with the formation of an enzyme-substrate complex at the vesicle membrane (BENFENATI et al. 1992), suggests that this regulatory mechanism can be activated very rapidly and mediate activity changes occurring in a very short time scale. In addition, synapsin I is an excellent substrate for transglutaminase, a potent protein-protein cross-linking enzyme present also in nerve terminals (FACCHIANO and LUINI 1992; FACCHIANO et al. 1993). Therefore, all stimuli leading to activation of transglutaminase (e.g., Ca^{2+}, tetanus toxin), might result an irreversible cross-linking of synaptic vesicles to actin filaments which prevents their release.

4 Molecular Mechanisms of Exocytosis: Targeting, Docking, and Fusion

The very short delay (200 ms) between the stimulus (Ca^{2+} influx) and the release of neurotransmitter (LLINAS et al. 1981) suggests that only few steps are involved in stimulus-secretion coupling and that only those vesicles which are already "docked" to the fusion sites, can undergo exocytosis (Fig. 1).

4.1 Targeting to the Release Sites

The population of docked vesicles is only a fraction of the population of available vesicles and the actual number of docked vesicles may be limited either by the availability of docking sites or by the kinetics of the docking process. Before docking vesicles must be targeted to the docking sites. Since synaptic vesicles appear to undergo continuous random mixing during their migration to the release sites on the plasma membrane, the targeting process is not likely to be driven by a directional molecular motor (e.g., a myosin I-like molecule; BETZ and BEWITZ 1992). A potential candidate for mediating the targeting step is the ras-like GTP-binding protein rab3a, which is specifically associated with the synaptic vesicle membrane (FISCHER VON MOLLARD et al. 1990), from which it has been reported to dissociate during the exoendocytotic cycle (FISCHER VON MOLLARD et al. 1991; MATTEOLI et al. 1991).

The rab proteins are the small G proteins essential for membrane traffic and exocytotic fusion in all mammalian cells. They are believed to achieve the directional targeting of donor membranes to the correct acceptor compartment by shuttling back and forth between a membrane-bound and a soluble form (ROTHMAN and ORCI 1992; SIMONS and ZERIAL 1993). At the donor membrane (the synaptic vesicle, in the case of rab3A), a guanine nucleotide exchange factor catalyzes the binding to rab of GTP, which replaces GDP. This promotes the association of rab-GTP with the donor membrane either via a geranylgeranyl posttranslational modification or via binding to a specific donor membrane-associated protein (identified in synaptic vesicles as rabphilin-3A, an extrinsic membrane protein homologous to synaptotagmin; SHIRATAKI et al. 1993). The rab proteins accompany the donor membrane to the acceptor compartment, where docking to specific sites occurs. In the presence of a GTPase-activating protein (GAP), this event leads to activation of the intrinsic rab GTPase and to the hydrolysis of rab-bound GTP. This event triggers a conformational change in rab which may play a role either in the activation of fusion or in the dissociation of the docking complex which is required for fusion to occur (see below). The last event of the cycle involves the dissociation of rab-GDP from the acceptor membrane in the presence of a GDP dissociation inhibitor, making the molecule available to carry another donor membrane to its destination (SIMONS and ZERIAL 1993). In analogy with what has been suggested for other secretory systems, it is possible that also other small G proteins may be involved in other steps of the synaptic vesicle life cycle (ROTHMAN and ORCI 1992; SIMONS and ZERIAL 1993).

4.2 Docking to the Release Sites

Docking involves a tight interaction between proteins specific for the synaptic vesicle and the presynaptic membrane. In addition to the voltage-dependent Ca^{2+} channel (ROBITAILLE et al. 1990; TORRI-TARELLI et al. 1991), in the last few years some proteins specifically associated with the presynaptic membrane and possibly involved in the docking/fusion of synaptic vesicles have been identified using complementary approaches. These proteins include the α-latrotoxin (α-LTX) receptor/neurexin I, HPC-1/syntaxin, and the synaptosomal-associated protein of M_r 25 000 (SNAP-25) (VALTORTA et al. 1984; USHKARYOV et al. 1992; INOUE et al. 1992; BENNETT et al. 1992; OYLER et al. 1989). However, for no protein has unequivocal proof been provided of its exclusive localization at active zones, and in addition the distribution of some of them (e.g., SNAP-25) is not exclusively restricted to the nerve terminal but also involves the plasma membrane of neuritic processes. This finding suggests that some of these proteins may also serve functions other than docking synaptic vesicles to the active zones. SNAP-25, for example, which is present in nerve terminals and throughout the axon but not in dendrites, has been implicated in the developmental axonal growth (Fig. 2), a process involving exocytosis and permanent incorporation of components of the vesicular membrane into the plasmalemma (OSEN-SAND et al. 1993).

In principle, any specific interaction between synaptic vesicle-associated and presynaptic membrane-associated proteins can be envisaged to mediate synaptic vesicle docking. Interactions involving distinct proteins have been described and well-characterized in vitro, although at the moment it is difficult to predict which one plays the major role in the interrelated processes of recognition

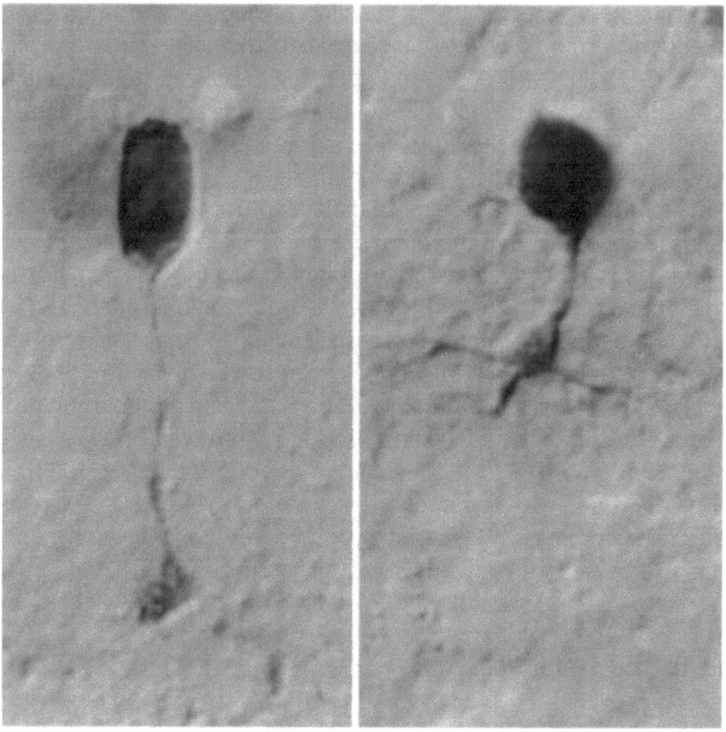

Fig. 2. Inhibition of axonal elongation in the retina after intraocular injections of SNAP-25 sense (*left*) or antisense (*right*) oligonucleotides. The figure shows two calbindin-positive amacrine cells located near the edge of the inner plexiform layer. (Courtesy of S. Catsicas)

of the fusion sites, docking, Ca^{2+} detection, and triggering of the fusion event, and whether they all cooperate in tethering the synaptic vesicle to the plasma membrane.

Synaptotagmin is an abundant protein of synaptic vesicles (MATTHEW et al. 1985), which has a single transmembrane domain and a cytoplasmic tail. The cytoplasmic tail exhibits two repeats homologous to the Ca^{2+}- and phospholipid-binding domain of protein kinase C and phospholipase A2 (C2 domain; PERIN et al. 1990). Indeed, purified synaptotagmin has been shown to bind Ca^{2+} and phospholipids (BROSE et al. 1992) and therefore has been hypothesized to represent the Ca^{2+} sensor preventing spontaneous fusion under resting conditions. Ca^{2+} flowing into the terminal through voltage-operated channels and binding to synaptotagmin would relieve the inhibition to fuse (POPOV and POO 1993; DEBELLO et al. 1993; see below). In addition, multiple interactions between synaptotagmin and proteins of the plasma membrane implicate this protein in mediating docking of synaptic vesicles to the active zone. Synaptotagmin has been reported to interact in vitro with HPC-1/syntaxin, an integral protein of the presynaptic membrane with a COOH terminal membrane anchor and an NH_2 terminal cytoplasmic domain (BENNETT et al. 1992; INOUE et al. 1992). HPC-1/syntaxin, in turn, interacts with the N-type Ca^{2+} channel, placing synaptotagmin in a strategic location to respond to the Ca^{2+} trigger (BENNETT et al. 1992). This is further confirmed by the observation that antibodies to synaptotagmin coprecipitate N-type Ca^{2+} channels (LEVEQUE et al. 1992).

Synaptotagmin also interacts with the cytoplasmic domain of the membrane receptor for α-LTX, the major active component of the black widow spider venom (PETRENKO et al. 1991; HATA et al. 1993a). Immunocytochemistry has shown that the receptor is localized in the nerve terminal plasma membrane (VALTORTA et al. 1984). The receptor for α-LTX has been purified to homogeneity and appears to be composed of a 160- to 220-kDa glycosylated polypeptide which might be complexed to a lower molecular weight subunit (SCHEER and MELDOLESI 1985). Cloning of the 160- to 220-kDa polypeptide has led to the identification of a family of neuronal cell surface proteins, the neurexins, which appear to span the membrane once and exhibit in their extracellular domain weak homology to proteins of the extracellular matrix (USHKARYOV et al. 1992). Its interaction with synaptotagmin implicates this receptor in the formation of the docking complex for synaptic vesicles at active zones, allowing a correct positioning of synaptotagmin for clamping spontaneous fusion. In addition to the formation of cationic channels (WANKE et al. 1986; HURLBUT et al. 1994), the binding of the toxin to its receptor may also remove the synaptotagmin clamp to exocytosis at resting Ca^{2+} levels.

These data suggest an important functional role of synaptotagmin in synaptic vesicle exocytosis, since the protein would be responsible both for docking (hence for the rapidity of release) and for the Ca^{2+} dependence of the process. In agreement with this hypothesis, antibodies to synaptotagmin and peptides corresponding to the C2 domains of the protein inhibit exocytosis from PC12 cells as well as from the squid giant synapse (ELFERINK et al. 1993; BOMMERT et al. 1993) and synaptotagmin-deficient mutants exhibit behavioral abnormalities attributable to an impairment of synaptic transmission (LITTLETON et al. 1993; NONET et al. 1993; GEPPERT et al. 1994).

HPC-1/syntaxin also interacts with the synaptic vesicle protein VAMP/synaptobrevin (TRIMBLE et al. 1988; BAUMERT et al. 1989) with dissociation constant in the micromolar range (4.7 μM; CALAKOS et al. 1994) and with SNAP-25, a protein associated with the cytoplasmic side of the axolemma (OYLER et al. 1989). The formation of such trimeric complex was indirectly suggested by results obtained from different approaches showing that the three proteins are capable of binding the soluble fusion machinery [*N*-ethylmaleimide sensitive factor (NSF) and α/β- and γ-soluble NSF-attachment proteins; see below] to the membrane (SÖLLNER et al. 1993a), and that they represent the specific and unique targets of tetanus and botulinum neurotoxins (SCHIAVO et al. 1992; 1993a,b; BLASI et al. 1993a,b; HUTTNER 1993). Immunoprecipitation experiments directly demonstrated that HPC-1/syntaxin, VAMP/synaptobrevin and SNAP-25 coprecipitate in roughly equimolar ratios (SÖLLNER et al. 1993b). Other nerve terminal proteins have been found to associate with this complex with various stoichiometries. These proteins include synaptotagmin (SÖLLNER et al. 1993b), rab3A (HORIKAWA et al. 1993) and a 67-kDa protein (called Munc-18/N-Sec 1) sharing homology with the yeast Sec-1 protein and with the product of the *C. elegans unc-18* gene (HATA et al. 1993b; GENGYO-ANDO et al. 1993; see O'CONNOR et al. 1994 for review). The latter protein is highly expressed in neurons (GENGYO-ANDO et al. 1993), and the mutation of the *unc-18* gene in *C. elegans* induces a paralytic phenotype consistent with an impaired synaptic transmission (HOSONO et al. 1992). Although Munc-18 is probably an important component of the docking/fusion complex, its precise function and subcellular localization is completely unknown.

An additional link between synaptic vesicles and the presynaptic membrane is represented by the recently described cysteine string protein (csp). The csp is a membrane protein containing an unusual cysteine repeat which was originally cloned as a putative subunit of the presynaptic Ca^{2+} channel (GUNDERSEN and UMBACH 1992) and subsequently found to be associated with synaptic vesicles (MASTROGIACOMO et al. 1994). It represents approximately 1% of the synaptic vesicle proteins and is presumably linked to the cytoplasmic side of the membrane via acylation of multiple cysteine residues. The interaction between csp and the presynaptic Ca^{2+} channel may cooperate with the complex Ca^{2+} channel syntaxin-synaptotagmin in docking the vesicle at the cytoplasmic mouth of the Ca^{2+} channel, so that it is exposed to the bulk of Ca^{2+} influx during stimulation. In addition, csp has been reported to be a Ca^{2+} channel modulator, converting the channel from a refractory to a voltage-sensitive state (GUNDERSEN and UMBACH 1992). It is therefore possible that Ca^{2+} channels at the active zones need to be primed by docked vesicles to open in response to the action potential, thus minimizing Ca^{2+} entry at release sites not occupied by synaptic vesicles. The important role played by csp in exocytosis is further strengthened by the temperature-dependent block of synaptic transmission followed by paralysis and early death observed in *Drosophila* after deletion of the csp gene (ZINSMAIER et al. 1994).

In summary, the docking of synaptic vesicles to the presynaptic membrane seems to involve a tangle of proteins belonging to both membranes and interacting with each other in a complex three-dimensional array which includes the Ca^{2+} channel (Fig. 3). Since there is no evidence, at present, that this multimeric protein complex forms a porelike structure (see below), the role of Ca^{2+} would then be that of disrupting this complex which keeps the two membranes fastened, while triggering fusion (see below).

4.3 Fusion with the Presynaptic Membrane

Membrane fusion is a ubiquitous process characterizing membrane traffic among the various intracellular compartments. Physicochemical analysis of pure phospholipid vesicles has revealed that membrane fusion is very unfavored and can be achieved only when a high free energy barrier is overcome. When the

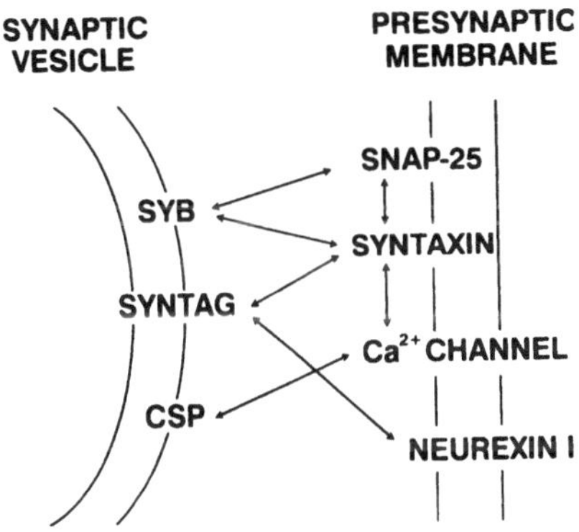

Fig. 3. Proteins involved in docking synaptic vesicles to the presynaptic membrane at the active zone. A tentative arrangement of the various proteins and of their respective interactions is shown. *SYB*, VAMP/ synaptobrevin; *SYNTAG*, synaptotagmin; *CSP*, cysteine string protein. For further details, see text

distance between the two membranes is less than 3 nm, an enormous repulsion due to the strong hydration of the polar phospholipid headgroups appears (WILSCHUT and HOEKSTRA 1984; PAPAHADJOPOULOS et al. 1990). Therefore the occurrence of a "hydrophobic contact" between apposed membranes involves the removal of the tight film of water lining the membrane surfaces and requires, in the absence of catalysis, a high energy expense.

The comparison between the concentrations of divalent cations necessary for inducing fusion of pure phospholipid vesicles and those reached by Ca^{2+} entering excitable cells during stimulation of exocytosis has led to the idea that fusion of biological membranes involves the intervention of catalytic proteins. The function of these "fusion proteins" would be that of decreasing the free energy barrier for fusion, thereby allowing this process to occur under physiological conditions. Regulated secretion would then need a further inhibitory control on the constitutive activity of fusion proteins which can be removed by an appropriate stimulus (WILSCHUT and HOEKSTRA 1984; WHITE 1992).

One possible mechanism for exocytosis overcoming the repulsive forces evoked by membrane fusion is the formation and regulated opening of a proteinaceous fusion pore. In mast cells and chromaffin cells, the formation of fusion pores as fundamental intermediates of membrane fusion during exocytosis of secretory granules has been analyzed at high resolution by whole-cell patch recordings. In these systems the single secretory event is characterized by an increase in membrane capacitance attributable to the incorporation of the granule membrane into the plasma membrane. The pore opens abruptly with a conductance of a large ion channel and then gradually expands with wide fluctuations in its conductance (flicker). This stage can be followed by either irreversible expansion or by closure of the pore (ALMERS 1990; MONCK and FERNANDEZ 1992). A pattern of secretion which parallels the changes in pore conductance has recently been observed by measuring electrochemically the release of serotonin from single mastcell granules (ALVAREZ DE TOLEDO et al. 1993).

Although the small size of synaptic vesicles does not allow the direct measurement of capacitance changes due to single exocytotic events, it is reasonable to assume that also these organelles can interact with the presynaptic membrane and release their content into the extracellular space through a pore-like structure. Given the small diameter of small synaptic vesicles and the small molecular weight of the secretory products, most if not all of the vesicle content can be released through the pore. This hypothesis can also explain the extremely short delay for the excitation-secretion coupling in neurons and the very efficient and precise retrieval of synaptic vesicles (see below; FESCE et al. 1994). A potential candidate for the formation of a fusion pore is the integral synaptic vesicle protein synaptophysin (JAHN et al. 1985; WIEDEMANN and FRANKE 1985). Similarly to the connexins (the gap junction monomers), synaptophysin monomers have four predicted membrane spanning regions (LEUBE et al. 1987; SÜDHOF et al. 1987; BETZ 1990) and, when reconstituted in planar lipid bilayers, form homo-oligomers with the morphological and functional features of channels (THOMAS et al. 1988). This observation has led to the hypothesis that synaptophysin may form an exocytotic fusion pore together with a synaptophysin-binding protein located on the plasma membrane. An interaction of synaptophysin with the synaptic vesicle protein VAMP/synaptobrevin participating in the docking fusion complex has recently been reported (CALAKOS and SCHELLER 1994; EDELMANN et al. 1995). Injection of anti-synaptophysin antibodies into neurons in culture or into *Xenopus* oocytes where secretion had been reconstituted by the injection of cerebellar mRNA has revealed that the presence of synaptophysin is required for both spontaneous and evoked neurotransmitter release (ALDER et al. 1992a,b).

Genetic and biochemical studies have demonstrated that in all cells membrane fusion is vectorial and requires the presence of cytoplasmic, peripheral and integral membrane proteins mediating organelle recognition and/or triggering fusion. The complex of such proteins necessary and sufficient to promote fusion includes a group of cytosolic factors, including the ATP-binding protein NSF and proteins which mediate the binding of NSF to the appropriate receptors present on the vesicle and target membranes (soluble NSF-attachment proteins). NSF and soluble NSF-attachment proteins appear to be highly conserved, essential components of a fusion machinery mediating all kinds of intracellular membrane traffic events, including constitutive and regulated secretion (ROTHMAN and ORCI 1992). A search for neuronal proteins interacting with the soluble NSF-attachment proteins and NSF complex revealed that three previously characterized nerve terminal proteins, VAMP/synaptobrevin, SNAP-25, and HPC-1/syntaxin are capable of binding specifically soluble NSF-attachment proteins and NSF in the presence of ATP (SÖLLNER et al 1993a). Because of these properties, the proteins were named soluble NSF-attachment protein receptors (SNAREs). Of the identified SNAREs, VAMP/synaptobrevin is specific for the donor compartment, the synaptic vesicle (v-SNARE), whereas SNAP-25 and HPC-1/syntaxin are specific for the target compartment, the presynaptic membrane (t-SNAREs).

It is noteworthy that all the identified SNAREs are proteolyzed by clostridial neurotoxins which irreversibly block neurotransmitter release (see above and elsewhere this book), thus emphasizing their important role in mediating the docking/fusion of secretory vesicles. If, as shown for intracellular vesicle traffic, the complex NSF and soluble NSF-attachment proteins turns out to be the fusion machine for synaptic vesicles, it means that a common, soluble device present in nonlimiting amounts within cells has been employed for mediating all fusion events in both the constitutive and regulated pathways (BENNETT and SCHELLER 1993). The specificity of the aggregation and fusion of the appropriate donor and acceptor compartments must therefore be guaranteed by the unique and non-overlapping distribution of specific v- and t-SNAREs. Only when complementary v- and t-SNAREs bind to each other, the NSF and soluble NSF-attachment proteins fusion machine can be recruited from the cytosol and activated to promote constitutive fusion (SÖLLNER et al. 1993a). An enhancement of neurotransmitter release from the squid giant synapse has been recently demonstrated after the microinjection of recombinant SNAPs (DEBELLO et al. 1995).

Regulation of fusion can be achieved by superimposing a negative regulator (so-called "clamp") which keeps the fusion machine inactive under resting conditions, and which can be relieved upon stimulation. For synaptic vesicles this clamp might be operated by the Ca^{2+}-binding protein synaptotagmin which, under resting conditions, binds to the comples of v- and t-SNAREs and masks the binding site(s) for the attachment of soluble NSF-attachment proteins and NSF. Binding of Ca^{2+} to synaptotagmin might induce a conformational change which would allow soluble NSF-attachment proteins and NSF to bind and trigger fusion. Indeed, it has been reported that in the absence of soluble NSF-attachment proteins and NSF, the synaptic SNAREs and synaptotagmin form a stable complex. Upon addition of α-SNAP, synaptotagmin is displaced and NSF can bind to the SNARE–soluble NSF-attachment proteins complex (SÖLLNER et al. 1993b). It is noteworthy that synaptotagmin mutants have a marked impairment of evoked neurotransmitter release, which was partially restored by elevation of extracellular Ca^{2+}, and a spontaneous release of neurotransmitter increased or unaffected (LITTLETON et al. 1993; GEPPERT et al. 1994). These results imply that, although synaptotagmin is not involved in the essential steps of vesicle fusion, it plays a fundamental regulatory role in synaptic vesicle exocytosis by preventing random, constitutive fusion of synaptic vesicles with the plasma membrane at resting Ca^{2+}. Upon transient elevation of Ca^{2+} levels to values higher than 50 μM in the microdomains near the voltage-dependent Ca^{2+} channel (VON GERSDORFF and MATTHEWS 1994), low-affinity binding of Ca^{2+} to synaptotagmin would remove the clamp and allow fusion to occur (Fig. 1). Exocytosis, however, is allowed for only a very short time due to the rapid diffusion of Ca^{2+} and to the subsequent rapid dissociation of the ion from the low-affinity binding site on synaptotagmin. Although other candidates for the clamp exist, this model is consistent with the current view that the rapidity of exocytosis is determined by the presence of ready-to-fuse synaptic vesicles with a fusion apparatus already primed but clamped at resting [Ca^{2+}].

How is fusion achieved by the fusion machine binding to the docked vesicles? NSF is an ATPase, and therefore it is possible that the energy obtained from ATP hydrolysis initiates fusion either by overcoming the repulsive forces between the two lipid bilayers, or by activating some other protein directly operating fusion. An attractive mechanism is represented by the dissociation of the stable complex between v- and t-SNAREs which occurs upon ATP hydrolysis by NSF, and which is a prerequisite for membrane fusion. The topological rearrangements of SNAREs after disruption of the complex may force membrane phospholipids towards reactive transition states leading to hemifusion and eventually full fusion (SÖLLNER et al. 1993b). Other proteins downstream the NSF activation might also be involved.

Alternatively, soluble NSF-attachment proteins and NSF may bind to the docking complex formed by HPC-1/syntaxin, SNAP-25 and VAMP-synaptobrevin under resting conditions. The subsequent ATP hydrolysis mediated by NSF may bring the docking complex to a reactive and unstable high energy state ("prefusion state"). Under resting conditions synaptotagmin would be required to stabilize the complex and prevent constitutive fusion. Upon stimulation, a Ca^{2+}-dependent conformational change in synaptotagmin would allow the docking complex to return to the ground state, releasing free energy for membrane fusion (O'CONNOR et al. 1994). In the latter model, the excitation-secretion coupling does not require an enzymatic reaction to occur but implies only a conformational change in a regulatory protein, consistent with the extremely short delay in the secretory response observed experimentally.

5 Endocytosis and Recycling of Synaptic Vesicles

The insertion of the vesicle membrane into the plasma membrane which accompanies exocytosis is only a transient event, since eventually the vesicle membrane is retrieved. The retrieval process appears essential for maintaining the surface area of the cell constant, for preserving membrane diversity and an intact population of synaptic vesicles even during periods of intense secretion. Although this mechanism operates in all secretory cells, endocytosis of neuronal small synaptic vesicles exhibits unique features. In fact, the nerve terminal possesses all the machinery necessary to synthetize and load neurotransmitter into synaptic vesicles, so that these organelles can undergo several cycles of exocytosis and endocytosis without being transported back to the cell body for refilling or resorting of their molecular components (CECCARELLI and HURLBUT 1980).

The retrieval process is quite selective. The lack of intermixing between vesicle and plasma membrane components which is observed during recycling (VALTORTA et al. 1988; TORRI-TARELLI et al. 1990) is likely to be due to the rapidity of retrieval and/or to the absence of complete lipid fusion between the two membranes. Only under extreme conditions (e.g., electrical stimulation at high frequency, where the rate of exocytosis exceeds the maximum rate of endocytosis), the vesicle membrane is permanently incorporated into the plasma membrane, intermixing does occur and vesicle antigens are found to be spread all over the plasma membrane of the nerve terminal, rather than being limited to the sites at which exocytosis occurs (CECCARELLI et al. 1972; TORRI-TARELLI et al. 1990, 1992).

The existence of synaptic vesicle recycling was initially demonstrated by showing that, when nerve terminals are stimulated to secrete in the presence of an extracellular tracer, the tracer is taken up by synaptic vesicles (CECCARELLI et al. 1972, 1973; HEUSER and REESE 1973). This phenomenon depends on the size of the tracer molecule, consistent with the idea that exocytosis does not necessarily involve a complete collapse of the vesicle. By loading synaptic vesicles with amphipathic fluorescent dyes and following the dye loss which occurs when exocytosis is stimulated, Betz and coworkers have been able to optically monitor vesicle recycling at the frog neuromuscular junction. Under these conditions dye loss and cumulative neurotransmitter release initially follow the same kinetics but then diverge after 60 sec of stimulation, providing an indication of the average recycling time for a vesicle (BETZ and BEWICK 1992). A similar kinetics has been calculated for synapses of the central nervous system. In this case the recycling time has been dissected into two time intervals: endocytosis and repriming of the endocytosed vesicle (RYAN et al. 1993). The high speed and completeness of the retrieval process was further confirmed by patch-clamp recordings performed in ribbon synaptic terminals of the retina and showing that, after strong electrical stimulation, capacitance returns to resting levels with a time constant of 30 s, which decreased to 2 s after weaker stimulations (VON GERSDORFF and MATTHEWS 1994).

When synaptic vesicle recycling was initially described, various mechanisms were proposed (CECCARELLI et al. 1972, 1973; HEUSER and REESE 1973). The most controversial issue concerns the problem of whether a synaptic vesicle is retrieved after fusion and complete collapse into the axolemma by a mechanism involving the assembly of clathrin to form a coated vesicle, or whether it opens only transiently through a fusion pore and is recycled directly, without ever losing its molecular identity (FESCE et al. 1994). The second mechanism offers the advantage of being considerably more rapid, allowing immediate reuse of the retrieved vesicle and may operate in the case of vesicles which have released their content through a fusion pore (see above).

Several lines of evidence indicate that direct retrieval may be the predominant mechanism under physiological conditions: (a) in stimulated preparations, no correlation exists between intensity of quantal turnover and density of coated structures (TORRI-TARELLI et al. 1987); (b) in nerve terminals frozen few milliseconds after a single stimulation, synaptic vesicles undergoing exocytosis were often observed in connection with the plasma membrane through a tubular neck but never collapsed into the axolemma (Fig. 4; TORRI-TARELLI et al. 1985); (c) under conditions of predominant endocytosis (e.g., during recovery after intense electrical stimulation) synaptic vesicle openings are still found virtually exclusively at the active zones, suggesting that endocytosis occurs at the same site of exocytosis (CECCARELLI et al. 1979); (d) the observation that the molecular components of the synaptic vesicle and plasma membranes do not intermix even when secretion is stimulated at high rates (VALTORTA et al. 1988; TORRI-TARELLI et al. 1990, 1992) suggests that the residence time of a vesicle in the membrane is very short. This collection of observations does not rule out the

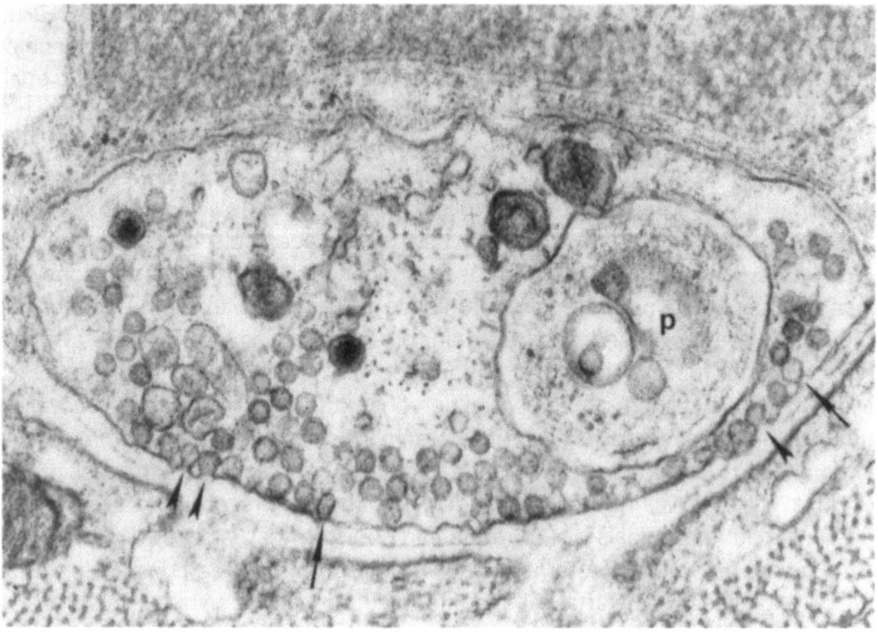

Fig. 4. Electron micrograph of a frog neuromuscular junction quickly frozen 2.5 ms after delivery of a single electrical stimulation to the nerve. *Arrows, arrowheads,* various degrees of association between synaptic vesicles and the presynaptic membrane. Synaptic vesicles undergoing exocytosis appear omega-shaped, with the vesicular profiles connected with the plasma membrane through a tubular neck. *P,* Schwann cell process. X 50 000 (Reprinted with permission from TORRI-TARELLI et al. 1985)

existence of clathrin-mediated endocytosis at the nerve terminal. This process certainly operates but is likely to represent a reserve mechanism which is activated at high rates of recycling, when the rate of exocytosis exceeds the maximum rate of direct recovery from the plasma membrane (FESCE et al. 1994).

Data obtained by studying the *shibire* mutants of *Drosophila melanogaster* which exhibit temperature-dependent paralysis as a consequence of a disfunction in endocytosis with depletion of synaptic vesicles (POODRY and EDGAR 1979) have implicated the GTPase dynamin in endocytosis in neuronal and non-neuronal cells (CHEN et al. 1991; VAN DER BLIEK and MEYEROWITZ 1991; HERSKOVITS et al. 1993; VAN DER BLIEK et al. 1993). In the mammalian brain dynamin appears to be at least partially membrane-bound, localized in both cell bodies and nerve terminals, but not specifically enriched in synaptic vesicles (SCAIFE and MARGOLIS 1990; MCPHERSON et al. 1994).

Dynamin has recently been found to be identical to dephosphin, a protein kinase C substrate which is dephosphorylated when synaptosomes are stimulated to release (ROBINSON et al. 1993). Interestingly, dephosphorylation is accompanied by a 12-fold increase in the GTPase activity of dynamin. Thus, a model can be envisaged according to which under resting conditions the state of phosphorylation of dynamin is high and the protein is inactive. When exocytosis is stimulated, dephosphorylation of the protein enhances its GTPase activity and possibly, as a consequence, its activation of endocytosis. Another interesting observation is that dynamin is able to bind to Src-homology 3 (SH3) domains of various proteins (PAWSON and SCHLESSINGER 1993) with its COOH terminal proline-rich domain, and that this binding activates dynamin GTPase activity (GOUT et al. 1993), opening the possibility that it is involved in other (ras-related?) signal transducing pathways (TROWNBRIDGE 1993).

It is interesting to note that, similarly to dynamin, synapsin I, a protein controlling synaptic vesicle availability, is also regulated by phosphorylation (see above) and binds SH3 domains with its proline-rich COOH terminal region (McPHERSON et al. 1994). However, while phosphorylation of synapsin I releases synaptic vesicles from the cytoskeletal constraint to undergo exocytosis, phosphorylation of dynamin inhibits its activity in endocytosis. Conversely, dephosphorylation may activate the dynamin-mediated endocytosis, a process which is accompanied by rapid re-embedding of the retrieved vesicles into the cytoskeletal meshwork mediated by the simultaneous dephosphorylation of synapsin I. These two phosphoproteins may therefore control the rate and extent of the processes of exo- and endocytosis depending on the activity of protein kinases and phosphatases within the nerve terminal.

6 Exocytosis of Large Dense-Core Vesicles

In contrast to small synaptic vesicles, which contain locally synthesized "classical" neurotransmitters, LDCVs, which store and release neuropeptides, resemble secretory granules of endocrine cells. The life cycle of these two kinds of secretory organelles, and consequently the mechanisms of release of classical neurotransmitters and of neuropeptides, differ in many respects. Assembly of LDCVs occurs at the *trans*-side of the golgi network, through a sorting mechanism which is believed to follow the condensation of secretory material. In addition to neuropeptides, other proteins are sorted into the budding vesicles, including enzymes necessary for the posttranslational processing of the secretory product and a family of acidic proteins, the secretogranins/chromogranins, presumably involved in the packaging process (KELLY 1991; HUTTNER et al. 1991).

LDCVs undergo fusion with the presynaptic membrane at morphologically nonspecialized sites. Fusion does not seem to be limited to the region of the plasma membrane facing the synaptic cleft and extrasynaptic exocytosis from these vesicles has also been described (THURESON-KLEIN and KLEIN 1990). Because of the proteinaceous nature of the molecules that are released from LDCVs, direct reloading of an endocytosed vesicle is not possible, with the result that LDCVs undergo only one round of exocytosis in the nerve terminal, and have to be transported back to the cell body for resorting and refilling.

Exocytosis from small synaptic vesicles and from LDCVs in the same nerve ending appears to be differentially regulated. Thus, exocytosis of LDCVs requires a higher frequency of stimulation (BARTFAI et al. 1988; LUNDBERG et al. 1990) and does not appear to be sensitive to α-LTX, which leads to a virtually complete depletion of small synaptic vesicles (MATTEOLI et al. 1988). These observations might be related to the different topological distribution and Ca^{2+} requirement of the two organelles (see below).

Although the fundamental components of the fusion machinery are probably shared by synaptic vesicles and LDCVs, no proteins specific for the membrane of LDCVs have been purified thus far; conversely, some of the proteins of small synaptic vesicles (e.g., synapsin I and synaptophysin) are not present at significant concentrations on LDCVs (NAVONE et al. 1984, 1986).

7 Dual Role of Ca^{2+} in the Nerve Terminal

Ca^{2+} plays a dual role in neuronal secretion. The influx of Ca^{2+} through voltage-dependent Ca^{2+} channels is the key event triggering synaptic vesicle exocytosis which follows the arrival of an action potential. On the other hand, Ca^{2+} appears also to play a complex modulatory role in neurotransmitter release, regulating the number of synaptic vesicles that become competent to fuse with the presynaptic plasma membrane following each action potential (ZUCKER 1989, 1993; GREENGARD et al. 1993).

Video-imaging analysis of intracellular Ca^{2+} in repetitively stimulated nerve terminals has revealed the occurrence of subplasmalemmal microdomains of high $[Ca^{2+}]$ (200–300 μ*M*), with a distribution resembling that of active zones (LLINAS et al. 1992). Recent studies aimed at verifying directly the Ca^{2+} levels necessary for eliciting secretion in chromaffin cells and in ribbon synapses of the retina showed that neurotransmitter release from nerve terminals starts to be stimulated at Ca^{2+} concentrations 50-fold higher that those required to drive maximal secretion in chromaffin cells (1.4 μ*M*; AUGUSTINE and NEHER 1993; VON GERSDORFF and METTHEWS 1994). These data suggest that low frequencies of stimulation lead to rapid transients of high $[Ca^{2+}]_i$ which are confined to the microenvironment surrounding Ca^{2+} channels. Since most LDCVs are located at sites distant from the active zones, they sense exclusively elevations in the average $[Ca^{2+}]$ of the axoplasm, which can be achieved only following repetitive stimulation (VERHAGE et al. 1991; PENG and ZUCKER 1993). This property also suggests that elevations in intracellular $[Ca^{2+}]$ subsequent to redistribution from the intracellular stores, rather than to influx from the channels, might stimulate the release of neuropeptides more efficiently than the release of classical neurotransmitters.

Ca^{2+} appears also to have a modulatory effect on neurotransmitter release. Some of the forms of short-term modulation of release such as paired-pulse facilitation (PPF) and posttetanic potentiation appear to depend linearly on a prolonged, diffuse elevation in the average intracellular [Ca^{2+}] rather than on transient localized increases in the concentration of the ion (ZUCKER 1989). This suggests the existence of at least another Ca^{2+} receptor of higher affinity (ZUCKER 1993). One such receptor might be represented by calmodulin, which then may activate CaM kinase II and lead to rapid phosphorylation of synapsin I and liberation of vesicles from the cytoskeletal constraint (VALTORTA et al. 1992; GREENGARD et al. 1993). Two recent reports obtained in CaM kinase II and synapsin I mutant mice support this hypothesis. In mice lacking the α-subunit of CaM kinase II there is an impaired ability to develop PPF in hippocampal nerve terminals (SILVA et al. 1992), whereas an enhanced PPF was observed in mice lacking synapsin I (ROSAHL et al. 1993), consistent with the idea that synapsin I repesents an inhibitory constraint for synaptic vesicle exocytosis and that its phosphorylation by CaM kinase II removes this inhibitory activity.

Ca^{2+} is the fundamental second messenger in neurotransmitter release. Rapid, localized, marked increases of Ca^{2+} levels by entry through the voltage-dependent Ca^{2+} channel promote the rapid, temporally and spatially focalized release of classical neurotransmitters from an already primed exocytosis machinery. Slow, diffuse, and quantitatively small increases in the intraterminal Ca^{2+} levels promote release of neuropeptides from LDCVs located far from active zones and at the same time modulate the efficiency of the release of classical neurotransmitters by regulating synaptic vesicle availability.

8 Conclusions

We have described above the fundamental features of neuroexocytosis. While release of neuropeptides from LDCVs exhibits essentially identical characteristics in neurons and endocrine cells, the release of classical neurotransmitters is the most efficient and plastic example of regulated secretion. Regulated exocytosis in neurons is characterized by high structural and functional complexity, with a series of interconnected steps which must occur in a well-defined temporal sequence. Complexity, although energetically expensive, has the great advantage of offering multiple levels of regulation by a cascade of intracellular and extracellular signals, and of allowing the integration of various inputs in a final plastic response. Where these remarkable characteristics come from? What are the molecules and the mechanisms which differentiate neurosecretion from constitutive and regulated exocytosis in other cells? Precise answers to these questions are still missing. The more we dissect the molecular machinery of neurotransmitter release, the more we find proteins which are highly conserved evolutionary and present also in nonneuronal cells (BENNETT and SCHELLER 1993). The basic machinery for constitutive fusion is shared by virtually all cells, and many isoforms of synaptic vesicle or presynaptic membrane proteins which were initially thought to be neurospecific have been identified in secretory and nonsecretory cells. Interestingly, the simple injection of acetylcholine into the cytoplasm of cultured myocytes or fibroblasts is sufficient to elicit quantal release (Popov and Poo 1993). This suggests that the machinery operating quantal neurotransmitter release in neurons has evolved from a general constitutive/recycling secretory pathway present in all cells. However, the poor excitation-secretion coupling observed in these nonneuronal cell emphasizes that neuron-specific molecular components are needed to transform a basic and nonspecialized secretory apparatus into the extremely sophisticated process of neurotransmitter release. Future research on the neuron-specific proteins involved will shed light on the fine mechanisms of neurosecretion.

Acknowledgments. We thank Drs. Franco Onofri and Giovanni Stefani for assistance in preparing the manuscript. This work was supported by grants from Telethon, C.N.R., and M.U.R.S.T.

References

Alder J, Xie ZP, Valtorta F, Greengard P, Poo M-M (1992a) Antibodies to synaptophysin interfere with transmitter secretion at neuromuscular synapses. Neuron 9: 759–768

Alder J, Lu B, Valtorta F, Greengard P, Poo M-M (1992b) Calcium-dependent transmitter secretion reconstituted in Xenopus oocytes. Science 257: 657–661

Almers W (1990) Exocytosis. Annu Rev Physiol 52: 607–624

Alvarez de Toledo G, Fernandez-Chacon R, Fernandez JM (1993) Release of secretory products during transient vesicle fusion. Nature 363: 554–558

Augustine GJ, Neher E (1992) Calcium requirements for secretion in bovine chromaffin cells. J Physiol (Lond) 450: 247–271

Barinaga M (1993) Secrets of secretion revealed. Science 260: 487–489

Bartfai T, Iverfeldt K, Fisone G, Serfözö B (1988) Regulation of the release of coexisting neurotransmitters. Annu Rev Pharmacol Toxicol 28: 285–310

Baumert M, Maycox PR, Navone F, De Camilli P, Jahn R (1989) Synaptobrevin: an integral membrane protein of 18,000 daltons present in small synaptic vesicles of rat brain. EMBO J 8: 379–384

Benfenati F, Valtorta F, Rubenstein JL, Gorelick F, Greengard P, Czernik AJ (1992) Synaptic vesicle-associated Ca^{2+}/calmodulin-dependent protein kinase II is a binding protein for synapsin I. Nature 359: 417–420

Benfenati F, Valtorta F, Rossi MC, Onofri F, Sihra T, Greengard P (1993) Interactions of synapsin I with phospholipids: possible role in synaptic vesicle clustering and in the maintenance of bilayer structures. J Cell Biol 123: 1845–1855

Bennett MK, Scheller RH (1993) The molecular machinery for secretion is conserved from yeast to neurons. Proc Natl Acad Sci USA 90: 2559–2563

Bennett MK, Calakos N, Scheller RH (1992) Syntaxin: a synaptic protein implicated in docking of synaptic vesicles at presynaptic active zones. Science 257: 255–259

Bernstein BW, Bamburg JR (1989) Cycling of actin assembly in synaptosomes and neurotransmitter release. Neuron 3: 257–265

Betz H (1990) Homology and analogy in transmembrane channel design: lessons from synaptic membrane proteins. Biochemistry 29: 3591–3599

Betz WJ, Bewick GS (1992) Optical analysis of synaptic vesicle recycling at the frog neuromuscular junction. Science 255: 200–203

Blasi J, Chapman ER, Link E, Binz T, Yamasaki S, De Camilli P, Südhof TC, Niemann H, Jahn R (1993a) Botulinum neurotoxin A selectively cleaves the synaptic protein SNAP-25. Nature 365: 160–162

Blasi J, Chapman ER, Yamasaki S, Binz T, Niemann H, Jahn R (1993b) Botulinum neurotoxin C1 blocks neurotransmitter release by means of cleaving HPC-1/syntaxin. EMBO J 12: 4821–4828
Bommert K, Charlton MP, DeBello WM, Chin GJ, Betz H, Augustine GJ (1993) Inhibition of neurotransmitter release by C2-domain peptides implicates synaptotagmin in exocytosis. Nature 363: 163–165
Brose N, Petrenko AG, Südhof TC, Jahn R (1992) Synaptotagmin: a calcium sensor on the synaptic vesicle surface. Science 256: 1021–1025
Calakos N, Bennett MK, Peterson KE, Scheller RH (1994) Protein-protein interactions contributing to the specificity of intracellular vesicular trafficking. Science 263: 1146–1149
Calakos N, Scheller RH (1994) Vesicle-associated membrane protein and synaptophysin are associated on the synaptic vesicle. J Biol Chem 269: 24534–24537
Ceccaldi P-E, Grohovaz F, Benfenati F, Chieregatti E, Greengard P, Valtorta F (1995) Dephosphorylated synapsin I anchors synaptic vesicles to actin cytoskeleton: an analysis by videomicroscopy. J Cell Biol 128: 905–912
Ceccarelli B, Hurlbut WP (1980) Vesicle hypothesis of the release of quanta of acetylcholine. Physiol Rev 60: 396–441
Ceccarelli B, Hurlbut WP, Mauro A (1972) Depletion of vesicles from frog neuromuscular junctions by prolonged tetanic stimulation. J Cell Biol 54: 30–38
Ceccarelli B, Hurlbut WP, Mauro A (1973) Turnover of transmitter and synaptic vesicles at the frog neuromuscular junction. J Cell Biol 57: 499–524
Ceccarelli B, Grohovaz F, Hurlbut WP (1979) Freeze-fracture studies of frog neuromuscular junctions during intense release of neurotransmitter. II. Effects of electrical stimulation and high potassium. J Cell Biol 81: 178–192
Chen MS, Obar RA, Schroeder CC, Austin TW, Poodry CA, Wadsworth SC, Vallee RB (1991) Multiple forms of dynamin are encoded by shibire, a Drosophila gene involved in endocytosis. Nature 351: 583–586
Couteaux R, Pecot-Dechavassine M (1970) Vesicles synaptiques et poches au niveau des "zones actives" de la jonction neuromusculaire. C.R Acad Sci D 271: 2346–2349
DeBello WM, Betz H, Augustine GJ (1993) Synaptotagmin and neurotransmitter release. Cell 74: 947–950
DeBello WM, O'Connor V, Dresbach T, Whiteheart SW, Wang SSH, Schweitzer FE, Betz H, Rothman JE, Augustine GJ (1995) SNAP-mediated protein-protein interactions essential for neurotransmitter release. Nature 373: 626–630
De Camilli P, Benfenati F, Valtorta F, Greengard P (1990) The synapsins. Annu Rev Cell Biol 6: 433–460
del Castillo J, Katz B (1954) Quantal components of the end-plate potential. J Physiol (Lond) 124: 560–573
Edelmann L, Hanson PI, Chapman ER, Jahn R (1995) Synaptobrevin binding to synaptophysin: a potential mechanism for controlling the exocytotic fusion machine. EMBO J 14: 224–231
Elferink LA, Peterson MR, Scheller RH (1993) A role for synaptotagmin (p65) in regulated exocytosis. Cell 72: 153–159
Facchiano F, Luini A (1992) Tetanus toxin potently stimulates tissue transglutaminase. J Biol Chem 267: 13267–13271
Facchiano F, Benfenati F, Valtorta F, Luini A (1993) Covcalent modification of synapsin I by a tetanus toxin-activated transglutaminase. J Biol Chem 268: 4588–4591
Fesce R (1990) Stochastic approaches to synaptic function. Prog Neurobiol 35: 85–133
Fesce R, Grohovaz F, Valtorta F, Meldolesi J (1994) Neurotransmitter release: fusion or kiss-and-run? Trends Cell Biol 4: 1–4
Fifkova E (1985) Actin in the nervous system. Brain Res Rev 9: 187–215
Fischer von Mollard G, Mignery GA, Baumert M, Perin MS, Hanson TJ, Burger PM, Jahn R, Südhof TC (1990) rab3 is a small GTP-binding protein exclusively localized to synaptic vesicles. Proc Natl Acad Sci USA 87: 1988–1992
Fischer von Mollard G, Südhof TC, Jahn R (1991) A small GTP-binding protein (rab3a) dissociates from synaptic vesicles during exocytosis. Nature 349: 79–81
Gengyo-Ando K, Kamiya Y, Yamakawa A, Kodaira K-I, Nishiwaki K, Miwa J, Hori I, Hosono R (1993) The C. elegans unc-18 gene encodes a protein expressed in motor neurons. Neuron 11: 703–711
Geppert M, Goda Y, Hammer RE, Li C, Rosahl TW, Stevens CF, Südhof TC (1994) Synaptotagmin I: a major Ca^{2+} sensor for transmitter release at a central synapse. Cell 79: 717–727
Gout I, Dhand R, Hiles ID, Fry MJ, Panayotou G, Das P, Truong O, Totty NF, Hsuan J, Booker GW, Campbell ID, Waterfield MD (1993) The GTPase dynamin binds to and is activated by a subset of SH3 domains. Cell 75: 25–36

Greengard P, Valtorta F, Czernik AJ, Benfenati F (1993) Synaptic vesicle phosphoproteins and regulation of synaptic function. Science 259: 780–785

Gundersen CB, Umbach JA (1992) Suppression cloning of the cDNA for a candidate subunit of a presynaptic calcium channel. Neuron 9: 527–537

Hata Y, Davletov B, Petrenko AG, Jahn R, Südhof TC (1993a) Interaction of synaptotagmin with the cytoplasmic domains of neurexins. Neuron 10: 307–315

Hata Y, Slaughter CA, Südhof TC (1993b) Synaptic vesicle fusion complex contains unc-18 homologue bound to syntaxin. Nature 366: 347–350

Herskovits JS, Burgess CC, Obar RA, Vallee RB (1993) Effects of mutant rat dynamin on endocytosis. J Cell Biol 122: 565–578

Heuser JE, Reese TS (1973) Evidence for recycling of synaptic vesicle membrane during transmitter release at the frog neuromuscular junction. J Cell Biol 57: 315–344

Hirokawa N, Sobue K, Kanda K, Harada A, Yorifuji H (1989) The cytoskeletal architecture of the presynaptic terminal and molecular structure of synapsin I. J Cell Biol 108: 111–126

Horikawa HPM, Saisu H, Ishizuka T, Sekine Y, Tsugita A, Odani S, Abe T (1993) A complex of rab3A, SNAP-25, VAMP/synaptobrevin-2, and syntaxins in brain presynaptic terminals. FEBS Lett 330: 236–240

Hosono R, Hekimi S, Kamiya Y, Sassa T, Murakami S, Nishiwaki K, Miwa J, Taketo A, Kodaira K-I (1992) The unc-18 gene encodes a novel protein affecting the kinetics of acetylcholine metabolism in the nematode Caenorhabditis elegans. J Neurochem 58: 1517–1525

Hurlbut WP, Chieregatti E, Valtorta F, Haimann C (1994) α-Latrotoxin channels in neuroblastoma cells. J Membr Biol 138: 91–102

Huttner WB (1993) Snappy exocytoxins. Nature 365: 104–105

Huttner WB, Gerdes HH, Rosa P (1991) The granin (chromogranin/secretogranin) family. Trends Biochem Sci 16: 27–30

Inoue A, Obata K, Akagawa K (1992) Cloning and sequence analysis of cDNA for a neuronal cell membrane antigen, HPC-1. J Biol Chem 267: 10613–10619

Jahn R, Schiebler W, Ouimet C, Greengard P (1985) A 38,000 dalton membrane protein (p38) present in synaptic vesicles. Proc Natl Acad Sci USA 82: 4137–4141

Kelly RB (1991) Secretory granule and synaptic vesicle formation. Curr Opin Cell Biol 3: 654–660

Kelly RB (1993) Storage and release of neurotransmitter. Cell 72/Neuron 19: 43–53

Landis DMD, Hall AK, Weinstein LA, Reese TS (1988) The organization of cytoplasm at the presynaptic active zone of a central nervous system synapse. Neuron 1: 201–209

Leube RE, Kaiser P, Seiter A, Zimbelmann R, Franke WW, Rehm HP, Knaus P, Prior P, Betz H, Reinke H, Beyreuther K, Wiedenmann B (1987) Synaptophysin: molecular organization and mRNA expression as determined from cloned cDNA. EMBO J 6: 3261–3268

Leveque C, Hoshino T, David P, Shoji-kasai Y, Leys K, Omori A, Lang B, E1 Far O, Sato K, Martin-Moutot N, Newsom-Davis J, Takahashi M, Seagar MJ (1992) The synaptic vesicle protein synaptotagmin associates with calcium channels and is a putative Lambert-Eaton myasthenic syndrome antigen. Proc Natl Acad Sci USA 89: 3625–3629

Littleton JT, Stern M, Schulze K, Perin M, Bellen HJ (1993) Mutational analysis of Drosophila synaptotagmin demonstrates its essential role in Ca^{2+}-activated neurotransmitter release. Cell 74: 1125–1134

Llinas R, Steinberg Z, Walton K (1981) Relationship between calcium current and post-synaptic potential in squid giant synapse. Biophys J 33: 323–352

Llinas R, McGuinness T, Leonard CS, Sugimori M, Greengard P (1985) Intraterminal injection of synapsin I or calcium/calmodulin-dependent protein kinase II alters neurotransmitter release at the squid giant synapse. Proc Natl Acad Sci USA 82: 3035–3039

Llinas R, Gruner JA, Sugimori M, McGuinness TL, Greengard P (1991) Regulation by synapsin I and Ca^{2+}/calmodulin-dependentprotein kinase II of transmitter release in squid giant synapse. J Physiol (Lond) 436: 257–282

Llinas R, Sugimori M, Silver RB (1992) Microdomains of high calcium concentration in a presynaptic terminal. Science 256: 677–679

Lundberg JM, Rudehill A, Sollevi A, Fried G, Wallin G (1990) Co-release of neuropeptide Y and noradrenaline from pig spleen in vivo: importance of subcellular storage, nerve impulse frequency and pattern, feedback regulation and resupply by axonal transport. Neuroscience 28: 475–486

Mastrogiacomo A, Parsons SM, Zampighi GA, Jenden DJ, Umbach JA, Gundersen CB (1994) Cysteine string proteins: a potential link between synaptic vesicles and presynaptic Ca^{2+} channels. Science 263: 981–982

Matteoli M, Haimann C, Torri-Tarelli F, Polak JM, Ceccarelli B (1988) Differential effect of α-latrotoxin on exocytosis from small synaptic vesicles and from large dense-core vesicles containing calcitonin gene-related peptide at the frog neuromuscular junction. Proc Natl Acad Sci USA 85: 7366–7370

Matteoli M, Takei K, Cameron R, Hurlbut P, Johnston PA, Südhof TC, Jahn R, De Camilli P (1991) Association of Rab3A with synaptic vesicles at late stages of the secretory pathway. J Cell Biol 115: 625–633

Matthew WD, Tsavaler L, Reichardt LF (1985) Identification of a synaptic vesicle-specific membrane protein with a wide distribution in neuronal and neurosecretory tissue. J Cell Biol 91: 257–269

McPherson PS, Czernik AJ, Chilcote T, Onofri F, Benfenati F, Greengard P, Schlessinger J, De Camilli P (1994) SH3 domains of Grb2 interact with several synaptic proteins including synapsin I. Proc Natl Acad Sci USA., 91: 6486–6490

Monck JR, Fernandez JM (1992) The exocytotic fusion pore. J Cell Biol 119: 1395–1404

Navone F, Greengard P, De Camilli P (1984) Synapsin I in nerve terminals: selective association with small synaptic vesicles. Science 226: 1209–1211

Navone F, Jahn R, Di Gioia G, Stukenbrok H, Greengard P, De Camilli P (1986) Protein p38: An integral membrane protein specific for small vesicles of neurons and neuroendocrine cells. J Cell Biol 103: 2511–2527

Nestler EJ, Greengard P (1984) Protein phosphorylation in the nervous system. Wiley, New York

Nonet ML, Grundahl K, Meyer BJ, Rand JB (1993) Synaptic function is impaired but not eliminated in C. elegans mutants lacking synaptotagmin. Cell 73: 1291–1305

O'Connor V, Augustine GJ, Betz H (1994) Synaptic vesicle exocytosis: molecules and models. Cell 76: 785–787

Osen-Sand A, Catsicas M, Staple JK, Jones KA, Ayala G, Knowles J, Grenningloh G, Catsicas S (1993) Inhibition of axonal growth by SNAP-25 antisense oligonucleotides in vitro and in vivo. Nature 364: 445–448

Oyler GA, Higgins GA, Hart RA, Battemberg E, Billingsley M, Bloom FE, Wilson MC (1989) The identification of a novel synaptosomal-associated protein, SNAP-25, differentially expressed by neuronal subpopulations. J Cell Biol 109: 3039–3052

Papahadjopoulos D, Nir S, Düzgünes N (1990) Molecular mechanisms of calcium-induced membrane fusion. J Bioenerg Biomembr 22: 157–179

Pawson T, Schlessinger J (1993) SH2 and SH3 domains. Curr Biol 3: 434–442

Peng Y, Zucker RS (1993) Release of LHRH is linearly related to the time integral of presynaptic Ca^{2+} elevation above a threshold level in bullfrog sympathetic ganglia. Neuron 10: 465–473

Perin MS, Fried VA, Mignery GA, Jahn R, Südhof TC (1990) Phospholipid binding by a synaptic vesicle protein homologous to the regulatory region of protein kinase C. Nature 345: 260–263

Petrenko AG, Perin MS, Davletov BA, Ushkaryov YA, Geppert M, Südhof TC (1991) Binding of synaptotagmin to the α-latrotoxin receptor implicates both in synaptic vesicle exocytosis. Nature 353: 65–68

Poodry CA, Edgar L (1979) Reversible alterations in the neuromuscular junctions of Drosophila melanogaster bearing a temperature-sensitive mutation shibire. J Cell Biol 81: 520–527

Popov SV, Poo M-M (1993) Synaptotagmin: a calcium-sensitive inhibitor of exocytosis? Cell 73: 1247–1249

Pumplin DW, Reese TS, Llinas R (1981) Are the presynaptic membrane particles the calcium channels? Proc Natl Acad Sci USA 78: 7210–7213

Robinson PJ, Sontag J-M, Liu J-P, Fykse EM, Slaughter C, McMahon H, Südhof TC (1993) Dynamin GTPase regulated by protein kinase C phosphorylation in nerve terminals. Nature 365: 163–166

Robitaille R, Adler EM, Charlton MP (1990) Strategic location of calcium channels at transmitter release sites of frog neuromuscular synapses. Neuron 5: 773–779

Rosahl TW, Geppert M, Spillane D, Herz J, Hammer RE, Malenka RC, Südhof TC (1993) Short-term synaptic plasticity is altered in mice lacking synapsin I. Cell 75: 861–870

Rothman JR, Orci L (1992) Molecular dissection of the secretory pathway. Nature 355: 409–415

Ryan TA, Reuter H, Wendland B, Schweizer FE, Tsien RW, Smith SJ (1993) The kinetics of synaptic vesicle recycling measured at single presynaptic boutons. Neuron 11: 713–724

Scaife R, Margolis RL (1990) Biochemical and immunochemical analysis of rat brain dynamin interaction with microtubules and organelles in vivo and in vitro. J Cell Biol 111: 3023–3033
Scheer H, Meldolesi J (1985) Purification of the putative α-latrotoxin receptor from bovine synaptosomal membranes in an active binding form. EMBO J 4: 323–327
Schiavo G, Benfenati F, Poulain B, Rossetto O, Polverino de Laurato P, DasGupta BR, Montecucco C (1992) Tetanus and botulinum-B neurotoxins block neurotransmitter release by proteolytic cleavage of synaptobrevin-2. Nature 359: 832–835
Schiavo G, Shone CC, Rossetto O, Alexander FCG, Montecucco C (1993a) Botulinum neurotoxin serotype F is a zinc endopeptidase specific for VAMP/synaptobrevin. J Biol Chem 268: 11516–11519
Schiavo G, Rossetto O, Catsicas S, Polverino de Laurato P, DasGupta BB, Benfenati F, Montecucco C (1993b) Identification of the nerve terminal targets of botulinum neurotoxin serotypes A, D, and E. J Biol Chem 268: 23784–23787
Shirataki H, Kaibuchi K, Sakoda T, Kishida S, Yamaguchi T, Wada K, Miyazaki M, Takai Y (1993) Rabphillin-3A, a putative target protein for smg p25A/rab3A p25 small GTP-binding protein related to synaptotagmin. Mol Cell Biol 13: 2061–2068
Silva AJ, Stevens CF, Tonegawa S, Wang Y (1992) Deficient hippocampal long-term potentiation in α-calcium-calmodulin kinase II mutant mice. Science 257: 201–206
Simons K, Zerial M (1993) Rab proteins and the road maps for intracellular transport. Neuron 11: 789–799
Söllner T, Whiteheart SW, Brunner M, Erdjument-Bromage H, Geromanos S, Tempst P, Rothman JE (1993a) SNAP receptors implicated in vesicle targeting and fusion. Nature 362: 318–324
Söllner T, Bennett MK, Whiteheart SW, Scheller RH, Rothman JE (1993b) A protein assembly-disassembly pathway in vitro that may correspond to sequential steps of synaptic vesicle docking, activation, and fusion. Cell 75: 409–418
Stevens CF (1993) Quantal release of neurotransmitter and long-term potentiation. Cell 72/Neuron 10: 55–63
Südhof TC, Lottspeich F, Greengard P, Mehl E, Jahn R (1987) A synaptic vesicle protein with a novel cytoplasmic domain and four transmembrane regions. Science 238: 1142–1144
Thomas L, Hartung K, Langosch D, Rehm H, Bamberg E, Franke WW, Betz H (1988) Identification of synaptophysin as a hexameric channel protein of the synaptic vesicle membrane. Science 242: 1050–1053
Thureson-Klein AK, Klein RL (1990) Exocytosis from neuronal large dense-cored vesicles. Int Rev Cytol 121: 67–126
Torri-Tarelli F, Grohovaz F, Fesce R, Ceccarelli B (1985) Temporal coincidence between synaptic vesicle fusion and quantal secretion of acetylcholine. J Cell Biol 101: 1386–1399
Torri-Tarelli F, Haimann C, Ceccarelli B (1987) Coated vesicles and pits during enhanced release of acetylcholine at the neuromuscular junction. J Neurocytol 16: 205–214
Torri-Tarelli F, Villa A, Valtorta F, De Camilli P, Greengard P, Ceccarelli B (1990) Redistribution of synaptophysin and synapsin I during α-latrotoxin-induced release of neurotransmitter at the neuromuscular junction. J Cell Biol 110: 449–459
Torri-Tarelli F, Passafaro M, Clementi F, Sher E (1991) Presynaptic localization of conotoxin-sensitive calcium channels at the frog neuromuscular junction. Brain Res 547: 331–334
Torri-Tarelli F, Bossi M, Fesce R, Greengard P, Valtorta F (1992) Synapsin I partially dissociates from synaptic vesicles during exocytosis induced by electrical stimulation. Neuron 9: 1143–1153
Trifaro J-M, Vitale ML (1993) Cytoskeleton dynamics during neurotransmitter release. Trends Neurosci 16: 466–472
Trimble WS, Cowan DM, Scheller RH (1988) VAMP-1: a synaptic vesicle-associated integral membrane protein. Proc Natl Acad Sci USA 85: 4538–4542
Trowbridge IS (1993) Dynamin, SH3 domains and endocytosis. Curr Biol 3: 773–775
Ushkaryov YA, Petrenko AG, Geppert M, Südhof TC (1992) Neurexins: synaptic cell surface proteins related to the α-latrotoxin receptor and laminin. Science 57: 50–56
Valtorta F, Madeddu L, Meldolesi J, Ceccarelli B (1984) Specific localization of the α-latrotoxin receptor in the nerve terminal plasma membrane. J Cell Biol 99: 124–132
Valtorta F, Jahn R, Fesce R, Greengard P, Ceccarelli B (1988) Synaptophysin (p38) at the frog neuromuscular junction: its incorporation into the axolemma and recycling after intense quantal secretion. J Cell Biol 107: 2719–2730

Valtorta F, Benfenati F, Greengard P (1992) Structure and function of the synapsins. J Biol Chem 267: 7195–7198

van der Bliek AM, Meyerowitz EM (1991) Dynamin-like protein encoded by the Drosophila shibire gene associated with vesicular traffic. Nature 351: 411–414

van der Bliek AM, Redelmeier TE, Damke H, Tisdale EJ, Meyerowitz EM, Schmid SL (1993) Mutations in human dynamin block an intermediate stage in coated vesicle formation. J Cell Biol 122: 553–563

Verhage M, McMahon HT, Ghijsen WEJM, Boomsma F, Scholten G, Wiegant VM, Nicholls DG (1991) Differential release of amino acids, neuropeptides, and catecholamines from isolated nerve terminals. Neuron 6: 517–524

von Gersdorff H, Matthews G (1994) Dynamics of synaptic vesicle fusion and membrane retrieval in synaptic terminals. Nature 367: 735–739

Wanke E, Ferroni A, Gattanini P, Meldolesi J (1986) α-Latrotoxin of the black widow spider venom opens a small, non-closing cation channel. Biochem Biophys Res Commun 134: 320–325

White JM (1992) Membrane fusion. Science 258: 917–923

Wiedenmann B, Franke W (1985) Identification and localization of synaptophysin, an integral membrane glycoprotein of M_r 38,000 characteristic of presynaptic vesicles. Cell 41: 1017–1028

Wilschut J, Hoekstra D (1984) Membrane fusion: from liposomes to biological membranes. Trends Biochem Sci 9: 479–483

Zinsmaier KE, Eberle KK, Buchner E, Walter N, Benzer S (1994) Paralysis and early death in cysteine string protein mutants of Drosophila. Science 263: 977–980

Zucker RS (1989) Short-term synaptic plasticity. Annu Rev Neurosci 12: 13–31

Zucker RS (1993) Calcium and transmitter release. J Physiol (Paris) 887: 25–36

Neurospecific Binding, Internalization, and Retrograde Axonal Transport

J.L. HALPERN[1] and E.A. NEALE[2]

1 Introduction

A number of pharmacologic agents which act on intracellular targets typically bind to specific cell surface receptors and are internalized in order to exert their effects on a cytosolic substrate (NEVILLE and CHANG 1978). Although many of the details remain to be determined, generally it is accepted that the clostridial neurotoxins (CNTs) interact with membrane receptors and are translocated into the cytosol where they act to block release of neurotransmitters (HABERMANN et al. 1980; SCHMITT et al. 1981; SIMPSON 1980, 1981). The active toxins consist of a light (L) chain and a heavy (H) chain that are linked by one disulfide bond. The neurotoxins comprise three functional domains, each of which is required for one of the steps in the intoxication process (Fig. 1). The carboxyl terminal half of the H chain (H_C) carries the receptor binding domain, the amino terminal half of the H chain (H_N)

[1] Division of Bacterial Products, Center for Biologics Evaluation and Research, Food and Drug Administration, Bldg 29, Room 103, 8800 Rockville Pike, Bethesda, MD 20892, USA
[2] Laboratory of Developmental Neurobiology, National Institute of Child Health and Human Development, Bldg 49, Room 5A38, NIH, Bethesda, MD 20892, USA

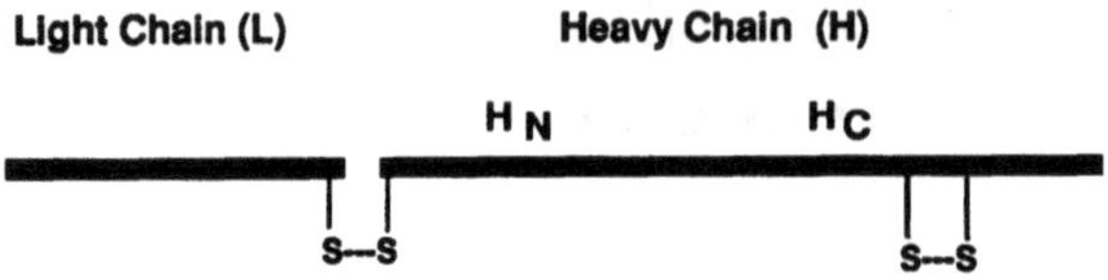

Fig. 1. Schematic structure of clostridial neurotoxins. Clostridial neurotoxins are synthesized as a single polypeptide chain which can be cleaved at a protease sensitive site to yield a light chain and a heavy chain linked by a disulfide bond. The heavy chain can be divided functionally into an amino terminal domain (H_N) and a carboxyl terminal domain (H_C)

constitutes the transmembrane domain, and the L chain is the catalytic domain (reviewed by NIEMANN 1991; WELLHÖNER 1992). This chapter summarizes current data on the binding of CNTs to sensitive cells and the mechanism whereby the toxins are thought to gain access to their cytoplasmic targets.

Botulinum neurotoxin (BoNT) reaches nerve terminals at the neuromuscular junction, where it binds to the neuronal membrane, moves into the cytoplasm of the axon terminal, and acts to block excitatory synaptic transmission, leading to flaccid paralysis (Fig. 2). Tetanus toxin (TeNT) also is taken into nerve endings at the neuromuscular junction but typically does not act at this site. Instead, TeNT is transported in a vesicular compartment within motor axons for a considerable distance, perhaps 1 m, to the soma of motor neurons within the spinal cord. Some of the toxin escapes lysosomal degradation in the motor neuron and moves "transsynaptically" out of this cell, through the synaptic cleft, and into impinging synaptic terminals. This process of retrograde (in the direction opposite to that of a nerve impulse) axonal transport and transsynaptic movement may be repeated, resulting in delivery of TeNT along neuron pathways to higher levels of the spinal cord. TeNT appears to act preferentially on inhibitory synapses, causing motor disinhibition, leading to spastic paralysis. In those synapses where it acts, the toxin moves out of its membrane compartment and acts in the cytosol to block the release of neurotransmitter. Thus, the initial uptake of TeNT at the periphery is separated, both temporally and spatially, from the translocation of the toxin to its cytoplasmic site of action, which occurs within the central nervous system.

2 Binding

2.1 Analysis of Clostridial Neurotoxin Binding Sites

Clostridial neurotoxins bind to plasma membrane receptors on target cells and undergo internalization. Because the receptor is fundamental for both recognition and uptake of a particular ligand, its identification and characterization are important for a complete understanding of CNT mechanism of action. TeNT and several of the BoNTs have identical or very similar intracellular mechanisms of action (SCHIAVO et al. 1992); however, the clinical symptoms associated with the two toxins are very different due to their relative specificity for different neuronal

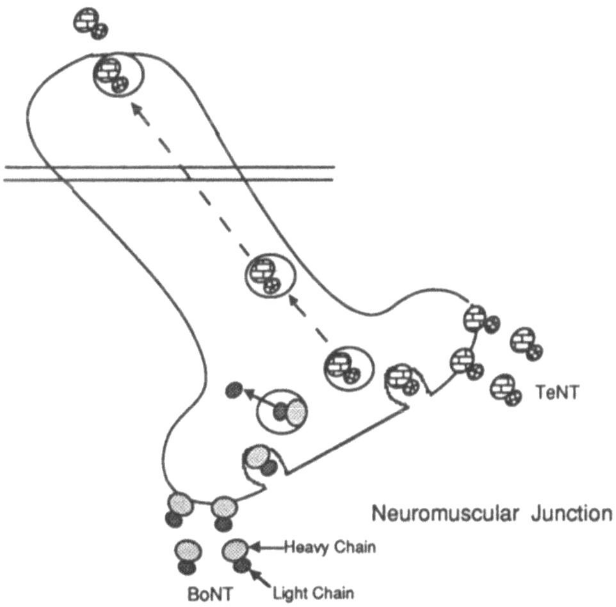

Fig. 2. Uptake of tetanus toxin and botulinum neurotoxins at the neuromuscular junction. This model shows the differential routing of TeNT and the BoNTs that accounts for their distinct clinical effects

subtypes. This issue has been discussed thoughtfully in previous reviews (MELLANBY 1984; HABERMANN and DREYER 1986; NIEMANN 1991), and it seems likely that the existence of specific receptors for these two toxins may in part explain their different effects in vivo.

Numerous studies have examined neurotoxin binding by both morphological and biochemical techniques. Although TeNT and BoNT bind primarily to neurons, there have been reports of binding sites on other cell types including macrophages (BLASI et al. 1990), kidney (HABERMANN and ALBUS 1986), thyroid (LEDLEY et al. 1977), and pancreas (EISENBARTH et al. 1982). Nonneuronal binding sites for CNTs have not been shown to be physiologically relevant. The pharmacology of TeNT and BoNT receptors has been detailed in several recent reviews (HABERMANN and DREYER 1986; NIEMANN 1991; WELLHÖNER 1992). Some general comments about these studies are applicable to both TeNT and BoNTs. The binding of these toxins has been studied in a variety of in vitro systems including primary neuronal cell cultures, cell lines, various membrane preparations, and isolated proteins or gangliosides (ROGERS and SNYDER 1981; WILLIAMS et al. 1983; EVANS et al. 1986; STAUB et al. 1986; WALTON et al. 1988). These studies have helped establish that TeNT and BoNT bind to membrane receptors and are internalized. Morphological studies have localized receptors for BoNT A and B to the unmyelinated area of motor nerve terminals (BLACK and DOLLY 1986a) and provided evidence for internalization of CNTs (BLACK and DOLLY 1986b; CRITCHLEY et al. 1985; PARTON et al. 1987, 1988). However, none of these studies has identified convincingly the receptors that mediate the biological activity of TeNT and BoNT. One difficulty is that binding has been measured under a wide range of conditions including variations in buffer ionic strength and pH, tissue or membrane concentration and method of ligand preparation. Each of these variables has been shown to influence the affinity and number of receptors (CRITCHLEY et al. 1986; OCHANDA et al. 1986; WELLER et al. 1986). Affinity constants ranging from 20 pM to 20 nM have been reported (WELLHÖNER 1992). Another problem is that under many conditions, the binding of CNTs appears to have a nonsaturable component. This high level of nonspecific binding makes it difficult to characterize a specific receptor that may be present in relatively low number. Because of these problems different values have been reported for both number and affinity of TeNT and BoNT receptors (GOLDBERG et al. 1981; ROGERS and SNYDER 1981; WILLIAMS et al. 1983; WELLHÖNER and NEVILLE 1987; BLASI et al. 1992), and it is difficult to draw any firm conclusions concerning these parameters.

2.2 Identification of Specific Clostridial Neurotoxin Receptors

The characterization of specific neuronal membrane components that bind CNTs has identified a number of different molecules proposed to function as receptors. Gangliosides are sialic acid containing glycosphingolipids found in neuronal membranes and were the first membrane component demonstrated to bind TeNT and BoNT (VAN HEYNINGEN and MILLER 1961; SIMPSON and RAPPORT 1971). A number of studies have reproduced these original findings, and it is widely accepted that CNTs bind to gangliosides . TeNT shows the highest affinity for polysialogangliosides G_{T1b} and G_{D1b} (VAN HEYNINGEN 1974; HOLMGREN et al. 1980) with approximately two- to threefold lower affinity for the monosialoganglioside G_{M1} and G_{D1a}. Botulinum toxin also has been reported to bind to G_{D1a} with high affinity (VAN HEYNINGEN 1974; HOLMGREN et al. 1980). Different BoNT serotypes seem to have somewhat different cell and ganglioside binding patterns (KITAMURA et al. 1980; BLACK and DOLLY 1986a; OCHANDA et al. 1986; KUROKAWA et al. 1987), and to interact with different domains of G_{T1b} (SCHENGRUND et al. 1991). The finding that CNTs bind similarly but not identically to gangliosides is not unexpected. There is a significant amount of divergence in the sequence of the receptor binding domains of CNTs (approximately 30%), and studies with antibodies have demonstrated little immunological cross-reactivity (TSUZUKI et al. 1988; HALPERN et al. 1989). Because of this lack of sequence homology it is likely that the region involved in ganglioside binding of different CNTs is not identical and individual CNTs would have slightly different binding specificities.

Although many reports have documented the interaction of CNTs with gangliosides, relatively few studies have attempted to correlate ganglioside binding with toxin action. The importance of gangliosides for TeNT and BoNT action was demonstrated in one study using bovine adrenal chromaffin cells

(MARXEN et al. 1989). Cultured chromaffin cells have no detectable level of G_{T1b} or G_{D1b} and very low levels of G_{D1a}. These cells are sensitive to BoNT A but not to TeNT. Chromaffin cells that were incubated with exogenous gangliosides (a mixture containing G_{D1a}, G_{D1b} and G_{T1b}) acquired sensitivity to TeNT and enhanced sensitivity to BoNT A. These data suggest that binding to gangliosides in sensitive cells is important for CNT action.

One difficulty with a model in which gangliosides are the sole receptor for CNTs is that it does not provide a basis for the very different effects of TeNT and BoNT. Many neuronal cells contain gangliosides to which both toxins bind; however, the two toxins act preferentially on certain neuronal subtypes in vivo. This relative specificity is seen also in vitro (BIGALKE et al. 1981; ALBUS and HABERMANN 1983; BERGEY et al. 1987; WILLIAMSON et al. 1992b). If binding to gangliosides were the only requirement for internalization, TeNT and BoNT might be expected to act similarly in the neuronal cells to which they both bind.

The clinical manifestations of the CNTs are due to the inhibition of neurotransmitter release and specifically by preferential action on different synapses. However, a body of evidence exists which demonstrates that TeNT, at high concentrations, can block the release of acetylcholine from neuromuscular terminals (DIAMOND and MELLANBY 1971; DUCHEN and TONGE 1973) and from central nervous system preparations (BIGALKE et al. 1978; RABASSEDA et al. 1988). The observation that TeNT works as well as BoNT at cholinergic synapses when the physiological binding and translocation mechanisms are by-passed (MOCHIDA et al. 1989; POULAIN et al. 1991; WELLER et al. 1991) also supports the existence of specific receptors that mediate appropriate entry. Similarly, at high concentration, several of the BoNT serotypes block central synaptic transmission in synaptosomal (BIGALKE et al. 1981; SANCHEZ-PRIETO et al. 1987; ASHTON and DOLLY 1988), homogenate (HABERMANN 1988), cell culture (BIGALKE et al. 1985; HABERMANN et al. 1988; L.C. WILLIAMSON, unpublished) and slice preparations (RABASSEDA et al. 1988) although physiologically, BoNT is not routed to central nervous system sites. These observations suggest that the preferential action of the toxins is related ultimately to specific receptors which determine toxin routing.

The presence of an additional receptor specific for TeNT or an individual serotype of BoNT that would explain the distinct actions of the two toxins has been investigated in several studies. Protease-sensitive TeNT binding sites have been reported on rat brain membranes and cultured neuronal cells (CRITCHLEY et al. 1986; PIERCE et al. 1986; YAVIN and NATHAN 1986), implying the existence of a protein receptor for TeNT. Protease sensitive binding sites for BoNT also have been demonstrated (EVANS et al. 1986; OGASAWARA et al. 1991).

Several proteins have been identified that may serve as the protein receptor for CNTs. In differentiated PC12 cells, cross-linking studies identified an approximately 20-kDa protein to which TeNT bound specifically (SCHIAVO et al. 1991). Botulinum toxin type A bound specifically to a 140-kDa presynaptic plasma membrane protein from the electric organ of *Torpedo marmorata* (BLASI et al. 1992). A detergent extract of rat brain synaptosomes was used to isolate a 58-kDa protein which bound to BoNT B (NISHIKI et al. 1994). This binding was enhanced by the presence of gangliosides. Amino acid sequence analysis identified this protein as synaptotagmin, a synaptic vesicle membrane protein. Each of these proteins is a candidate for a protein receptor; however, additional work is needed to demonstrate their exact role in toxin binding and uptake.

2.3 A Model for Neurotoxin Binding

From the above discussion it is apparent that there are still fundamental questions regarding the role of gangliosides as CNT receptors, the identification of additional receptors specific for individual toxins, and the importance of each of these receptors in mediating toxin binding and internalization. A "dual receptor" model proposes that CNTs interact with both ganglioside and protein (MONTECUCCO 1986). In this model TeNT and BoNT initially interact with gangliosides in the plasma membrane. The high concentration of gangliosides in neuronal membranes should ensure efficient binding of CNT molecules in the proximity of nerve endings. The complex of CNT and ganglioside can then interact with a protein receptor specific for an individual toxin. The formation of this trimeric complex would result in high affinity, productive binding, and internalization. The dual receptor model of MONTECUCCO (1986) has been extended by NIEMANN (1991), relating the receptor to the mechanism for differential routing of the CNTs. Interpreted simply, toxin bound to gangliosides is routed for transport; that bound to its specific protein receptor is translocated into the cytosol. These models appear consistent with results obtained by a number of investigators, and provide useful hypotheses for future experimental testing.

2.4 Structure of the Receptor Binding Domain

The CNTs may be considered A-B toxins in which specific functions can be assigned to discrete structural domains. The receptor binding domain of CNTs is thought to be located in the carboxyl terminal half of the H chain. The H chain of TeNT can be cleaved by papain (HELTING and ZWISLER 1977) and other proteases to generate amino terminal (H_N) and carboxyl terminal domains (H_C) which are each approximately 50 kDa. The H_C domain of TeNT (traditionally referred to as fragment C) retains the receptor binding activity of holotoxin (MORRIS et al. 1980; WELLER et al. 1986). A study to analyze the structural requirements for the interaction of H_C with receptor identified the carboxyl terminal ten amino acids as an essential region for binding (HALPERN and LOFTUS 1993). Although H_C clearly is important for binding, differences in the binding of H_C and holotoxin have been noted in direct comparisons. One interpretation of these data is that additional domains in H_N also contribute to binding.

Preparation of H_C domain from BoNT has been reported only for BoNT A (KOZAKI et al. 1989); this fragment retained the ability to bind gangliosides. The preparation and characterization of H_C from other BoNT serotypes may have been hampered by the increased protease sensitivity of BoNT relative to TeNT (SHONE et al. 1985). Although direct binding studies with H_C from BoNT have not been performed, a number of indirect experiments provide good evidence that the receptor binding domain is most likely conserved in this region (KOZAKI et al. 1986; SHONE et al. 1985).

3 Internalization

Several lines of evidence have strengthened the argument that the CNTs act within the neuronal cytosol rather than at the neuronal surface: first, that there is a time interval after which the toxins are inaccessible to neutralization by antibodies; secondly, that toxicity is antagonized by agents known to interfere with receptor-mediated endocytosis; and thirdly, that toxins have been localized morphologically within the neuronal cytoplasm.

3.1 Endocytosis and Membrane Translocation

Electron microscopic autoradiography localized ^{125}I-labeled BoNT A on or within the axon terminal membrane in mouse hemidiaphragms. Importantly, the relative distribution of silver grains was shifted by treatment of preparations with metabolic inhibitors, reduced temperature, and lysosomotropic agents (BLACK and DOLLY 1986a,b), providing morphologic evidence for receptor-mediated endocytosis of BoNT. The resolution of this technique was not sufficient, however, to allow precise localization of toxin relative to endocytic structures.

Primary spinal cord neurons in culture serve as a model for understanding toxin trafficking and the electrophysiologic and neurochemical expression of toxicity (BIGALKE et al. 1978; BERGEY et al. 1983; BIGALKE et al. 1985; HABIG et al. 1986; WILLIAMSON et al. 1992a,b, 1993, 1995; WILLIAMSON and NEALE 1994a,b). The binding and uptake of TeNT have been examined in spinal cord cultures, and surface-bound TeNT was shown to be internalized rapidly within vesicular structures (CRITCHLEY et al. 1985). With electron microscopy, gold-labeled TeNT was found in these neurons associated with coated pits, coated vesicles, and endosomes at the level of the cell body and in coated and uncoated vesicles at synaptic terminals and in axons (PARTON et al. 1987). In contrast, entry of toxin-gold into ganglioside-treated BALB/c 3T3 cells was mediated by noncoated structures, with the toxin-gold complex visualized at later times in lysosomes and multivesicular bodies (Parton et al. 1988). In liver cells gold-labeled TeNT was shown to be associated with uncoated structures and this interaction was proposed to be ganglioside mediated. However, this study used extremely high concentrations of TeNT, and the specificity of this binding was not demonstrated (Montesano et al. 1982). It was proposed that TeNT associated with clathrin-coated structures and endosomes is bound to high-affinity protein receptors involved in productive sites of entry, whereas toxin associated with gangliosides enters the cell in uncoated structures and is degraded without producing neurotoxicity. This model, however, requires definitive proof. The recent availability of a multicompartment culture preparation of striated muscle innervated by spinal cord neurons promises to provide a relevant system for a comparative study of the uptake and action versus transport of the BoNTs and TeNT (Parfitt et al. 1994).

While endosomes might mediate toxin internalization, the expression of toxicity requires that at least the catalytic domain of the toxin exit this membrane-limited compartment. There are numerous reports of interactions of the CNTs with artificial membranes, leading to the formation of ion-conducting pores. In most of these studies, although the specific biophysical properties of the pores vary somewhat, pore formation generally is acid-dependent and appears to be mediated by the H_N of the toxin. The pores typically are permeable to small cations, but have been shown to allow the diffusion of fluorescent markers up to a molecular weight of 700 Da (Menestrina et al. 1989). These artificial membrane systems have been either planar lipid bilayers or lipid vesicles, sometimes with added gangliosides. Toxin-induced channel formation in planar lipid bilayers has been reported for TeNT (Borochov-Neori et al. 1984; Hoch et al. 1985; Rauch et al. 1990; Montal et al. 1992), BoNT A (Blaustein et al. 1987; Montal et al. 1992), BoNT B (Hoch et al. 1985), BoNT C (Donovan and Middlebrook 1986), and BoNT D (Hoch et al. 1985). Tetanus toxin forms pores in lipid vesicles (Boquet and Duflot 1982; Boquet et al. 1984; Roa and Boquet 1985; Menestrina et al. 1989), as does BoNT (Shone et al. 1987; Schmid et al. 1993). It has been more difficult to study pore formation in biologic membranes. For diphtheria toxin (DT) movement of the A chain (catalytic domain) out of purified endosomes has been demonstrated (Beaumelle et al. 1992); similar studies on isolated endosomal preparations have not been reported for the CNTs. However, pore formation induced by TeNT in physiologic membranes (in which the putative protein receptor should be present) has been studied. Rat cortical synaptosomes have been shown to lose lactate dehydrogenase and potassium ions upon exposure to complete H chain of TeNT but not to its H_N fragment or to holotoxin (Högy et al. 1992). More recently, using membrane patches obtained from spinal cord neurons in culture, Beise et al. (1994) have reported that TeNT forms pores at pH 5.0 but not at pH 7.4. The H chain of TeNT, as well as H_N, form pores at both pH 5.0 and 7.4, whereas the L chain alone does not form pores in neuronal membranes at either pH 5.0 or 7.4. Because TeNT does not form pores at neutral pH, the authors believe that extracellular TeNT (or its L chain) does not penetrate the plasma membrane directly and hypothesize that, for productive entry, TeNT passes through an acidic intracellular compartment following receptor-mediated endocytosis.

The acid dependence of toxin translocation in intact cells also has been inferred by the ability to block toxin action using agents that neutralize or dissipate pH gradients in acidic compartments or that block the acidification of endosomes. Lysosomotropic amines and proton ionophores such as monensin and nigericin act to accomplish the former. Bafilomycin A1, a specific inhibitor of the vacuolar H^{+} –ATPase, achieves the latter. In studies with DT these agents blocked the expression of cytotoxicity (LEPPLA et al. 1980; MARNELL et al. 1982; PAPINI et al. 1993). The lysosomotropic amines chloroquine and hydroxychloroquine were found to antagonize the onset of paralysis caused at the neuromuscular junction by BoNT A and B, but neither drug antagonized TeNT (SIMPSON 1982). Ammonium chloride and methylamine hydrochloride both antagonized BoNT A and B and TeNT (SIMPSON 1983). Monensin and nigericin blocked the action of BoNT A and B at the neuromuscular junction (SHERIDAN 1993; ADLER et al. 1994) and bafilomycin delayed the onset of paralysis (SIMPSON et al. 1994), while monensin and bafilomycin blocked the action of TeNT in spinal cord neuronal cultures (WILLIAMSON et al. 1992a, 1993; WILLIAMSON and NEALE 1994a,b). In the latter studies the drugs did not interfere with toxin binding. In experiments analogous to those performed using DT (reviewed by SANDVIG and OLSNES 1991) pre-bound TeNT entered spinal cord neurons directly through the plasma membrane when the cultures were pulsed briefy, in the presence of monensin, with extracellular medium at low pH (WILLIAMSON et al. 1992a; WILLIAMSON and NEALE 1994a). Furthermore, TeNT injected directly into chromaffin cells (PENNER et al. 1986) or into *Aplysia* neurons (MOCHIDA et al. 1989) acts to block stimulated secretion, indicating that a direct modification of the toxin by low pH is not required for the toxin to act on its substrate. Thus, for the expression of toxicity the CNTs must be exposed to conditions of low pH, and it is concluded that this requirement operates at the level of membrane translocation.

Some bacterial protein toxins appear to be endocytosed and then trafficked through the Golgi apparatus prior to translocation, for example, cholera toxin (ORLANDI et al. 1993) and *Pseudomonas* exotoxin A (YOSHIDA et al. 1991). Movement through a Golgi pathway is demonstrated by blocking toxin activity with brefeldin A, a drug which causes redistribution of Golgi membranes into the endoplasmic reticulum (FUJIWARA et al. 1988; LIPPINCOTT-SCHWARTZ et al. 1989). Tetanus toxin action in spinal cord cultures clearly is not affected by brefeldin A (WILLIAMSON and NEALE 1994b), suggesting that CNTs are not trafficked through Golgi compartments.

A source of controversy for the pore or channel theory of entry is that the size of toxin-induced pores, as inferred from the single pore conductance, is too small to allow passage of the catalytic domain of a toxin. It might be argued that there is no clear correlation between pore conductance and pore size, and that the single pore conductances reported may have been measured with the toxin inserted within the lumen of the pore (see SIMON and BLOBEL 1991). However, pores formed in neuronal membranes by intact TeNT exhibit conductances very similar to those formed by H_N (see Table 1 in BEISE et al. 1994). MONTECUCCO et al. (1991, 1992) have hypothesized recently that the ion-conducting channel formed by DT differs in its properties from the toxin-translocating channel and have offered a model which conceives the channel as a hydrophilic cleft rather than as a tunnel. A cleft could spread to accommodate the catalytic domain but would tighten up after its passage to a low conductance state allowing the flow of only small ions. Alternatively, it has been claimed that the toxin molecule "unfolds" and passes through the pore in an extended state (HOCH et al. 1985).

Total elimination of the membrane barrier has been posited as a mechanism for release of toxin from endosomes. BEISE et al. (1994) have suggested that toxin-induced pores, rather than providing for the direct exit of the catalytic domain of the toxin, may cause the loss of physiologic gradients and result in lysis of the endosome. While such lysis would be effective, it remains to be demonstrated. The literature on DT translocation has not implicated endosomal lysis as the mechanism for delivery of the catalytic domain from the endosome to the cytosol. Furthermore, acid-pulse experiments which subject intact cells to toxin in the presence of monensin, have demonstrated toxin translocation directly across the plasma membrane into the cytosol. By analogy, pores formed in the plasma membrane might be expected to cause a similar loss of physiologic gradients, leading to cell lysis. Lysis of Vero cells has not been reported for DT nor was it observed with spinal cord neurons and TeNT (L.C. WILLIAMSON, unpublished).

3.2 Structure of the Translocation Domain

The crystal structure of DT shows that its transmembrane domain is composed of nine α-helices (CHOE et al. 1992). A hydrophobic α-helix motif commonly is associated with pore-forming ability (SANSOM 1991). Whereas the primary sequences of the CNTs are known, their crystal structure has not been determined.

Recent studies have provided evidence that the DT-induced pore is oligomeric (SILVERMAN et al. 1994). The data of DONOVAN and MIDDLEBROOK (1986) suggested that channel formation by BoNT C in lipid membranes involved aggregates of the toxin, possibly dimers. MONTAL et al. (1992) employed sequence analysis and molecular modeling to identify potential channel-forming regions in the primary structure of TeNT and BoNT A. They synthesized peptides of these regions and evaluated their single channel properties in lipid bilayers. A 23 amino acid peptide was shown to be a transmembrane domain and an amphipathic α-helix. Pores formed by this peptide in planar lipid bilayers showed channel properties identical to those of intact TeNT. These authors proposed four amphipathic α-helices as a possible structure for the pore-forming entity of both TeNT and BoNT A. BEISE et al. (1994) identified three regions of the N-terminal region of the TeNT H chain with sufficiently high amphiphilic character to be considered candidates for pore formation. LEBEDA (LEBEDA and OLSON 1995; LEBEDA et al. 1994) applied a combination of hydrophobic moment and neural net analyses to TeNT and to six serotypes of BoNT and predicted a complementary pattern of four adjacent amphipathic, possibly transmembrane, regions that might participate in forming ion channels. Further theoretical evaluation of a number of models led these authors to conclude that a model having four transmembrane segments could account for the optimal pH values for channel formation and for the known ionic selectivities of TeNT- and BoNT-induced channels (F.J. LEBEDA et al., personal communication). Channels formed by BoNT B in ganglioside G_{T1b}/phosphatidylcholine lipid vesicles and visualized by electron cryomicroscopy and image processing had an estimated diameter of 14×25 Å and appeared to arise from the interaction of four neurotoxin molecules (SCHMID et al. 1993). The channels appeared obstructed partially by toxin protein, prompting the speculation that a low conductance, partially "blocked" channel forms when the neurotoxin interacts with the cell surface. Following endocytosis, exposure to low pH might lead to the "uncovering" (rather than creation) of the channel and a shift to a high conductance state.

Although acid-dependent pore formation has been demonstrated amply for the CNTs, it remains to be proven definitively that the pores play an essential role in toxin translocation. Site-directed mutations of a DT-based fusion toxin provide evidence that although pores may be required for the expression of toxicity, they are not of themselves sufficient (VANDERSPEK et al. 1993).

The disulfide bond linking the H and L chains has been shown to be important for the membrane translocation of the CNTs. Extracellular reduction of the interchain disulfide bond of TeNT did not affect toxin binding, but did eliminate toxicity, suggesting that this bond plays a role in nerve cell penetration (SCHIAVO et al. 1990). This suggestion has been strengthened by studies on the effect of BoNT A on identified *Aplysia* neurons which are sufficiently large to allow presynaptic intracellular injection, and which form cholinergic synapses whose activity can be monitored with intracellular recordings. These studies demonstrated clearly that reduction of the interchain disulfide bond of BoNT A decreased its paralytic potency when applied extracellularly. Reduction of this linkage did not eliminate either the toxin's ability to bind to its productive ectoacceptors or its ability to block neurotransmitter release when injected intracellularly, suggesting that it is important in the membrane translocation step (DE PAIVA et al. 1993).

4 Movement of Tetanus Toxin in the Central Nervous System

It is well established that the primary site of action of TeNT is within the spinal cord. The disease tetanus is characterized by symptoms of muscle hyperactivity including increased muscle tone, trismus, muscle rigidity, and generalized spasms (BLECK 1989). These symptoms result from the inhibition of inhibitory neurotransmitter release from spinal cord interneurons. Since the synthesis and release of TeNT from *Clostridium tetani* typically occur at a site distant from the central nervous system, TeNT must have a transport mechanism for reaching the spinal cord. This mechanism has been demonstrated to involve retrograde axonal transport and transsynaptic transport. After synthesis of TeNT at a peripheral site the toxin binds to receptors present on nerve endings and is internalized in endocytic vesicles. These vesicles migrate from the nerve endings to the neuronal cell body. At the cell body the toxin is released and transferred across the synapse to nerve endings that impinge upon the neuron from which the toxin was released. The process of retrograde axonal transport is not unique to TeNT but has been described for a variety of substances, including BoNT A (HABERMANN 1974; WEIGAND et al. 1976). In contrast, TeNT appears somewhat unique in its ability to undergo efficient transsynaptic transport (SCHWAB et al. 1979).

4.1 Retrograde Axonal Transport

Neurons are highly polarized cells whose axons can extend over a long distance. Retrograde axonal transport might be thought of as a modification of a normal cellular endocytic pathway which has been specialized for neurons. Material that is endocytosed at the nerve endings must be transported back to the cell body to be processed or degraded. In neurons material is transported continually both away from and toward the cell body by distinct anterograde and retrograde pathways. Several anterograde pathways and one retrograde pathway have been identified (VALLEE and BLOOM 1991). Molecules undergo retrograde axonal transport after binding to receptor sites on nerve endings and internalization. Alternatively, molecules that do not bind to nerve endings can be internalized by a pinocytic mechanism if present in high enough concentration. Both retrograde and anterograde transport vesicles have been demonstrated to move along microtubules. Two microtubule-associated proteins, dynein (also known as MAP 1C) and kinesin appear to be the molecular motors responsible for movement of vesicles along microtubules in the retrograde and anterograde directions, respectively (VALE et al. 1985; PASCHAL and VALLEE 1987; PASCHAL et al. 1987). Anterograde transport vesicles are small, with diameters of less than 150 nm, whereas retrograde transport vesicles are larger, multivesicular, and more heterogeneous in appearance (TSUKITA and ISHIKAWA 1980).

After identification of the spinal cord as a physiologically important site of TeNT action many studies were undertaken to define its pathway of entry into the central nervous system (DIMPFEL and HABERMANN 1973). The use of ^{125}I-labeled TeNT and autoradiography allowed several investigators to demonstrate the existence of the retrograde axonal pathway (ERDMANN et al. 1975; PRICE et al. 1975; STÖCKEL et al. 1975). In each of these studies radiolabeled toxin was found inside membrane-bound vesicles. These studies demonstrated that in addition to α–motoneurons, a retrograde transport pathway for TeNT also existed in peripheral sensory and adrenergic neurons. Both autoradiographic and, more recently, immunohistochemical techniques have been used to visualize toxin following its injection into an anatomically distinct site. These studies have consistently localized TeNT within neurons in membrane-bound vesicles. One study examined the retrograde transport of TeNT, cholera toxin, nerve growth factor, and several lectins (SCHWAB and THOENEN 1976), molecules known to bind to nerve endings. There were no apparent morphologic differences among vesicles containing TeNT and the other ligands. The receptor binding domain (H_C) of TeNT retains the ability to undergo retrograde axonal transport (BIZZINI et al. 1977) although it may be transported less efficiently (see below).

Only a few papers have characterized the quantitative aspects of neuronal transport of TeNT. DUMAS et al. (1979b) studied the transport of TeNT, nerve growth factor, wheat germ agglutinin, and cholera toxin in adrenergic nerve terminals. They reported that the transport of each of these ligands is saturable, indicating a specific receptor or transport system. The extent of transport has been measured by quantifying the toxin accumulated at a ligation site on the nerve after injection into a muscle. In one paper, it was reported that the retrograde transport of H_C was comparable to that of TeNT (MORRIS et al. 1980). However in another study H_C was transported with significantly less efficiency than holotoxin (WELLER et al. 1986). The reason for this discrepancy is not clear.

4.2 Transsynaptic Transport

The transsynaptic movement of TeNT was demonstrated first by autoradiographic localization of ^{125}I-labeled TeNT in spinal cord interneurons after injection into a muscle (SCHWAB and THOENEN 1976). Since then a number of other papers have described the transsynaptic transport of both TeNT and H_C (DUMAS et al. 1979a; SCHWAB et al. 1979; EVINGER and ERICHSEN 1986; FISHMAN and CARRIGAN 1987; MANNING et al., 1990; CABOT et al. 1991). These studies have shown that transsynaptic transport of TeNT is not selective for motoneurons but is seen also in peripheral sensory and adrenergic neurons. There has been no convincing demonstration of the transsynaptic transfer of BoNT.

The nontoxic nature of H_C has made it an especially attractive molecule to neurobiologists interested in the use of TeNT as a neuronal cell marker (DIMPFEL et al. 1975; NEALE et al. 1989). Many of the papers that report on the transsynaptic transport of TeNT and H_C are primarily neuroanatomic studies focusing on characterization of neuronal pathways and synaptic connections rather than on the molecular mechanism of the transsynaptic transport process. Transsynaptic transport must involve release of TeNT from the first neuron, movement across the synapse, and entry into the presynaptic terminal of the second neuron. ERDMANN et al. (1981) demonstrated that anti-TeNT antibodies injected into the intrathecal space could protect an animal against intramuscular injections of TeNT, and receptors for TeNT have been identified on presynaptic nerve endings (PRICE et al. 1977). These results imply that TeNT is not inside a vesicle in the synaptic cleft but diffuses to the presynaptic membrane after release from the first neuron and enters a second neuron by receptor-mediated endocytosis. The mechanism for this process has not been characterized. One difficulty with studying transcellular and transsynaptic movement in neurons is the absence of a convenient in vitro system similar to those available for studying transcellular transport in other types of polar cells.

Since it seems unlikely that nerve cells would have developed a transsynaptic pathway specifically for a bacterial toxin, it is probable that TeNT is using a pathway that exists for the movement of endogenous molecules between neurons. Such a pathway has not been described, however a transcytotic pathway was recently demonstrated in cultured hippocampal neurons that had been transfected with the polymeric immunoglobulin receptor (IKONEN et al. 1994). This receptor transported immunoglobulin molecules from a somatodendritic to an axonal domain or in an anterograde direction. The authors suggested that the movement of the polymeric immunoglobulin receptor in neurons represents a transcytotic pathway analogous to the pathway described in epithelial cells. TeNT and H_C may be useful tools for further characterization of the mechanism and function of transsynaptic pathways.

From the above discussion, it is clear that delivery of TeNT to the spinal cord involves a complex routing mechanism and a number of questions remain unanswered: Does the trafficking for all retrograde transneuronal transfer coincide with sites of synaptic connectivity? What is the underlying mechanism for a transfer of substances across the synapse in the direction opposite to the conventional flow of information between neurons? Does this transfer serve a physiologically relevant purpose? The basis for the ability of TeNT to bypass a degradative route in α-motoneurons (SCHWAB et al. 1979) and to enter the cytosol only in specific neurons is unclear. The specificity may lie in the nature of the receptor or receptors for TeNT on different populations of neurons. Alternatively, TeNT may have some internal signal sequence which specifies its routing pathway. These interesting questions may be answered by characterization of the toxin receptor and by site-directed mutagenesis of TeNT and H_C.

4.3 Use of H_C as a CNS Delivery System

The ability of nontoxic H_C fragment to undergo retrograde axonal and transsynaptic transport has prompted interest in the possibility that it could be a useful system for the delivery of other molecules to the central nervous system. BIZZINI et al. (1980) chemically conjugated the $L.H_N$ fragment of TeNT to H_C and demonstrated that this complex is transported retrogradely. When a noncovalent complex of TeNT and a monoclonal antibody against the toxin was injected into the rat gastrocnemius muscle, both toxin and antibody were transported efficiently in the sciatic nerve (HABIG et al. 1983). There was no transport of the monoclonal antibody in the absence of TeNT. In a more recent study an H_C enzyme conjugate was able to repair an enzyme defect in cultured neuronal cells

(DOBRENIS et al. 1992). H_C was conjugated chemically to the lysosomal enzyme B-*N*-acetylhexosaminidase A (Hex A) to determine whether this construct would be useful for replacement of neuronal lysosomal enzymes. Lysosomal storage diseases which lead to serious neurodegeneration result from a genetic defect in lysosomal enzymes such as Hex A. The H_C-Hex A construct bound to and was internalized by cerebral cortex cell cultures. In feline cerebral cortical cells in which there is a buildup of ganglioside G_{M2} due to lysosomal enzyme defects, H_C-Hex A was able to reduce levels of lysosomal G_{M2}. These experiments suggest that H_C might be useful for specific routing of therapeutic or diagnostic molecules to neuronal cells.

5 Conclusions

The molecular model for CNT action includes at least three distinct steps: (a) binding; (b) uptake/internalization; and (c) intoxication. The recent clarification that CNTs are zinc endoproteases and the identification of their respective substrates (reviewed in this volume) have significantly advanced our understanding of the basis of neuronal intoxication. These exciting discoveries on the mode of action and intracellular targets of the CNTs have contributed further to our understanding of the cell biology of both regulated and constitutive exocytosis. Additional research will almost certainly result in similar advances in our understanding of the binding and internalization of these toxins. Important areas of research include (but are not limited to) identification of specific cell surface receptors, intracellular trafficking and sorting pathways, and mechanisms of membrane translocation and transsynaptic transport. The intriguing differences among the clostridial neurotoxins undoubtedly reside in the distribution of their respective receptors. Elucidation of the molecular bases for the internalization and routing of the CNTs promises to provide yet another significant contribution to cell biology with respect to our understanding of membrane trafficking, and may be expected to be of significance for the clinical treatment of tetanus and botulism.

References

Adler M, Deshpande SS, Sheridan RE, Lebeda FJ (1994) Evaluation of captopril and other potential therapeutic compounds in antagonizing botulinum toxin-induced muscle paralysis . In: Jankovic J, Hallett M (eds) Therapy with botulinum toxin. Dekker, New York, pp 63–70

Albus U, Habermann E (1983) Tetanus toxin inhibits the evoked outflow of an inhibitory (GABA) and an excitatory (D-aspartate) amino acid from particulate brain cortex. Toxicon 21: 97–110

Ashton AC, Dolly JO(1988) Characterization of the inhibitory action of botulinum neurotoxin type A on the release of several transmitters from rat cerebrocortical synaptosomes. J Neurochem 50: 1808–1816

Beaumelle B, Bensammar L, Bienvenüe A (1992) Selective translocation of the A chain of diphtheria toxin across the membrane of purified endosomes. J Biol Chem 267: 11525–11531
Beise J, Hahnen J, Andersen-Beckh B, Dreyer F (1994) Pore formation by tetanus toxin, its chain and fragments in neuronal membranes and evaluation of the underlying motifs in the structure of the toxin molecule. Naunyn Schmiedebergs Arch Pharmacol 349: 66–73
Bergey GK, Macdonald RL, Habig WH, Hardegree MC, Nelson PG (1983) Tetanus toxin: convulsant action on mouse spinal cord neurons in culture: J Neurosci 3: 2310–2323
Bergey GK, Bigalke H, Nelson PG (1987) Differential effects of tetanus toxin on inhibitory and excitatory synaptic transmission in mammalian spinal cord neurons in culture: a presynaptic locus of action for tetanus toxin. J Neurophysiol 57: 121–131
Bigalke H, Dimpfel W, Habermann E (1978) Suppression of ^{3}H-acetylcholine release from primary nerve cell cultures by tetanus and botulinum-A toxin. Naunyn Schmiedebergs Arch Pharmacol 303: 133–138
Bigalke H, Heller I, Bizzini B, Habermann E (1981) Tetanus toxin and botulinum A toxin inhibit release and uptake of various transmitters, as studied with particulate preparations from rat brain and spinal cord. Naunyn Schmiedebergs Arch Pharmacol 316: 244-251
Bigalke H, Dreyer F, Bergey G (1985) Botulinum A neurotoxin inhibits non-cholinergic synaptic transmission in mouse spinal cord neurons in culture. Brain Res 360: 318–324
Bizzini B, Stoeckel K, Schwab M (1977) An antigenic polypeptide fragment from tetanus toxin: chemical characterization, binding to gangliosides and retrograde axonal transport in various neuron systems. J Neurochem 28: 529–542
Bizzini B, Grob P, Glicksman MA, Akert K (1980) Use of the B-II_b tetanus toxin derived fragment as a specific neuropharmacological transport agent. Brain Res 193: 221–227
Black JD, Dolly JO (1986a) Interaction of ^{125}I-labeled botulinum neurotoxins with nerve terminals. I. Ultrastructural autoradiographic localization and quantitation of distinct membrane acceptors for types A and B on motor nerves. J Cell Biol 103: 521–534
Black JD, Dolly JO (1986b) Interaction of ^{125}I-labeled botulinum neurotoxins with nerve terminals. II. Autoradiographic evidence for its uptake into motor nerves by acceptor-mediated endocytosis. J Cell Biol 103: 535–544
Blasi E, Pitzurra L, Burhan-Fuad AM, Marconi P, Bistoni F (1990) Gamma interferon-induced specific binding of tetanus toxin on the GG2EE macrophage cell line. Scand J Immunol 32: 289–292
Blasi J, Egea G, Castiella MJ, Arribas M, Solsona C, Richardson PJ, Marsal J (1992) Binding of botulinum neurotoxin to pure cholinergic nerve terminals isolated from the electric organ of Torpedo. J Neural Transm Gen Sect 90: 87–102
Blaustein RO, Germann WJ, Finkelstein A, DasGupta BR (1987) The N-terminal half of the heavy chain of botulinum type A neurotoxin forms channels in planar phospholipid bilayers. FEBS Lett 226: 115–120
Bleck T (1989) Clinical aspects of tetanus. In: Simpson LL (ed) Botulinum neurotoxin and tetanus toxin. Academic, San Diego, pp 379–398
Boquet P, Duflot E (1982) Tetanus toxin fragment forms channels in lipid vesicles at low pH. Proc Natl Acad Sci USA 79: 7614–7618
Boquet P, Duflot E, Hauttecoeur B (1984) Low pH induces a hydrophobic domain in the tetanus toxin molecule. Eur J Biochem 144: 339–344
Borochov-Neori H, Yavin E, Montal M (1984) Tetanus toxin forms channels in planar lipid bilayers containing gangliosides. Biophys J 45: 83–85
Cabot JB, Mennone A, Bogan N, Carroll J, Evinger C, Erichsen JT (1991) Retrograde, transsynaptic and transneuronal transport of fragment C of tetanus toxin by sympathetic preganglionic neurons. Neuroscience 40: 805–823
Choe S, Bennett MJ, Fujii G, Curmi PM, Kantardjieff KA, Collier RJ, Eisenberg D (1992) The crystal structure of diphtheria toxin. Nature 357: 216–222
Critchley DR, Nelson PG, Habig WH, Fishman PH (1985) Fate of tetanus toxin bound to the surface of primary neurons in culture: evidence for rapid internalization. J Cell Biol 100: 1499–1507
Critchley DR, Habig WH, Fishman PH (1986) Reevaluation of the role of gangliosides as receptors for tetanus toxin. J Neurochem 47: 213–222
de Paiva A, Poulain B, Lawrence GW, Shone CC, Tauc L, Dolly JO (1993) A role for the interchain disulfide or its participating thiols in the internalization of botulinum neurotoxin A revealed by a toxin derivative that binds to ecto-acceptors and inhibits transmitter release intracellularly. J Biol Chem 268: 20838–20844

Diamond J, Mellanby J (1971) The effect of tetanus toxin in the goldfish. J Physiol (Lond) 215: 727–741

Dimpfel W, Habermann E (1973) Histoautoradiographic localization of ^{125}I-labelled tetanus toxin in rat spinal cord. Naunyn Schmiedebergs. Arch Pharmacol 280: 177–182

Dimpfel W, Neale J, Habermann E (1975) ^{125}I-labeled tetanus toxin as a neuronal marker in tissue cultures derived from embryonic CNS. Naunyn Schmiedebergs Arch Pharmacol 290: 329–333

Dobrenis K, Joseph A, Rattazzi MC (1992) Neuronal lysosomal enzyme replacement using fragment C of tetanus toxin. Proc Natl Acad Sci USA 89: 2297–2301

Donovan JJ, Middlebrook JL (1986) Ion-conducting channels produced by botulinum toxin in planar lipid membranes. Biochemistry 25: 2872–2876

Duchen LW, Tonge DA (1973) The effects of tetanus toxin on neuromuscular transmission and on the morphology of motor end-plates in slow and fast skeletal muscle of the mouse. J Physiol (Lond) 228: 157–172

Dumas M, Schwab ME, Baumann R, Thoenen H (1979a) Retrograde transport of tetanus toxin through a chain of two neurons. Brain Res 165: 354–357

Dumas M, Schwab ME, Thoenen H (1979b) Retrograde axonal transport of specific macromolecules as a tool for characterizing nerve terminal membranes. J Neurobiol 10: 179–197

Eisenbarth GS, Shimizu K, Bowring MA, Wells S (1982) Expression of receptors for tetanus toxin and monoclonal antibody A2B5 by pancreatic islet cells. Proc Natl Acad Sci USA 79: 5066–5070

Erdmann G, Wiegand H, Wellhöner HH (1975) Intraaxonal and extraaxonal transport of ^{125}I-tetanus toxin in early local tetanus. Naunyn Schmiedebergs Arch Pharmacol 290: 357–373

Erdmann G, Hanauske A, Wellhöner HH (1981) Intraspinal distribution and reaction in the grey matter with tetanus toxin of intracisternally injected anti-tetanus toxoid $F(ab')_2$ fragments. Brain Res 211: 367–377

Evans DM, Williams RS, Shone CC, Hambleton P, Melling J, Dolly JO (1986) Botulinum neurotoxin type B. Eur J Biochem 154: 409–416

Evinger C, Erichsen JT (1986) Transsynaptic retrograde transport of fragment C of tetanus toxin demonstrated by immunohistochemical localization. Brain Res 380: 383–388

Fishman PS, Carrigan DR (1987) Retrograde transneuronal transfer of the C-fragment of tetanus toxin. Brain Res 406: 275–279

Fujiwara T, Oda K, Yokota S, Takatsuki A, Ikehara Y (1988) Brefeldin A causes disassembly of the Golgi complex and accumulation of secretory proteins in the endoplasmic reticulum. J Biol Chem 263: 18545–18552

Goldberg R, Costa T, Habig W, Kohn L, Hardegree M (1981) Characterization of fragment C and tetanus toxin binding to rat brain membranes. Mol Pharmacol 20: 565–570

Habermann E (1974) ^{125}I-labeled neurotoxin from Clostridium botulinum A: preparation binding to synaptosomes and ascent to the spinal cord. Naunyn Schmiedebergs Arch Pharmacol 281: 47–56

Habermann E (1988) Inhibition by tetanus and botulinum A toxin of the release of [^{3}H]noradrenaline and [^{3}H]GABA from rat brain homogenate. Experientia 44: 224–226

Habermann E, Albus U (1986) Interaction between tetanus toxin and rabbit kidney: a comparison with rat brain preparations. J Neurochem 46: 1219–1226

Habermann E, Dreyer F (1986) Clostridial neurotoxins: handling and action at the cellular and molecular level. In: Compans RW, Cooper M, Koprowski H et al. (eds) Current topics in microbiology and immunology, Vol 129. Springer, Berlin Heidelberg New York, pp 93–179

Habermann E, Dreyer F, Bigalke H (1980) Tetanus toxins blocks the neuromuscular transmission in vitro like botulinum A toxin. Naunyn Schmiedebergs Arch Pharmacol 311: 33–40

Habermann E, Müller H, Hudel M (1988) Tetanus toxin and botulinum A and C neurotoxins inhibit noradrenaline release from cultured mouse brain. J Neurochem 51: 522–527

Habig WH, Kenimer JG, Hardegree MC (1983) Retrograde axonal transport of tetanus toxin: toxin mediated antibody transport. In: Liu T-Y, Sakakibara S, Schechter AN, Yagi K, Yajima H, Yasunobu KT (eds) Frontiers in biochemical and biophysical studies of proteins and membranes. Elsevier, New York, pp 463–473

Habig WH, Bigalke H,Bergey GK, Neale EA, Hardegree MC, Nelson PG (1986) Tetanus toxin in dissociated spinal cord cultures: long-term characterization of form and action. J Neurochem 47: 930–937

Halpern JL, Loftus A (1993) Characterization of the receptor-binding domain of tetanus toxin. J Biol Chem 268: 11188–11192

Halpern JL, Smith LA, Seamon KB, Groover KA, Habig WH (1989) Sequence homology between tetanus and botulinum toxins detected by an antipeptide antibody. Infect Immun 57: 18–22

Helting TB, Zwisler O (1977) Structure of tetanus toxin. I. Breakdown of the toxin and discrimination between polypeptide fragments. J Biol Chem 252: 187–193

Hoch DH, Romero-Mira M, Ehrlich BE, Finkelstein A, DasGupta BR, Simpson LL (1985) Channels formed by botulinum, tetanus, and diphtheria toxins in planar lipid bilayers: relevance to translocation of proteins across membranes. Proc Natl Acad Sci USA 82: 1692–1696

Högy B, Dauzenroth M-E, Hudel M, Weller U, Habermann E (1992) Increase of permeability of synaptosomes and liposomes by the heavy chain of tetanus toxin. Toxicon 30: 63–76

Holmgren J, Elwing H, Fredman P, Svennerholm L (1980) Polystyrene-absorbed gangliosides for investigation of the structure of the tetanus toxin receptor. Eur J Biochem 106: 371–379

Ikonen E, Parton RG, Hunziker W, Simons K, Dotti CG (1994) Transcytosis of the polymeric immunoglobulin receptor in cultured hippocampal neurons. Curr Biol 3:635–644

Kitamura M, Iwamori M, Nagai Y (1980) Interaction between Clostridium botulinum neurotoxin and gangliosides. Biochim Biophys Acta 628: 328–335

Kozaki S, Kamata Y, Nagai T, Ogasawara J, Sakaguchi G (1986) The use of monoclonal antibodies to analyze the structure of Clostridium botulinum type E derivative toxin. Infect Immun 52: 786–791

Kozaki S, Miki A, Kamata Y, Ogasawara J, Sakaguchi G (1989) Immunological characterization of papain-induced fragments of Clostridium botulinum type A neurotoxin and interaction of the fragments with brain synaptosomes. Infect Immun 57: 2634–2639

Kurokawa Y, Oguma K, Yokosawa N, Syuto B, Fukatsu R, Yamashita I (1987) Binding and cytotoxic effects of Clostridium botulinum type A, C_1 and E toxins in primary neuron cultures from foetal mouse brains. J Gen Microbiol 133: 2647–2657

Lebeda FJ, Olson MA (1995) Structural predictions of the channel-forming region of botulinum neurotoxin heavy chain. Toxicon (in press)

Lebeda FJ, Hack DC, Gentry MK (1994) Theoretical analyses of the functional regions of the heavy chain of botulinum neurotoxin. In: Jankovic J, Hallett M (eds) Therapy with botulinum toxin. Dekker, New York, pp 51–61

Ledley FD, Lee G, Kohn LD, Habig WH, Hardegree MC (1977) Tetanus toxin interactions with thyroid plasma membranes. J Biol Chem 252: 4049–4055

Leppla S, Dorland RB, Middlebrook JL (1980) Inhibition of diphtheria toxin degradation and cytotoxic action by chloroquine. J Biol Chem 255: 2247–2250

Lippincott-Schwartz J, Yuan LC, Bonifacino JS, Klausner RD (1989) Rapid redistribution of Golgi proteins into the ER in cells treated with brefeldin A: evidence for membrane cycling from Golgi to ER. Cell 56: 801–813

Manning KA, Erichsen JT, Evinger C (1990) Retrograde transneuronal transport properties of fragment C of tetanus toxin. Neuroscience 34: 251–263

Marnell MH, Stookey M, Draper RK (1982) Monensin blocks the transport of diphtheria toxin to the cell cytoplasm. J Cell Biol 93: 57–62

Marxen P, Fuhrmann U, Bigalke H (1989) Gangliosides mediate inhibitory effects of tetanus and botulinum A neurotoxins on exocytosis in chromaffin cells. Toxicon 27: 849–859

Mellanby J (1984) Comparative activities of tetanus and botulinum toxins. Neuroscience 11: 29–34

Menestrina G, Forti S, Gambale F (1989) Interaction of tetanus toxin with lipid vesicles. Effects of pH, surface charge, and transmembrane potential on the kinetics of channel formation. Biophys J 55: 393–405

Mochida S, Poulain B, Weller U, Habermann E, Tauc L (1989) Light chain of tetanus toxin intracellularly inhibits acetylcholine release at neuro-neuronal synapses, and its internalization is mediated by heavy chain. FEBS Lett 253: 47-51

Montal MS, Blewitt R, Tomich JM, Montal M (1992) Identification of an ion channel-forming motif in the primary structure of tetanus and botulinum neurotoxins. FEBS Lett 313: 12–18

Montecucco C (1986) How do tetanus and botulinum toxins bind to neuronal membranes? Trends Biochem Sci 11: 314–317

Montecucco C, Papini E, Schiavo G (1991) Molecular models of toxin membrane translocation. In: Alouf JE, Freer JH (eds) Sourcebook of bacterial protein toxins. Academic, London, pp 45–56

Montecucco C, Papini E, Schiavo G, Padovan E, Rossetto O (1992) Ion channel and membrane translocation of diphtheria toxin. FEMS Microbiol Immunol 105: 101–112

Montesano R, Roth J, Robert A, Orci L (1982) Non-coated invaginations are involved in binding and internalization of cholera and tetanus toxin. Nature 296: 651–653

Morris NP, Consiglio E, Kohn LD, Habig WH, Hardegree MC, Helting TB (1980) Interaction of fragments B and C of tetanus toxin with neural and thyroid membranes and with gangliosides. J Biol Chem 255: 6071–6076

Neale EA, Habig WH, Schrier BK, Bergey GK, Bowers LM, Koh J (1989) Applications of tetanus toxin for structure-function studies in neuronal cell cultures. In: Nisticó G, Bizzini B, Bytchenko B, Triau R (eds) 8th international conference on tetanus. Pythagora, Rome, pp 58–65
Neville DM, Chang TM (1978) Receptor-mediated protein transport into cells. Entry mechanisms for toxins, hormones, antibodies, viruses, lysosomal hydrolases, asialoglycoproteins, and carrier proteins. Curr Top Membr Transport 10: 65–150
Niemann H (1991) Molecular biology of clostridial neurotoxins . In: Alouf JE, Freer JH (eds) Sourcebook of bacterial protein toxins. Academic, London, pp 303–348
Nishiki T-i, Kamata Y, Nenoto Y, Omori A, Ito T, Takahashi M, Kozaki S (1994) Identification of protein receptor for Clostridium botulinum type B neurotoxin in rat brain synaptosomes. J Biol Chem 269: 10498–10503
Ochanda JO, Syuto B, Ohishi I, Naiki M, Kubo S (1986) Binding of Clostridium botulinum neurotoxin to gangliosides. J Biochem 100: 27–33
Ogasawara J, Kamata Y, Sakaguchi G, Koraki S (1991) Properties of a protease-sensitive acceptor component in mouse brain synaptosomes for Clostridium botulinum type B neurotoxin. FEMS Microbiol Lett 79: 351–356
Orlandi PA, Curran PK, Fishman PH (1993) Brefeldin A blocks the response of cultured cells to cholera toxin. Implications for intracellular trafficking in toxin action. J Biol Chem 268: 12010–12016
Papini E, Rappuoli R, Murgia M, Montecucco C (1993) Cell penetration of diphtheria toxin. Reduction of the interchain disulfide bridge is the rate-limiting step of translocation in the cytosol. J Biol Chem 268: 1567–1574
Parfitt A, Neale EA, Fitzgerald SC, Nelson PG (1994) Homonymous neuromuscular preparation using the Campenot chamber. Soc Neurosci Abstr 20: 1089
Parton RG, Ockleford CD, Critchley DR (1987) A study of the mechanism of internalisation of tetanus toxin by primary mouse spinal cord cultures. J Neurochem 49: 1057–1068
Parton RG, Ockleford CD, Critchley DR (1988) Tetanus toxin binding to mouse spinal cord cells: an evaluation of the role of gangliosides in toxin internalization. Brain Res 475: 118–127
Paschal BM, Vallee RB (1987) Retrograde transport by the microtubule-associated protein MAP 1C. Nature 330: 181–183
Paschal BM, Shpetner HS, Valle RB (1987) MAP 1C is a microtubule-activated ATPase that translocates microtubules in vitro and has dynein-like properties. J Cell Biol 105: 1273–1282
Penner R, Neher E, Dreyer F (1986) Intracellularly injected tetanus toxin inhibits exocytosis in bovine adrenal chromaffin cells. Nature 324: 76–78
Pierce EJ, Davison MD, Parton RG, Habig WH, Critchley DR (1986) Characterization of tetanus toxin binding to rat brain membranes. Biochem J 236: 845–852
Poulain B, Mochida S, Weller U, Högy B, Habermann E, Wadsworth JDF, Shone CC, Dolly JO, Tauc L (1991) Heterologous combinations of heavy and light chains from botulinum neurotoxin A and tetanus toxin inhibit neurotransmitter release in Aplysia. J Biol Chem 266: 9580–9585
Price DL, Griffin J, Young A, Peck K, Stocks A (1975) Tetanus toxin: direct evidence for retrograde intraaxonal transport. Science 188: 945–947
Price DL, Griffin JW, Peck K (1977) Tetanus toxin: evidence for binding at presynaptic nerve endings. Brain Res 121: 379–384
Rabasseda X, Blasi J, Marsal J, Dunant Y, Casanova A, Bizzini B (1988) Tetanus and botulinum toxins block the release of acetylcholine from slices of rat striatum and from the isolated electric organs of Torpedo at different concentrations. Toxicon 26: 329–336
Rauch G, Gambale F, Montal M (1990) Tetanus toxin channel in phosphatidylserine planar bilayers: conductance states and pH dependence. Eur Biophys J 18: 79–83
Roa M, Boquet P (1985) Interaction of tetanus toxin with lipid vesicles at low pH. J Biol Chem 260: 6827–6835
Rogers TB, Snyder SH (1981) High-affinity binding of tetanus toxin to mammalian brain membranes. J Biol Chem 256: 2402–2407
Sanchez-Prieto J, Sihra TS, Evans D, Ashton A, Dolly JO, Nicholls DG (1987) Botulinum toxin A blocks glutamate exocytosis from guinea-pig cerebral cortical synaptosomes. Eur J Biochem 165: 675–681
Sandvig K, Olsnes S (1991) Membrane translocation of diphtheria toxin. In: Alouf JE, Freer JH (eds) Sourcebook of bacterial protein toxins. Academic, London, pp 57–73
Sansom M (1991) The biophysics of peptide models of ion channels. Prog Biophys Mol Biol 55: 139–235
Schengrund C-L, DasGupta BR, Ringler NJ (1991) Binding of botulinum and tetanus neurotoxins to ganglioside GT1b and derivatives thereof. J Neurochem 57: 1024–1032
Schiavo G, Papini E, Genna G, Montecucco C (1990) An intact interchain disulfide bond is required for the neurotoxicity of tetanus toxin. Infect Immun 58: 4136–4141

Schiavo G, Ferrari G, Rossetto O, Montecucco C (1991) Tetanus toxin receptor. Specific cross-linking of tetanus toxin to a protein of NGF-differentiated PC 12 cells. FEBS Lett 290: 227–230

Schiavo G, Benfenati F, Poulain B, Rossetto O, Polverino de Laureto P, DasGupta BR, Montecucco C (1992) Tetanus and botulinum-B neurotoxins block neurotransmitter release by proteolytic cleavage of synaptobrevin. Nature 359: 832–835

Schmid SL, Fuchs R, Kielian M, Helenius A, Mellman R (1989) Acidification of endosome subpopulations in wild-type Chinese hamster ovary cells and temperature-sensitive acidification-defective mutants. J Cell Biol 108: 1291–1300

Schmid MF, Robinson JP, DasGupta BR (1993) Direct visualization of botulinum neurotoxin-induced channels in phospholipid vesicles. Nature 364: 827–830

Schmitt A, Dreyer F, John C (1981) At least three sequential steps are involved in the tetanus toxin-induced block of neuromuscular transmission. Naunyn Schmiedebergs Arch Pharmacol 317: 326–330

Schwab ME, Thoenen H (1976) Electron microscopic evidence for a transsynaptic migration of tetanus toxin in spinal cord motorneurons: an autoradiographic and morphometric study. Brain Res 105: 213–227

Schwab ME, Suda K, Thoenen H (1979) Selective retrograde transsynaptic transfer of a protein, tetanus toxin, subsequent to its retrograde axonal transport. J Cell Biol 82: 798–810

Sheridan RE (1993) Proton ionophores delay onset of botulinum paralysis in isolated mouse skeletal muscle. Soc Neurosci Abstr 19: 1125

Shone CC, Hambleton P, Melling J (1985) Inactivation of Clostridium botulinum type A neurotoxin by trypsin and purification of two tryptic fragments. Eur J Biochem 151: 75-82

Shone CC, Hambleton P, Melling J (1987) A 50-kDa fragment from the NH_2-terminus of the heavy subunit of Clostridium botulinum type A neurotoxin forms channels in lipid vesicles. Eur J Biochem 167: 175–180

Silverman JA, Mindell JA, Zhan H, Finkelstein A, Collier RJ (1994) Structure-function relationships in diphtheria toxin channels. I. Determining a minimal channel-forming domain. J Membr Biol 137: 17–28

Simon SM, Blobel G (1991) A protein-conducting channel in the endoplasmic reticulum. Cell 65: 371–380

Simpson LL (1980) Kinetic studies of the interaction between botulinum toxin type A and the cholinergic neuromuscular junction. J Pharmacol Exp Ther 212: 16–21

Simpson LL (1981) The origin, structure and pharmacological activity of botulinum toxin. Pharmacol Rev 33: 155–188

Simpson LL (1982) The interaction between aminoquinolines and presynaptically acting neurotoxins. J Pharmacol Exp Ther 222. 34–48

Simpson LL (1983) Ammonium chloride and methylamine hydrochloride antagonize clostridial neurotoxins. J Pharmacol Exp Ther 225: 546–552

Simpson LL, Coffield JA, Bakry N (1994) Inhibition of vacuolar adenosine triphosphatase antagonizes the effects of clostridial neurotoxins but not phospholipase A2 neurotoxins. J Pharmacol Exp Ther 269: 256–262

Simpson LL, Rapport MM (1971) Ganglioside inactivation of botulinum toxin. J Neurochem 18: 1341–1343

Staub GC, Walton KM, Schnaar RL, Nichols T, Baichwal R, Sandberg K, Rogers TB (1986) Characterization of the binding and internalization of tetanus toxin in a neuroblastoma hybrid cell line. J Neurosci 6: 1443–1451

Stöckel K, Schwab M, Thoenen H (1975) Comparison between the retrograde axonal transport of nerve growth factor and tetanus toxin in motor, sensory and adrenergic neurons. Brain Res. 99: 1–16

Tsukita S, Ishikawa H (1980) The movement of membranous organelles in axons. J Cell Biol 84: 513–530

Tsuzuki K, Yokosawa N, Syuto B, Ohishi I, Fujii N, Kimura K, Oguma K (1988) Establishment of a monoclonal antibody recognizing an antigenic site common to Clostridium botulinum types B, C_1, D, and E toxins and tetanus toxin. Infect Immun 56: 898–902

Vale RD, Reese TS, Sheetz MP (1985) Identification of a novel force-generating protein, kinesin, involved in microtubule-based motility. Cell 42: 39–50

Vallee RB, Bloom GS (1991) Mechanisms of fast and slow axonal transport. Annu Rev Neurosci 14: 59–92

van Heyningen WE (1974) Gangliosides as receptors for tetanus toxin, cholera toxin and serotonin. Nature 249: 415–417

van Heyningen WE, Miller P (1961) The fixation of tetanus toxin by ganglioside. J Gen Microbiol 24: 107–119
vanderSpek JC, Mindell JA, Finkelstein A, Murphy JR (1993) Structure/function analysis of the transmembrane domain of DAB_{389}-interleukin-2, an interleukin-2 receptor-targeted fusion toxin. J Biol Chem 268: 12077–12082
Walton KM, Sandberg K, Rogers TB, Schnaar RL (1988) Complex ganglioside expression and tetanus toxin binding by PC12 pheochromocytoma cells. J Biol Chem 263: 2055–2063
Weigand H, Erdmann G, Wellhöner HH (1976) ^{125}I-labelled botulinum A neurotoxin: pharmacokinetics in cats after intramuscular injection. Naunyn Schmiedebergs Arch Pharmacol 292: 161–165
Weller U, Taylor CF, Habermann E (1986) Quantitative comparison between tetanus toxin, some fragments, and toxoid for binding and axonal transport in the rat. Toxicon 24: 1055–1063
Weller U, Dauzenroth M-E, Gansel M, Dreyer F (1991) Cooperative action of the light chain of tetanus toxin and the heavy chain of botulinum toxin type A on the transmitter release of mammalian motor endplates. Neurosci Lett 122: 132–134
Wellhöner HH (1992) Tetanus and botulinum neurotoxins. In: Herken H, Hucho F (eds) Handbook of experimental pharmacology. Springer, Berlin Heidelberg New York, pp 357–417
Wellhöner HH, Neville D Jr (1987) Tetanus toxin binds with high affinity to neuroblastoma X glioma hybrid cells NG 108–15 and impairs their stimulated acetylcholine release. J Biol Chem 262: 17374–17378
Williams RS, Tse C-K, Dolly JO, Hambleton P, Melling J (1983) Radioiodination of botulinum neurotoxin type A with retention of biological activity and its binding to brain synaptosomes. Eur J Biochem 131: 437–445
Williamson LC, Neale EA (1994a) Tetanus toxin enters neurons through acidic endosomes: studies in spinal cord cell cultures. In: Freer J, Alouf JE (eds) Bacterial protein toxins. Fischer, Stuttgart, pp 410–411
Williamson LC, Neale EA (1994b) Bafilomycin A1 inhibits the action of tetanus toxin in spinal cord neruons in cell culture. J Neurochem 63: 2342–2345
Williamson LC, Brown JE, Neale EA (1995) Differential effects of the seven serotypes of botulinum neurotoxin on mammalian central synapses. Soc Neurosci Abstr (in press)
Williamson LC, Clarke WY, Fitzgerald SC, Neale EA (1992a) Tetanus toxin enters neurons through acidic endosomes. Soc Neurosci Abstr 18: 1601
Williamson LC, Fitzgerald SC, Neale EA (1992b) Differential effects of tetanus toxin on inhibitory and excitatory neurotransmitter release from mammalian spinal cord cells in culture. J Neurochem 59: 2148–2157
Williamson LC, Fitzgerald SC, Neale EA (1993) Bafilomycin A1, a specific-vacuolar type H^+–ATPase inhibitor, blocks the action of tetanus toxin in spinal cord neurons in cell culture. Mol Biol Cell 4: 325a
Yavin E, Nathan A (1986) Tetanus toxin receptors on nerve cells contain a trypsin-sensitive component. Eur J Biochem 154: 403–407
Yoshida T, Chen CC, Zhang MS, Wu HC (1991) Disruption of the Golgi apparatus by brefeldin A inhibits the cytotoxicity of ricin, modeccin, and Pseudomonas toxin. Exp Cell Res 192: 389–395

Quantal Neurotransmitter Release and the Clostridial Neurotoxins' Targets

B. POULAIN[1], J. MOLGÓ[1], and S. THESLEFF[2]

1 Introduction

The eight clostridial neurotoxins so far known, tetanus toxin (TeNT) and botulinum neurotoxins (BoNTs) types A–G, have been extensively studied, not only because they are responsible for the clinical manifestations of tetanus and botulism, but also because they have the unique ability to effectively block neurotransmitter release from nerve terminals in a long-lasting manner (for reviews see MELLANBY and GREEN 1981; SIMPSON 1986; HABERMANN and DREYER 1986; SELLIN 1987; DREYER 1989; NIEMANN 1991; POULAIN and MOLGÓ 1992; DOLLY 1992; WELLHÖNER 1992).

BoNTs and TeNT act on nerve terminals through a sequence of three main steps (reviewed by SIMPSON 1986). There is an initial binding step during which the neurotoxins attach to specific acceptors located on presynaptic membranes. The subsequent step involves the internalization and translocation of the toxins into the cytosol of the nerve terminal. The final step results in the disruption of the transmitter release process. Detailed analysis of the binding and internalization processes of BoNTs and TeNT in nerve terminals as well as identification of the toxin-molecular domain(s) implicated in these action have been reviewed (see NIEMANN 1991; WELLHÖNER 1992; DOLLY 1992; POULAIN et al. 1993a; Halpern and

[1]Laboratoire de Neurobiologie Cellulaire et Moléculaire, Centre National de la Recherche Scientifique, 91198-Gif sur Yvette cedex, France
[2]Department of Pharmacology, University of Lund, Solvegatan 10, Lund 223 62, Sweden

Neale, this volume). To summarize, the heavy chain of TeNT and BoNTs (approx. 100 kDa) is responsible for binding and internalization, whereas the light chain (approx. 50 kDa) is implicated in the intracellular impairment of neurotransmitter release.

In vitro TeNT has been shown to bind and activate transglutaminase a Ca^{2+} -dependent enzyme that can form stable cross-links between substrate proteins (Facchiano and Luini 1992). Further studies revealed that transglutaminase present in nerve terminals recognizes synapsin I, a major protein of synaptic vesicle membranes (Facchiano et al. 1993). It has been proposed that TeNT, by stimulating transglutaminase, covalently cross-links synapsin I to microfilaments. This in turn prevents fusion of synaptic vesicles with the nerve terminal membrane (see Benfenati and Valtorta, this volume).

Recent findings have definitively established that TeNT and BoNTs act as Zn-proteases (Schiavo et al. 1992a, b; De Paiva et al. 1993; Simpson et al. 1993; Schiavo et al., this volume). The target of TeNT, BoNT/B, /D, /F, and /G is VAMP/synaptobrevin (Schiavo et al. 1992a, b; 1993a, b, c; 1994; Link et al. 1992; Poulain et al. 1993b; Yamasaki et al. 1994). BoNT/A and /E cleave SNAP-25 (Blasi et al. 1993a; Schiavo et al. 1993b, c; Binz et al. 1994), and the target of BoNT/C1 is HPC1/syntaxin (Blasi et al. 1993b). These results suggest that all three nerve terminal proteins are directly involved in neurotransmitter release.

This chapter analyzes the modifications caused on quantal transmitter release by BoNTs and TeNT revealed by electrophysiological techniques which allow a fast-time resolution of the events involved in synaptic transmission and neurotransmitter release. Since analysis of neurotransmission can be easily performed at vertebrate skeletal neuromuscular junctions, most of the data reviewed below are from studies involving these model synapses. In addition, we try to correlate some characteristic features of the neuronal action of TeNT and BoNTs with the cleavage of the synaptic proteins targeted by these neurotoxins with the final goal of understanding elementary processes involved in neurotransmission.

2 Quantal Aspects of Neurotransmission as Revealed by Electrophysiological Methods

When recording electrical activity at a single vertebrate neuromuscular junction (NMJ) with an intracellular microelectrode, one observes spontaneous quantal acetylcholine (ACh) release manifested as miniature endplate potential (MEPP; Katz 1969; for instance, see Fig. 1c). Each MEPP is the postsynaptic response resulting from the presynaptic release of a multimolecular packet, or quantum, of ACh contained in a single small clear synaptic vesicle. Occasionally large spontaneous potentials (Fig. 1c), so-called "giant" miniature endplate potentials (G-MEPPs), have also been observed (Liley 1957). Under normal conditions when a nerve terminal is stimulated by nerve impulses, there is an influx of Ca^{2+} which

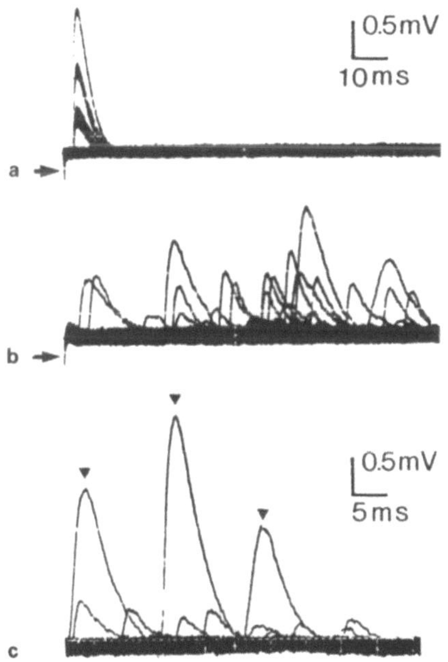

Fig. 1. Intracellular recordings obtained from two different rat nerve extensor digitorum longus muscle preparations poisoned in vivo with BoNT/A (**a**) or BoNT/D (**b**, **c**) and in the presence of 8 m*M* Ca^{2+}. **a** Oscilloscope tracings showing superimposed endplate potentials and failures of release recorded following nerve stimulation at 1 Hz. Notice that a great number of nerve stimuli induced only failures of release, while EPPs of various quantal contents (1, 2, or 3 quanta) were synchronized after nerve stimulation (*arrow*, stimulus artifact). **b** Superimposed tracings showing asynchronous evoked MEPPs in response to nerve stimulation (1 Hz). Notice the temporal dispersion of events appearing after the nerve stimuli (*arrow*, stimulus artifact). **c** Superimposed oscilloscope tracings showing the presence of spontaneous MEPPs and the so-called G-MEPPs (*inverted triangles*). Same calibration as in **b**. (Modified from Molgó et al. 1989a, 1990)

in less than 500 µs triggers the synchronous release of hundreds of quanta giving rise to a phasic post-synaptic response called, at the NMJ, the endplate potential (EPP).

3 Changes in Nerve-Impulse Evoked Quantal Release Caused by Clostridial Neurotoxins

BoNTs and TeNT block neurotransmitter release by a mechanism which apparently affects neither transmitter synthesis and storage (Gundersen 1980) nor the propagation of the nerve action potential to the terminal and the ensuing Ca^{2+} influx across the presynaptic membrane (Gundersen et al. 1982; Dreyer et al. 1983; Mallart et al. 1989; Molgó et al. 1989a). In addition, neither obvious structural abnormalities nor degenerative changes are observed at motor nerve terminals examined at various times after TeNT or BoNTs poisoning (see Dreyer et al. 1987; Molgó et al. 1990). In fact, the number of small clear synaptic vesicles docked at the presynaptic membrane appeared to be slightly elevated (Mellanby et al. 1988; Neale et al. 1989; Pécot-Dechavassine et al. 1991).

When BoNTs or TeNT are administered either in vivo or in vitro they markedly reduce nerve-impulse evoked quantal neurotransmitter release. At the NMJ this is manifested by the disappearance of the EPP or a reduction of its amplitude so that a single or only a few quanta of ACh are released (Harris and Miledi 1971; Mellanby and Thompson 1972; Duchen and Tonge 1973; Boroff et al. 1974; Cull-Candy et al. 1976; Dreyer and Schmitt 1981, 1983; Sellin et al. 1983;Bevan and Wendon 1984; Kauffman et al. 1985; Gansel et al. 1987; Molgó et al. 1989a, b). Similar observations have been made at inhibitory or excitatory vertebrate central synapses (Kanda and Takano 1983), mouse spinal cord neurones in culture (Bigalke et al. 1985) and *Aplysia* cholinergic and non-cholinergic neuro-neuronal synapses (Poulain et al. 1988; Mochida et al. 1989). It is worth noting that at NMJs poisoned in vitro evoked monoquantal EPPs are of similar amplitude to those of spontaneous MEPPs recorded in untreated preparations (Dreyer and Schmitt 1983; Dolly et al. 1987). At neuro-neuronal synapses of *Aplysia*, noise analysis of long duration postsynaptic responses revealed no change in quantal size during the action of BoNT/A (Poulain et al. 1988). These findings indicate that the decrease in neurotransmitter release is due to a reduction in the number of quanta released by the nerve impulse and not to a decreased amount of transmitter per quantum liberated.

Some important differences were detected when studying the blockade of quantal release by the different serotypes of clostridial neurotoxins (Table 1). In contrast to the results observed with BoNT/A (Fig. 1a) or /E, the synchronization of evoked release by a nerve impulse is greatly altered by TeNT, BoNT/B or /D, as detected by the temporal dispersion of the quanta released (Fig. 1b) after the arrival of the action potential at the terminal (Harris and Miledi 1971; Dreyer and Schmitt 1983; Bevan and Wendon 1984; Gansel et al. 1987; Molgó et al. 1989a).

The low probability of release from BoNT- or TeNT-treated terminals may be enhanced by increasing extracellular [Ca^{2+}], by increasing the frequency of nerve stimulation, by blockade of presynaptic potassium channels, or by reducing the temperature of the preparation (Boroff et al. 1974; Cull-Candy et al. 1976; Dreyer

Table 1. Quantal acetylcholine release evoked by nerve stimulation at neuromuscular junctions poisoned with clostridial neurotoxins and their identified molecular targets

Neurotoxin	Quantal release evoked by nerve impulses	Synchronous release after aminopyridines	Target proteins
TeNT	Asynchronous [1,2]	No [1,2]	VAMP/Synaptobrevin [3,4,5]
BoNT/B	Asynchronous [2,6]	No [2,6]	VAMP/Synaptobrevin [4]
BoNT/D	Asynchronous [7,8]	No [8]	VAMP/Synaptobrevin [9]
BoNT/F	Asynchronous [10]	No [10]	VAMP/Synaptobrevin [11]
BoNT/G	Not known	Not known	VAMP/Synaptobrevin [12]
BoNT/A	Synchronous [13,14]	Yes [1,2,14]	SNAP-25 [9,15,16]
BoNT/E	Synchronous [17]	Yes [17]	SNAP-25 [9,16]
BoNT/C1	Not known	Not known	Syntaxin [18]

Numbers in brackets indicate references quoted:
[1] Dreyer and Schmitt (1983); [2] Gansel et al. (1987); [3–4] Schiavo et al. (1992a,b); [5] Link et al. (1992); [6] Sellin et al. (1983); [7] Harris and Miledi (1971); [8] Molgó et al. (1989a); [9] Schiavo et al. (1993b); [10] Kauffman et al. (1985); [11] Schiavo et al. (1993a); [12] Schiavo et al. (1994); Yamasaki et al. (1994); [13] Cull-Candy et al. (1976); [14] Molgó et al. (1980); [15] Blasi et al. (1993a); [16] Schiavo et al. (1993c); Binz et al. (1994); [17] Molgó et al. (1989b); [18] Blasi et al. (1993b). For further details, see text and review by Schiavo et al. (this volume) or by Niemann et al. (1994).

and SCHMITT 1981, 1983; LUNDH 1983; MOLGÓ and THESLEFF 1984; GANSEL et al. 1987). However, the resultant increase in ACh release may differ with toxin serotypes and treatments used. This is particularly evident with aminopyridines. These drugs block fast potassium currents in nerve terminals and thus enhance phasic Ca^{2+} influx and thereby increase evoked quantal neurotransmitter release. Aminopyridines were potent antagonists of the neuromuscular block produced by BoNT/A or /E, provided ACh release is not completely abolished. Notably quantal release remained phasic and synchronized (LUNDH et al. 1977; MOLGÓ et al. 1980; 1989a,b; GANSEL et al. 1987). At TeNT, BoNT/B, /D or /F poisoned nerve terminals, similar treatments with aminopyridines also increased the probability of quantal release. However, in marked contrast to BoNT/A or /E, the release observed with TeNT and BoNT /B, /D, /F after aminopyridine treatment still exhibited a high temporal dispersion of evoked MEPPs after the arrival of an action potential to the nerve terminal. Thus, with these neurotoxins even when Ca^{2+} influx is enhanced by aminopyridines the asynchrony of release prevents the build-up of EPPs and thereby recovery of transmission (DREYER and SCHMITT 1981, 1983; SELLIN et al. 1983; KAUFFMAN et al. 1985; GANSEL et al. 1987; MOLGÓ et al. 1989a).

As highlighted in Table 1, there is a clear correlation between the toxin-induced modifications of quantal release, the ability of aminopyridines to counteract toxin effects, and the identity of the toxins' targets. Thus, cleavage of SNAP-25 by BoNT/A or /E seems to inhibit quantal release without altering synchronization, whereas, cleavage of VAMP/synaptobrevin by TeNT, BoNT/B, /D or /F not only blocks quantal release but also strongly affects synchronization. It remains to be determined whether cleavage of HPC1/syntaxin by BoNT/C1 affects quantal synchronization in response to nerve impulses.

4 Alterations in Spontaneous Quantal Release Caused by Clostridial Neurotoxins

Following in vitro or in vivo administration of various BoNTs or TeNT, the frequency of spontaneous MEPPs is strongly reduced but not completely abolished (Table 2; see reviews by THESLEFF 1986; HABERMANN and DREYER 1986; SELLIN 1987; MOLGÓ et al. 1990). Although comparative studies with equimolar concentrations of BoNTs or of TeNT have not been performed, it appears that MEPP frequency is less affected by TeNT, BoNT/B, /D, or /F than by BoNT/A or /E (HARRIS and MILEDI 1971; SELLIN et al. 1983; GANSEL et al. 1987; MOLGÓ et al. 1989a,b). However, it is not clear whether these discrepancies are related to toxin doses, the animal species used, or to the different target proteins cleaved by the toxins.

The spontaneous MEPPs recorded after in vivo poisoning with clostridial neurotoxins or after in vitro administration of the neurotoxins have, in general, a lower amplitude and slower time course than normal MEPPs (HARRIS and MILEDI 1971; DUCHEN and TONGE 1973; CULL-CANDY et al. 1976; KRIEBEL et al. 1976; DREYER

Table 2. Alterations of spontaneous quantal acetylcholine release at neuromuscular junctions poisoned with clostridial neurotoxins

Neurotoxin serotype	Target	Frequency of MEPPs	MEPPs frequency after α-LTX	Frequency of G-MEPPs
TeNT	VAMP/ Synaptobrevin[a]	Decreased [1,2]	Moderately increased [3]	–
BoNT/B		Decreased [2,4]	Moderately increased [2]	Increased [4]
BoNT/D		Decreased [5,6]	Moderately increased [6]	Increased [7]
BoNT/F		Decreased [8]	–	–
BoNT/G		–	–	–
BoNT/A	SNAP-25[a]	Strongly decreased [1,2,4, 9,10]	Strongly increased up to depletion of synaptic vesicles [3,9, 11]	Increased [12]
BoNT/E		Strongly decreased [13]	–	–
BoNT/C1	Syntaxin[a]	–	–	–

Numbers in brackets indicate references quoted:
[1] Dreyer and Schmitt (1983); [2] Gansel et al. (1987); [3] Dreyer et al. (1987); [4] Sellin et al. (1983); [5] Harris and Miledi (1971); [6] Molgó et al. (1989a); [7] Molgó et al. (1990); [8] Kauffman et al. (1985); [9] Cull-Candy et al. (1976); [10] Dolly et al. (1987); [11] Kao et al. (1976); [12] Kim et al. (1984); [13] Molgó et al. (1989b). For further details, see text.
[a]For references see Table 1.

and Schmitt 1981, 1983; Sellin et al. 1983; Thesleff et al.1983; Kim et al. 1984; Kauffman et al. 1985; Dolly et al. 1987; Molgó et al.1989a; Dunant et al. 1987). It has been suggested that either MEPP amplitudes become progressively smaller following toxin administration (Harris and Miledi 1971) or that the MEPPs of normal amplitude disappear leaving behind a small-amplitude MEPP population unaffected by the toxins. These so-called sub-MEPPs have been proposed to make-up the normally sized MEPPs (Kriebel et al. 1976). However, this view has been disputed (see Van der Kloot 1989). It is not yet clear whether the proportion of small-amplitude MEPPs, relative to that of normal sized MEPPs, remains unaffected (TeNT, Dreyer and Schmitt 1983), is diminished (BoNT/A, Dreyer and Schmitt 1983) or is increased while spontaneous normally sized MEPPs disappear (BoNT/A, Dunant et al. 1987). In the absence of comparative studies between the different serotypes of clostridial neurotoxins, it seems impossible, at present, to correlate such observations with specific toxin-induced cleavage of a given presynaptic protein.

4.1 Effects of Ca^{2+}, La^{3+}, and α-Latrotoxin

The spontaneous quantal release from TeNT or BoNTs poisoned motor terminals is relatively insensitive to treatments that enhance the intraterminal Ca^{2+} concentration (increases in external K^+ and Ca^{2+} concentration, actions of calcium ionophores A-23187 or X-537, inhibition of the Na^+/K^+ ATPase by ouabain, administration of metabolic inhibitors (reviewed in Dreyer 1989; Molgó et al. 1990).

Collectively these findings indicate that the release process in poisoned terminals exhibits a reduced sensitivity to intraterminal Ca^{2+} concentrations. La^{3+} and Cd^{2+}, which within minutes cause a large and sustained increase in MEPP frequency at normal junctions, are by far less effective in promoting quantal release from terminals poisoned with TeNT, BoNT/A, /D and E (the only toxin types studied; MELLANBY et al. 1988; MOLGÓ et al. 1990). Accordingly, no striking depletion of clear synaptic vesicles was seen contrary to the situation found at unpoisoned junctions (MELLANBY et al. 1988; MOLGÓ et al. 1990).

Black widow spider venom or its major component, α-latrotoxin (α-LTX), induces an asynchronous release of ACh quanta within minutes of administration to vertebrate NMJs poisoned with TeNT, BoNT/A, /B and /D (CULL-CANDY et al. 1976; KAO et al. 1976; DREYER et al. 1987; MOLGÓ et al. 1989a, GANSEL et al. 1987; Table 2). However, black widow spider venom was found to be much more effective in stimulating quantal release at BoNT/A-poisoned nerve terminals as compared to BoNT/B or TeNT-treated ones (DREYER et al. 1987; GANSEL et al. 1987). It is worth noting that while black widow spider venom causes a marked depletion of synaptic vesicles at BoNT/A-blocked nerve terminals, no such action has ever been observed with TeNT (DREYER et al. 1987). It is known that synaptotagmin, a synaptic vesicle protein, interacts both with the receptor of α-LTX (neurexin Iα; HATA et al. 1993) and with a molecular complex comprised of VAMP/synaptobrevin, SNAP-25, and HPC1/syntaxin (SÖLLNER et al. 1993). Thus, the above findings indicate that the blockade of release ensuing from SNAP- 25 cleavage by BoNT/A can be actively bypassed by α-LTX induced clustering of synaptotagmin with neurexins. This does not seem to be the case when VAMP/synaptobrevin is cleaved (for instance by either TeNT or BoNT/B). Further studies on the ability of α-LTX to evoke quantal release in BoNTs or TeNT treated nerve terminals are needed for better understanding the protein-protein interactions in the release process.

5 Giant MEPPs Are Not Inhibited by Clostridial Neurotoxins

Analysis of the amplitude distribution of spontaneous MEPPs at unpoisoned normal rat, mouse, and frog NMJs revealed that a small proportion of events have an amplitude exceeding twice the modal amplitude of MEPPs; these are called G-MEPPs (LILEY 1957; KIM et al. 1984; THESLEFF 1986; THESLEFF et al. 1990). The relative proportion of G-MEPPs is about 1%–3% of the total number of events. The amplitude of a G-MEPP may exceed 10 mV and is therefore sufficiently large to trigger an action potential in the muscle fiber. The time-to-peak of a G-MEPP is generally longer than that of MEPPs, and they have therefore also been called slow MEPPs. Amplitude histograms of G-MEPPs reveal no peaks corresponding to multiples of MEPPs, suggesting that they result from the release of a single large packet of ACh and not from multiple packets.

In contrast to the blockade of EPPs and the strong diminution in the occurrence of MEPPs caused by clostridial neurotoxins, the frequency of G-MEPPs is unaffected (note that only BoNT/A, /B, /D were studied); in fact, during long-lasting paralysis in vivo, G-MEPP frequency is increased (Fig. 1, Table 2; SELLIN et al. 1983; KIM et al. 1984; THESLEFF 1986; MOLGÓ et al. 1990). Upon recovery from BoNT-induced paralysis, the frequency of G-MEPPs decreases to the low level seen at unpoisoned junctions while MEPP occurrence reaches near normal values (THESLEFF et al. 1983). The observation that only MEPPs and not G-MEPPs are affected by various BoNTs indicates that part of the set of proteins underlying these two types of neurosecretion may be different. That G-MEPPs are observed after BoNT/A and after BoNT/B or /D treatments means that the respective targets of these toxins, SNAP-25 and VAMP/synaptobrevin, are not implicated in the release of ACh giving rise to G-MEPPs. It should be emphasized that the enhancement of G-MEPP frequency observed at BoNTs-paralyzed muscles is not a direct effect of the toxins but is presumably secondary to the prolonged synaptic inactivity. Indeed, an increased frequency of G-MEPPs is also observed during prolonged muscle paralysis caused by tubocurarine or tetrodotoxin and at regenerated NMJs (reviewed by THESLEFF et al. 1990).

It is therefore of interest to examine some of the properties of G-MEPPs as compared to those of MEPPs. As with MEPPs, G-MEPPs are produced by ACh since they are both blocked by tubocurarine and α-bungarotoxin. In addition, cholinesterase inhibitors increase their amplitude and prolong their duration. The ACh originates from the motor nerve terminal since denervation resulting in the degeneration of the terminals abolishes both MEPPs and G-MEPPs (reviewed by THESLEFF 1986). Both MEPPs and G-MEPPs presumably result from synaptic vesicle exocytosis of ACh since blockade of ACh uptake into synaptic vesicles by the drug vesamicol abolishes both types of events (LUPA et al. 1986). However, a major difference between MEPPs and G-MEPPs is that the latter do not enter into the composition of the EPPs since they are not evoked by nerve stimulation (MOLGÓ and THESLEFF 1982). In addition, the frequency of G-MEPPs is less sensitive to an increase of intraterminal Ca^{2+} concentration and has a much higher dependence on temperature than MEPPs frequency (reviewed by THESLEFF 1986; THESLEFF et al. 1990).

The mechanism whereby G-MEPPs are generated remains an enigma. The neuron, as all eukaryotic cells, has multiple constitutive and regulated pathways of vesicle exocytosis implicated in the delivery of membrane components or/and secretory products (reviewed by DECAMILLI and JAHN 1990; KELLY 1993). It is possible that G-MEPPs results from the release of the ACh content of vesicles participating in these exocytosis pathways. Indeed, if there are vesicles other than the small clear synaptic vesicles, which irrespective of their main content also contain ACh, their exocytosis at sites of the presynaptic membrane facing ACh receptor rich postsynaptic regions will give rise to different types of spontaneous MEPPs (small and giant of variable time-course). Given this, it is worth noting that ACh has been shown to be stored in synaptic or like vesicles of endosomal origin which share some of the membrane proteins characteristic of the small clear synaptic vesicles (BAUERFEIND et al. 1993). A recent commentary raises the possibility that early endosomes may release neurotransmitters in quantal manner (BAUERFEIND et al. 1994). However, it remains to be determined whether such type of quantal release is affected by any of the clostridial neurotoxins. Because the so-called large dense core vesicles may accumulate ACh and do not release their content in response to single nerve stimuli but only following high frequency stimulation, it has been proposed that these organelles are involved in the generation of G-MEPPs (THESLEFF et al. 1990).

6 Conclusions

Evidence summarized above indicates that BoNTs and TeNT block both the spontaneous and nerve-impulse evoked quantal release of neurotransmitter. However, the toxins do not affect some of the quantal events (those generating G-MEPPs and a population of small-MEPPs). Because SNAP-25, VAMP/synaptobrevins 1 and 2, and HPC1/syntaxin have been shown to interact together and were identified as targets for clostridial neurotoxins, it is likely that the same protein complex is responsible for evoked and spontaneous quantal release. Although a G-MEPP is the result of the liberation of a multimolecular packet of neurotransmitter, this release mechanism seems to proceed with a set of proteins distinct to those targeted by clostridial neurotoxins. Supporting this view it has been recently suggested that the incomplete inhibition of transferrin receptor exocytosis produced by the TeNT light chain may involve TeNT-insensitive proteins of the VAMP/synaptobrevin family (GALLI et al. 1994).

Due to the observation that those clostridial neurotoxins which cleave VAMP/synaptobrevin also desynchronize evoked quantal release, it is tempting to ascribe to this protein a major role in the process of quantal synchronisation after Ca^{2+} entry triggered by the nerve action potential. Alternatively, it is possible that the diminution (by cleavage) in the number of VAMP/synaptobrevin molecules present on each vesicle leads to such a slowing in the kinetics of the release process that the quanta liberated appear dispersed in time.

Although the identification of the toxins' targets has led to a breakthrough in our thinking of the neurotransmitter release process, we must be aware that some important points still await clarification. Since it is plausible that the turnover of synaptic proteins implicated in the release process is shorter than the weeks or months needed for recovery from paralysis, how can the cleavage of a given synaptic protein account for the long-lasting blockade of quantal release? Is it possible that the toxin remains at the nerve ending for a very long time, or that secondary activities of the toxins other than the cleavage of VAMP/synaptobrevin, SNAP-25, or syntaxin are involved?

The fact that the number of vesicles in close proximity of the presynaptic membrane does not diminish and the possibility to induce quantal release by α-LTX at synapses blocked by clostridial neurotoxins (mainly BoNT/A) are indicative that neither the docking of clear synaptic vesicles at the active zone, nor the release machinery is completely disrupted by clostridial neurotoxins. Thus, we conclude that clostridial neurotoxins act after Ca^{2+} entry at a step of the release process downstream to the docking of synaptic vesicles but upstream to the final fusion process.

Acknowledgments. This study was supported in part by grants from Association Française contre les Myopathies (AFM) and Direction des Recherches Etudes et Techniques (94/67 to J.M. and 92/102). Support was also obtained from grants from the Swedish Medical Research Council, Stockholm (14x-3112 to S.T.), Sweden.

References

Bauerfeind R, Régnier-Vigouroux A, Flatmark T, Huttner WB (1993) Selective storage of acetylcholine, but not catecholamines, in neuroendocrine synaptic-like microvesicles of early endosomal origin. Neuron 11: 105–121

Bauerfeind R, Huttner WB, Almers W, Augustine GJ (1994) Quantal neurotransmitter release from early endosomes. Trends Cell Biol 4: 155–156

Bevan S, Wendon LM (1984) A study of the action of tetanus toxin at rat soleus neuromuscular junctions. J Physiol (Lond) 348: 1–17

Bigalke H, Dreyer F, Bergey G (1985) Botulinum A neurotoxin inhibits non-cholinergic synaptic transmission in mouse spinal cord neurons in culture. Brain Res 360: 318–324

Binz T, Blasi J, Yamasaki S, Baumeister A, Link E, Südhof TC, Jahn R, Niemann H (1994) Proteolysis of SNAP-25 by types E and A botulinal neurotoxins. J Biol Chem 269: 1617–1620

Blasi J, Chapman ER, Link E, Binz T, Yamasaki S, De Camilli P, Südhof TC, Niemann H, Jahn R (1993a) Botulinum neurotoxin A selectively cleaves the synaptic protein SNAP-25. Nature 365: 160–163

Blasi J, Chapman ER, Yamasaki S, Binz T, Niemann H, Jahn R (1993b) Botulinum neurotoxin C1 blocks neurotransmitter release by means of cleaving HPC-1/syntaxin. EMBO J 12: 4821–4828

Boroff DA, del Castillo J, Evoy WH, Steinhardt RA (1974) Observations on the action of type A botulinum toxin on frog neuromusuclar junctions. J Physiol (Lond) 240: 227–253

Cull-Candy SG, Lundh H, Thesleff S (1976) Effects of botulinum toxin on neuromuscular transmission in the rat. J Physiol (Lond) 260: 177–203

DeCamilli P, Jahn R (1990) Pathways to regulated exocytosis in neurons. Annu Rev Physiol 52: 625–645

De Paiva A, Ashton AC, Foran P, Schiavo G, Montecucco C, Dolly JO (1993) Botulinum A like type B and tetanus toxin fulfils criteria for being a zinc-dependent protease. J Neurochem 61: 2338–2341

Dolly JO (1992) Peptide toxins that alter neurotransmitter release. In: Herken H, Hucho F (eds) Selective neurotoxicity. Springer, Berlin Heidelberg New York, pp 681–717 (Handbook of experimental pharmacology, vol 102)

Dolly JO, Lande S, Wray W (1987) The effects of in vitro application of purified botulinum neurotoxin at mouse motor nerve terminals. J Physiol (Lond) 386: 475–484

Dreyer F (1989) Peripheral actions of tetanus toxin. In: Simpson LL (ed) Botulinum neurotoxin and tetanus toxin. Academic, San Diego, pp 179–202

Dreyer F, Schmitt A (1981) Different effects of botulinum A toxin and tetanus toxin on the transmitter releasing process at the mammalian neuromuscular junction. Neurosci Lett 26: 307–311

Dreyer F, Schmitt A (1983) Transmitter release in tetanus and botulinum A toxin-poisoned mammalian motor end-plates and its dependence on nerve stimulation and temperature. Pflugers Arch 399: 228–234

Dreyer F, Mallart A, Brigant JL (1983) Botulinum A toxin and tetanus toxin do not affect presynaptic membrane currents in mammalian motor nerve endings. Brain Res 270: 373–375

Dreyer F, Rosenberg F, Becker C, Bigalke H, Penner R (1987) Differential effects of various secretagogues on quantal transmitter release from mouse motor nerve terminals treated with botulinum A and tetanus toxin. Naunyn Schmiedeberg's Arch Pharmacol 335: 1–7

Duchen LW, Tonge DA (1973) The effects of tetanus toxin on neuromuscular transmission and on the morphology of motor endplates in slow and fast skeletal muscle of the mouse. J Physiol (Lond) 228: 157–172

Dunant Y, Esquerda JE, Loctin F, Marsal J, Müller D (1987) Botulinum toxin inhibits quantal acetylcholine release and energy metabolism in Torpedo electric organ. J Physiol (Lond) 385: 677–692

Facchiano F, Luini A (1992) Tetanus toxin potently stimulates tissue transglutaminase a possible mechanism of neurotoxicity. J Biol Chem 267: 13267–13271

Facchiano F, Benfenati F, Valtorta F, Luini A (1993) Covalent modification of synapsin I by a tetanus toxin-activated transglutaminase. J Biol Chem 268: 4588–4591

Galli T, Chilcote T, Mundigl O, Binz T, Niemann H, De Camilli P (1994) Tetanus toxin-mediated cleavage of cellubrevin impairs exocytosis of transferrin receptor-containing vesicles in CHO cells. J Cell Biol 125: 1015–1024

Gansel M, Penner R, Dreyer F (1987) Distinct sites of action of clostridial neurotoxins revealed by double poisoning of mouse motor nerve terminals. Pflugers Arch 409: 533–539

Gundersen CB (1980) The effects of botulinum toxin on the synthesis, storage and release of acetylcholine. Prog Neurobiol 4: 99–119

Gundersen CB, Katz B, Miledi R (1982) The antagonism between botulinum toxin and calcium in motor nerve terminals. Proc R Soc Lond [Biol] 216: 369–376

Habermann E, Dreyer F (1986) Clostridial neurotoxins: handling and action at the cellular and molecular level. In: Capron A, Compans RW, Cooper M et al. (eds) Current topics in microbiology and immunology, vol 129. Springer, Berlin Heidelberg New York, pp 93–179

Harris AJ, Miledi R (1971) The effect of type D botulinum toxin on frog neuromuscular junctions. J Physiol (Lond) 217: 497–515

Hata Y, Davletov B, Petrenko AG, Jahn R, Südhof TC (1993) Interaction of synaptotagmin with the cytoplasmic domain of neurexins. Neuron 10: 307–315

Kanda K, Takano K (1983) Effect of tetanus toxin on the excitatory and the inhibitory post-synaptic potentials in the cat motoneurone. J Physiol (Lond) 335: 319–333

Kao I, Drachman DB, Price DL (1976) Botulinum toxin: mechanism of presynaptic blockade. Science 193: 1256–1258

Katz B (1969) The release of neural transmitter substances. Sherrington Lectures X. Liverpool University Press, Liverpool

Kauffman JA, Way JF, Siegel LS, Sellin LC (1985) Comparison of the action of types A and F botulinum toxin at the rat neuromuscular junction. Toxicol Appl Pharmacol 79: 211–217

Kelly RB (1993) Storage and release of neurotransmitters. Cell 72/Neuron 10 [Suppl]: 42–53

Kim YI, Lömo T, Lupa MT, Thesleff S (1984) Miniature end-plate potentials in rat skeletal muscle poisoned with botulinum toxin. J Physiol (Lond) 356: 587–599

Kriebel ME, Llados F, Matteson DR (1976) Spontaneous subminiature end-plate potentials in mouse diaphragm muscle: evidence for synchronous release. J Physiol (Lond) 262: 553–581

Liley AW (1957) Spontaneous release of transmitter substance in multiquantal units. J Physiol (Lond) 136: 595–605

Link E, Edelmann L, Chou JH, Binz T, Yamasaki S, Eisel U, Baumert M, Südhof TC, Niemann H, Jahn R (1992) Tetanus toxin action: inhibition of neurotransmitter release linked to synaptobrevin proteolysis. Biochem Biophys Res Commun 189: 1017–1023

Lundh H (1983) Antagonism of botulinum toxin paralysis by low temperature. Muscle Nerve 6: 56–60

Lundh H, Leander S, Thesleff S (1977) Antagonism of the paralysis produced by botulinum toxin in the rat. J Neurol Sci 32: 29–43

Lupa MT, Tabti N, Thesleff S, Vyskocil F, Yu SP (1986) The nature and origin of calcium-insensitive miniature end-plate potentials at rodent neuromuscular junctions. J Physiol (Lond) 381: 607–618

Mallart A, Molgó J, Angaut-Petit D, Thesleff S (1989) Is the internal calcium regulation altered in type A botulinum toxin poisoned motor endings? Brain Res 479: 167–171

Mellanby J, Green J (1981) How does tetanus toxin act? Neuroscience 6: 281–300

Mellanby J, Thompson PA (1972) The effects of tetanus toxin at the neuromuscular junction in the goldfish. J Physiol (Lond) 224: 407–419
Mellanby J, Beaumont MA, Thompson PA (1988) The effect of lanthanum on nerve terminals in goldfish muscle after paralysis with tetanus toxin. Neuroscience 25: 1095–1106
Mochida S, Poulain B, Weller U, Habermann E, Tauc L (1989) Light chain of tetanus toxin intracellularly inhibits acetylcholine release at neuro-neuronal synapses, and its internalization is mediated by heavy chain. FEBS Lett 253: 47–51
Molgó J, Thesleff S (1982) 4-Aminoquinoline induced "giant" miniature end-plate potentials at mammalian neuromuscular junctions. Proc R Soc Lond [Biol] 214: 229–247
Molgó J, Thesleff S (1984) Studies on the mode of action of botulinum toxin type A at the frog neuromuscular junction. Brain Res 296: 309–316
Molgó J, Lundh H, Thesleff S (1980) Potency of 3,4-diaminopyridine and 4-aminopyridine on mammalian neuromuscular transmission and the effects of pH changes. Eur J Pharmacol 61: 25–34
Molgó J, Siegel LS, Tabti N, Thesleff S (1989a) A study of synchronization of quantal transmitter release from mammalian motor endings by the use of botulinal toxins type A and D. J Physiol (Lond) 411: 195–205
Molgó J, DasGupta BR, Thesleff S (1989b) Characterization of the actions of botulinum neurotoxin type E at the rat neuromuscular junction. Acta Physiol Scand 137: 497–501
Molgó J, Comella JX, Angaut-Petit D, Pécot-Dechavassine M, Tabti N, Faille L, Mallart A, Thesleff S (1990) Presynaptic actions of botulinal neurotoxins at vertebrate neuromuscular junctions. J Physiol (Paris) 84: 152–166
Neale EA, Habig WH, Schrier BK, Bergey GK, Bowers LM, Koh J (1989) Applications of tetanus toxin for structure-function studies in neuronal cell cultures. In: Nisticò G, Bizzini B, Bytchenko B, Triau R (eds) 8th international conference on tetanus. Pythagora, Rome pp 58–65
Niemann H (1991) Molecular biology of clostridial neurotoxins. In: Alouf JE, Freer JH (eds) Sourcebook of bacterial protein toxins. Academic, San Diego, pp 303–348
Niemann H, Blasi J, Jahn R (1994) Clostridial neurotoxins: new tools for dissecting exocytosis. Trends Cell Biol 4: 179–185
Pécot-Dechavassine M, Molgó J, Thesleff S (1991) Ultrastructure of botulinum type-A poisoned frog motor nerve terminals after enhanced quantal transmitter release caused by carbonyl cyanide m-chlorophenylhydrazone. Neurosci Lett 130: 5–8
Poulain B, Molgó J (1992) Botulinal neurotoxins: mode of action on neurotransmitter release. In: Conn PM (ed) Methods in neuroscience: neurotoxins, vol 8. Academic, San Diego, pp 38–54
Poulain B, Tauc L, Maisey EA, Wadsworth JDF, Mohan PM, Dolly JO (1988) Neurotransmitter release is blocked intracellularly by botulinum neurotoxin, and this requires uptake of both toxin polypeptides by a process mediated by the larger chain. Proc Natl Acad Sci USA 85: 4090–4094
Poulain B, Weller U, Binz T, Niemann H, de Paiva A, Dolly JO, Leprince C, Tauc, L (1993a) Functional roles of domains of Clostridial neurotoxins: the contribution from studies on Aplysia. In: DasGupta BR (ed) Botulinum and tetanus neurotoxins: neurotransmission and biochemical aspects. Plenum, New York, pp 345–360
Poulain B, Rossetto O, Deloye F, Schiavo G, Tauc L, Montecucco C (1993b) Antibodies against rat brain VAMP/synaptobrevin prevent inhibition of acetylcholine release by tetanus toxin or botulinum neurotoxin type B. J Neurochem 61: 1175–1178
Schiavo G, Poulain B, Rossetto O, Benfenati F, Tauc L, Montecucco C (1992a) Tetanus toxin is a zinc protein and its inhibition of neurotransmitter release and protease activity depend on zinc. EMBO J 11: 3577–3583
Schiavo G, Benfenati F, Poulain B, Rossetto O, Polverino de Laureto P, DasGupta BR, Montecucco C (1992b) Tetanus and botulinum-B neurotoxins block neurotransmitter release by proteolytic cleavage of synaptobrevin. Nature 359: 832–835
Schiavo G, Shone CC, Rossetto O, Alexander FCG, Montecucco C (1993a) Botulinum neurotoxin serotype F is a zinc endopeptidase specific for VAMP/synaptobrevin. J Biol Chem 268: 11516–11519
Schiavo G, Rossetto O, Catsicas S, Polverino de Laureto P, DasGupta BR, Benfenati F, Montecucco C (1993b) Identification of the nerve terminal targets of botulinum neurotoxin serotypes A, D, and E. J Biol Chem 268: 23784–23787
Schiavo G, Santucci A, DasGupta BR, Metha PP, Jontes J, Benfenati F, Wilson MC, Montecucco C (1993c) Botulinum neurotoxins serotypes A and E cleave SNAP-25 at distinct COOH-terminal peptide bonds. FEBS Lett, 335: 99–103
Schiavo G, Malizio C, Trimble WS, Polverino de Laureto P, Milan G, Sugiyama H, Johnson EA, Montecucco C (1994) Botulinum G neurotoxin cleaves VAMP/synaptobrevin at a single Ala/Ala peptide bond. J Biol Chem 269: 20213–20216

Sellin LC (1987) Botulinum toxin and the blockade of transmitter release. Asia Pac J Pharmacol 2: 203–222

Sellin LC, Thesleff S, DasGupta BR (1983) Different effects of types A and B botulinum toxin on transmitter release at the neuromuscular junction. Acta Physiol Scand 119: 127–133

Simpson LL (1986) Molecular pharmacology of botulinum toxin and tetanus toxin. Annu Rev Pharmacol Toxicol 26: 427–453

Simpson LL, Coffield JA, Bakry N (1993) Chelation of zinc antagonizes the neuromuscular blocking properties of the seven serotypes of botulinum neurotoxin as well as tetanus toxin. J Pharmacol Exp Ther 267: 720–727

Söllner T, Bennett MK, Whiteheart SW, Scheller RH, Rothman JE (1993) A protein assembly-diassembly pathway in vitro that may correspond to sequential steps of synaptic vesicle docking, activation and fusion. Cell 75: 409–418

Thesleff S (1986) Different kinds of acetylcholine release from the motor nerve. Int Rev Neurobiol 28: 59–88

Thesleff S, Molgó J, Lundh H (1983) Botulinum toxin and 4-aminoquinoline induce a similar abnormal type of spontaneous quantal transmitter release at the rat neuromuscular junction. Brain Res 264: 89–97

Thesleff S, Molgó J, Tågerud S (1990) Trophic interrelations at the neuromuscular junction as revealed by the use of botulinal neurotoxins. J Physiol (Paris) 84: 167–173

Van der Kloot W (1989) Statistical and graphical methods for testing the hypothesis that quanta are made up of subunits. J Neurosci Methods 27: 81–89

Wellhöner HH (1992) Tetanus and botulinum neurotoxins. In: Herken H, Hucho F (eds) Selective neurotoxicity. Springer, Berlin Heidelberg New York, pp 357–417 (Handbook of experimental pharmacology, vol 102)

Yamasaki S, Binz T, Hayashi T, Szabo E, Yamasaki N, Eklund M, Jahn R, Niemann H (1994) Botulinum type G peoteolyses Ala^{81}-Ala^{82} bond of rat synaptobrevin 2. Biochem Biophys Res Commun 200: 829–835

Intracellular Targets and Metalloprotease Activity of Tetanus and Botulism Neurotoxins

G. Schiavo, O. Rossetto, F. Tonello, and C. Montecucco

1 Introduction

Tetanus and botulism neurotoxins (TeNT and BoNT, respectively) produced by clostridia are the most toxic substances known: the mouse LD_{50} of highly purified preparations is between 0.1 and 1 ng/kg. They block the release of neurotransmitters either at the peripheral (BoNT) or central (TeNT) nervous system. This tremendous potency derives from two essential features of these bacterial toxins: (a) their absolute neurospecificity and (b) their intracellular catalytic activity. By concentrating their action on a limited number of cells, whose complete functionality is essential to the survival of very complex animals such as the vertebrates, neurotoxins lead to animal death with a minimal amount of toxic molecules. The basis of this cell selectivity resides on receptors uniquely present on neuronal cells (see Halpern and Neale, this volume). Although questioned by a number of researchers, it was not unexpected that clostridial neurotoxins are enzymes, acting in the neuron cytosol (Schiavo et al. 1993a). In fact, an enzyme can modify one after another all the target molecules present in the system and

Centro CNR Biomembrane and Dipartimento di Scienze Biomediche, Università di Padova, Via Trieste 75, 35121 Padova, Italy

hence one single molecule of an enzymic toxin is able to intoxicate a synapse. TeNT and the seven BoNTs, A–G, are zinc endopeptidases specific for protein components of the neuroexocytosis apparatus. This enzymatic activity is the subject of this chapter.

2 Structural Aspects of Clostridial Neurotoxins

All clostridial neurotoxins are synthesized as a single inactive polypeptide chain of 150 kDa without leader sequence and hence are presumably released by bacterial lysis (NIEMANN 1991; Minton, this volume). Bacterial or tissue proteases cleave these toxins within an exposed highly protease-sensitive loop (DASGUPTA 1989; KRIEGLSTEIN et al. 1991) and generate the active di-chain neurotoxins composed of a heavy chain (H, 100 kDa) and a light chain (L, 50 kDa) joined by a disulfide bond. This interchain S-S bond plays a critical role in cell penetration, and its cleavage by reduction abolishes toxicity (SCHIAVO et al. 1990; DE PAIVA et al. 1993a). TeNT and BoNTs are purified from bacterial culture filtrates by precipitation and successive chromatographic steps (Shone and Tranter, this volume; SCHIAVO and MONTECUCCO 1995). A very convenient and simple final purification step that it is advisable to be used with commercial preparations of TeNT and BoNTs is immobilized metal ion affinity chromatography (IMAC) with bound zinc (ROSSETTO et al. 1992).

Available structural and biochemical data (ROBINSON et al. 1988; DASGUPTA 1989; WELLHONER 1992; MONTECUCCO and SCHIAVO 1993; SCHIAVO et al. 1993a; ANDERSON et al. 1993) indicate that TeNT and BoNTs, similarly to diphtheria toxin and exotoxin A (ALLURED et al. 1986; CHOE et al. 1992), are folded into three functionally distinct domains, which play different roles in cell intoxication (Fig. 1). This structural organization and its significance with respect to the process of cell penetration has recently been discussed in detail (MONTECUCCO et al. 1994). Domain L is the catalytic part, responsible for the blockade of neuroexocytosis; domain H_N, the 50-kDa amino terminal half of the H chain, appears to govern cell penetration; and domain H_C, the 50-kDa carboxyl terminal half of the H chain, is responsible mainly for the neurospecific binding.

3 Structure of the L Chain

The L chain is composed of about 450 amino acid residues: the precise number depends on serotypes and the exact site of proteolytic nicking. The L chains of TeNT and BoNTs show homologous segments, dispersed along the sequence (NIEMANN 1991; Minton, this volume). The most conserved segment is located in the middle of the L chain and comprises 19 residues (Fig. 2; SCHIAVO et al.

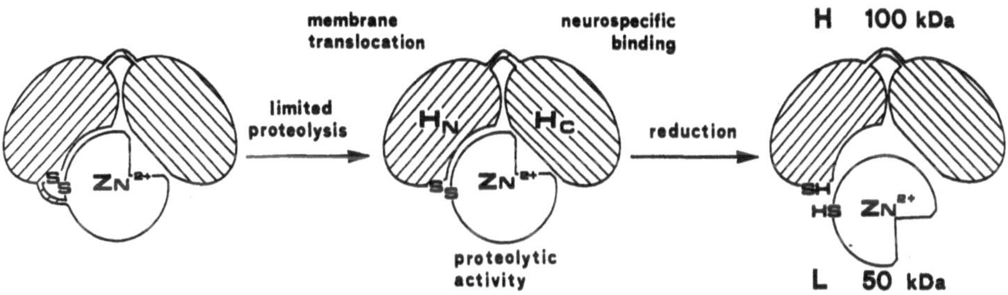

Fig. 1. Scheme of the structure and mechanism of activation of tetanus and botulinum neurotoxins. These toxins are synthetized as an 150-kda inactive single polypeptide chain composed of three 50-kDa domains, which play different roles in nerve cell intoxication. The three domains are connected by protease-sensitive loops. The toxin becomes active upon selective proteolytic cleavage, performed by various proteases, that generates two disulfide-linked chains. H_C is responsible for cell binding and H_N for cell penetration. Reduction takes place inside the nerve cells and liberates the activity of the L chain, which blocks neuroexocytosis via a zinc endopeptidase activity specific for three components of the neuroexocytosis apparatus.

1992a,c; 1993b; WRIGHT et al. 1992; MONTECUCCO and SCHIAVO 1993, 1994). It contains the His-Glu-Xaa-Xaa-His zinc-binding motif of zinc endopeptidases (JONGENEEL et al. 1989; VALLEE and AULD 1990; JIANG and BOND 1992; SCHIAVO et al. 1992a,c; WRIGHT et al. 1992; BODE et al. 1993; YAMASAKI et al. 1994a). Due to the limited amount of purified toxins available, zinc content of clostridial neurotoxins has been determined directly only on TeNT and on serotypes A, B, E, and F of BoNT. TeNT and BoNT/A, /B, and /F have one atom of zinc bound to the L chain (SCHIAVO and MONTECUCCO 1994; SCHIAVO et al. 1992a,c, 1993a). The affinity of zinc binding to TeNT and BoNT/A, /B and /E, measured by flow dialysis with $^{65}Zn^{2+}$, is comprised in the 50–150 nM range, at the lower limit of the known range of affinities among metalloproteases (SCHIAVO et al. 1992c). Zinc can be removed upon incubation with heavy metal chelators, thus generating an aponeurotoxin. The most efficient chelator is orthophenantroline, which must be used in the cold to avoid denaturating effects. The metal atom can be reacquired upon incubation in zinc-containing buffers with the regeneration of the fully active holoneurotoxin.

The zinc atom of zinc endopeptidases is coordinated either by two or three histidines (JIANG and BOND 1992; BODE et al. 1993). In thermolysinlike enzymes the zinc atom is at the center of a tetrahedron formed by the imidazole rings of the two histidines of the motif, a water molecule bound to the glutamic acid of the motif and the carboxylate of another glutamate residue (MATTHEWS et al. 1972; PAUPTIT et al. 1988; THAYER et al. 1991). On the contrary, astacin-like proteinases pentacoordinate zinc via three histidines and one tyrosine, while adamalysin, alkaline protease, and collagenases adopt a tetrahedral coordination via three histidines and the glutamate-bound water molecule (BODE et al. 1992, 1993, 1994; BAUMANN et al. 1993; GOMIS-RUTH et al. 1993; LOVEJOY et al. 1994). The active site has a cleft shape with zinc in the center and the residues of the motif organized as α-helix. The orientation and volume of the residue forming the cleft determine the peptide bond specificity (see below). The water molecule bound to the glutamate of the motif is involved in the hydrolysis of the peptide bond with a

BoNT/A	D	P	A	V	T	L	A	H	E	L	I	H	A	G	H	R	L	Y	G
BoNT/A infant	D	P	A	V	T	L	A	H	E	L	I	H	A	E	H	R	L	Y	G
BoNT/B GpI	D	P	A	L	I	L	M	H	E	L	I	H	V	L	H	G	L	Y	G
BoNT/B GpII	D	P	A	L	I	L	M	H	E	L	I	H	V	L	H	G	L	Y	G
BoNT/C	D	P	I	L	I	L	M	H	E	L	N	H	A	M	H	N	L	Y	G
BoNT/D	D	P	V	I	A	L	M	H	E	L	T	H	S	L	H	Q	L	Y	G
BoNT/E	D	P	A	L	T	L	M	H	E	L	I	H	S	L	H	G	L	Y	G
BoNT/E butyricum	D	P	A	L	T	L	M	H	E	L	I	H	S	L	H	G	L	Y	G
BoNT/F	D	P	A	I	S	L	A	H	E	L	I	H	A	L	H	G	L	Y	G
BoNT/F baratii	D	P	A	I	S	L	A	H	E	L	I	H	V	L	H	G	L	Y	G
BoNT/F Langeland	D	P	A	I	S	L	A	H	E	L	I	H	A	L	H	G	L	Y	G
BoNT/G	D	P	A	L	T	L	M	H	E	L	I	H	V	L	H	G	L	Y	G
TeNT	D	P	A	L	L	L	M	H	E	L	I	H	V	L	H	G	L	Y	G
	*	*				*		*	*	*		*			*		*	*	*
Zincins								H	E	x	x	H							
						*		*	*			*							
Metzincins					x	h	x	H	E	x	h	H	x	h	G	h	x	H	x

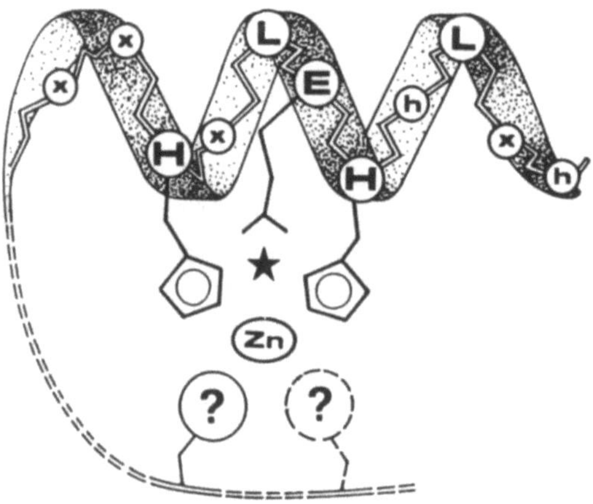

Fig. 2. The conserved segment of clostridial neurotoxin L chains contain the zinc endopeptidase signature. *Above*, sequence of the longest conserved segment among the L chains of tetanus and botulinum neurotoxins, present in the central part of the chain. It contains the zinc-binding motif of metalloproteinases (zincins), and it is dissimilar from the consensus sequence of the metzincins metalloproteinase family. *Below*, that which is known about the zinc coordination of clostridial neurotoxins: zinc is bound via the two histidines and the glutamate-bound water molecule (*star*) of the motif arranged in an α-helical configuration. Additional ligand(s) is still to be identified

mechanism that has been studied in chemical detail only for thermolysin (Matthews 1988).

The L chains of TeNT and BoNT/A, /B and /E were chemically modified with diethylpyrocarbonate (DEPC), a histidine-specific reagent (Miles 1977) in their holo- and apo-forms. Two additional histidines are available to DEPC modification in the apo- with respect to the holoneurotoxins (Schiavo et al. 1992a,c), as it is found with thermolysin (Tonello et al. unpublished results). This result indicates that the zinc atom of clostridial neurotoxins is coordinated via two histidines, as in the case of thermolysin. Yamasaki et al. (1994a) have mutagenized the two histidine residues of the motif and found that the mutated proteins are inactive and bind zinc poorly. They also mutated the conserved Glu-271–Glu-272 couple of residues with a resulting decrease of zinc binding and loss of proteolytic activity. On this basis it was suggested that one of these two residues is the fourth zinc ligand, as in the case of thermolysinlike proteases (YAMASAKI et al. 1994a). However the result of single Glu mutation was not reported, thus leaving open the possibility that the double mutation has lead to a gross folding alteration with ensuing loss of function.

The metal atom of TeNT can be exchanged with other transition metals of similar size with preservation of toxic activity (HÖHNE-ZELL et al. 1993; TONELLO et al., in preparation). The cobalt-substituted TeNT shows an absorption specturm more similar to that of Co-thermolysin than to that of Co-astacin (TONELLO et al., in preparation). While three zinc ligands have been identified, further experiments are needed to identify the additional ligand(s) and to establish definitively the spatial orientation of residues around the zinc atom of the clostridial neurotoxins.

4 Metalloproteinase Activity

TeNT and BoNTs were recently shown to be zinc endopeptidases specific for protein components of the neuroexocytosis apparatus (reviewed in MONTECUCCO and SCHIAVO 1994; Fig. 3). TeNT and BoNT/B, /D, /F, and /G act on the same intraneuronal target: they cleave VAMP, also called synaptobrevin, a membrane protein of synaptic vesicles (SCHIAVO et al. 1992a,b, 1993b,c, 1994b; YAMASAKI et al. 1994a–c). At variance, BoNT/A, /C, and /E act on proteins of the presynaptic membrane: BoNT/A and /E cleave SNAP-25, while serotype C cleaves syntaxin (BLASI et al. 1993a,b; SCHIAVO et al 1993c,d, 1994c; BINZ et al. 1994).

Except for TeNT and BoNT/B each one of the different clostridial neurotoxins catalyses the hydrolysis of different peptide bonds (Table 1). No common features of the amino acid sequences around the cleavage sites of the various neurotoxins in terms of amino acid lateral chain charge, hydrophilicity/hydrophobicity, and volume can be identified. This difference is at the basis of a different geometry of the active sites of these proteinases which accounts for the binding and hydrolysis of the different peptide segments listed in Table 1.

A characteristic feature of the clostridial neurotoxins protease activity is that they do not cleave short peptides or peptide analogues, spanning the respective cleavage sites (SCHIAVO et al. 1992a; SHONE et al. 1993; YAMASAKI et al. 1994a, b). On the other hand, it is clear that short peptides encompassing the cleavage sites do bind to the respective neurotoxins because they are effective inhibitors of the protease activity in vitro and in vivo (SCHIAVO et al. 1992b, 1993b, 1994a; HOHNE-ZELL et al. 1993; DAYANITHI et al. 1994; HUNT et al. 1994). Proteolysis of VAMP peptides of different length by BoNT/B, /D, and /F was appreciable only with 40-residue-long peptides and maximal cleavage rates were only observed with segment 33–96 of VAMP (SHONE et al. 1993; YAMASAKI et al. 1994a,b). BINZ et al. (1994) found that the minimal segment of SNAP-25 that is cleaved by BoNT/A and

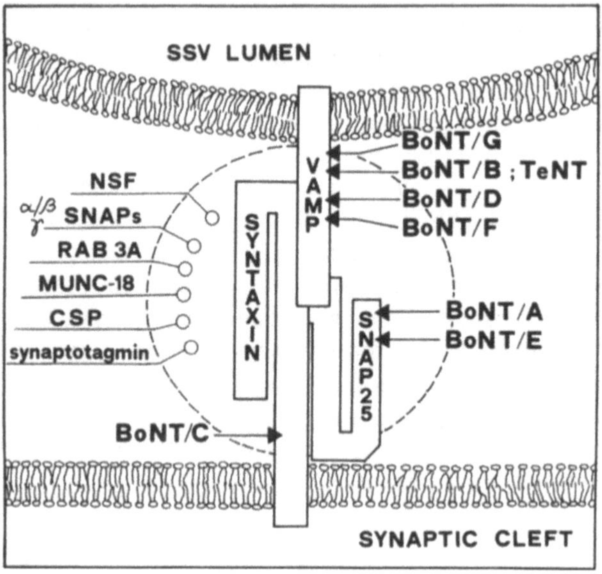

Fig. 3. Scheme of the intracellular zinc endopeptidase activity of tetanus and botulinum neurotoxins. VAMP, SNAP-25, and syntaxin are essential parts of the 20S multiprotein complex which mediates synaptic vesicle docking and fusion with the presynaptic membrane. VAMP is located on the synaptic vesicle membrane, while SNAP-25 and syntaxin are located on the target membrane. Zinc-dependent proteolytic cleavage of VAMP by TeNT and BoNT/B, /D, /F, or /G releases the amino terminal part of VAMP in the cytosol and impairs the function of the neuroexocytosis apparatus. Each toxin cleaves a different peptide bond. BoNT/A and /E cleave SNAP-25 at the carboxyl terminus with the release of 9 and 26 residues, respectively. BoNT/C cleaves the two syntaxin isoforms present in the nervous tissue at a single site near the carboxy terminus. Thus, the action of TeNT and BoNT/B, /C, /D, /F, and /G causes the release of a large portion of the cytosolic domain of VAMP and syntaxin. At variance, only a small part of SNAP-25 is released by the selective proteolysis of BoNT/A and /E. A set of additional proteins, indicated here with acronyms, have been suggested to constitute the neuroexocytosis apparatus (BENNETT and SCHELLER 1994)

/E is the 133–206 fragment. Moreover, in serveral cases clostridial neurotoxins cleave only a single peptide bond and leave intact identical bonds located in other parts of the sequence of the target protein. These results indicate that TeNT and BoNTs are proteases that recognize the tertiary structure of their targets rather than the primary sequence around the peptide bond to be cleaved, as most proteases do. The structural basis of the specificity of the proteolytic activity of TeNT and BoNTs is still unknown.

5 VAMP/Synaptobrevin

VAMP was first identified by cloning the gene from a *Torpedo* electric organ expression library (TRIMBLE et al. 1988). This protein is present in different isoforms in all eukaryotes where it has been looked at and primary sequences are available

Table 1. Tetanus and Botulism Neurotoxins: Target and peptide bond specificities

Toxin type	Target	Peptide bond cleaved $P3–P_2–P_1–P_1'–P_2'–P3'$
TeNT	VAMP	Ala-Ser-Gln—Phe-Glu-Thr
BoNT/A	SNAP-25	Ala-Asn-Gln—Arg-Ala-Thr
BoNT/B	VAMP	Ala-Ser-Gln—Phe-Glu-Thr
BoNT/C	Syntaxin	Thr-Lys-Lys—Ala-Val-Lys
BoNT/D	VAMP	Asp-Gln-Lys—Leu-Ser-Glu
BoNT/E	SNAP-25	Ile-Asp-Arg—Ile-Met-Glu
BoNT/F	VAMP	Arg-Asp-Gln—Lys-Leu-Ser
BoNT/G	VAMP	Thr-Ser-Ala—Ala-Lys-Leu

from rat, bovine, human, squid, *Drosophila*, *Aplysia*, and yeast (Fig. 4, upper panel; ELFERINK et al. 1989; ARCHER et al. 1990; SÜDHOF et al. 1989; TRIMBLE et al. 1991; DI ANTONIO et al. 1993; HUNT et al. 1994; YAMASAKI et al. 1994a; BENNETT and SCHELLER 1994). The VAMP molecule (about 120 residues; the exact number varies in different species and isoforms) is anchored to the vesicle membrane via a carboxyl terminal part, likely to span the membrane in an α-helical configuration, as depicted in the lower part of Fig. 4 (TRIMBLE et al. 1988; BAUMERT et al. 1989). Hence, the majority of the molecule is exposed to the cytosol. A 30-residue-long amino terminal part is highly enriched in prolines and its sequence diverges considerably among species and isoforms. By contrast, the central part (residues 30–96) of VAMP is highly conserved and is enriched in charged residues. The intraluminal portion displays the maximum heterogeneity both in terms of size and sequence: this domain spans from 75-residues in *Aplysia* to few residues in most mammalian species.

The nervous tissue of vertebrates contain two isoforms, termed VAMP-1 and VAMP-2, of 118 and 116 amino acids respectively, which differ almost exclusively in the proline-rich amino terminal part (ELFERINK et al. 1989; ARCHER et al. 1990; TRIMBLE et al. 1991). Recent evidence indicate that VAMP-1 and -2 (or very closely related proteins) are present in all tissues, although in different amounts and ratios in specialized portions and cells. VAMPs are present on the glucose transporter containing vesicles of adipocytes and muscle cells and on malpighian cells (CORLEY CAIN et al. 1992; CHIN et al. 1994; RALSTON et al. 1994). VAMP-1 is expressed more than VAMP-2 in the exocrine pancreas, while VAMP-2 is predominant in the islets of Langerhans and in kidney glomerular cells (BRAUN et al. 1994; ROSSETTO et al. 1995). Another TeNT-cleavable isoform, termed cellubrevin (VAMP-3), is ubiquitously distributed in non neuronal cells (MCMAHON et al. 1993). These findings indicate that clostridial neurotoxins restrict their activity to the nerve tissue only because of the presence of toxin receptors and not because of lack of cleavable targets.

Clostridial neurotoxins cleave those VAMP isoforms that contain an appropriate cleavage site. This is the case of all the variants known so far, except for chicken and rat neuronal VAMP-1 which carries a Val residue in place of the Gln

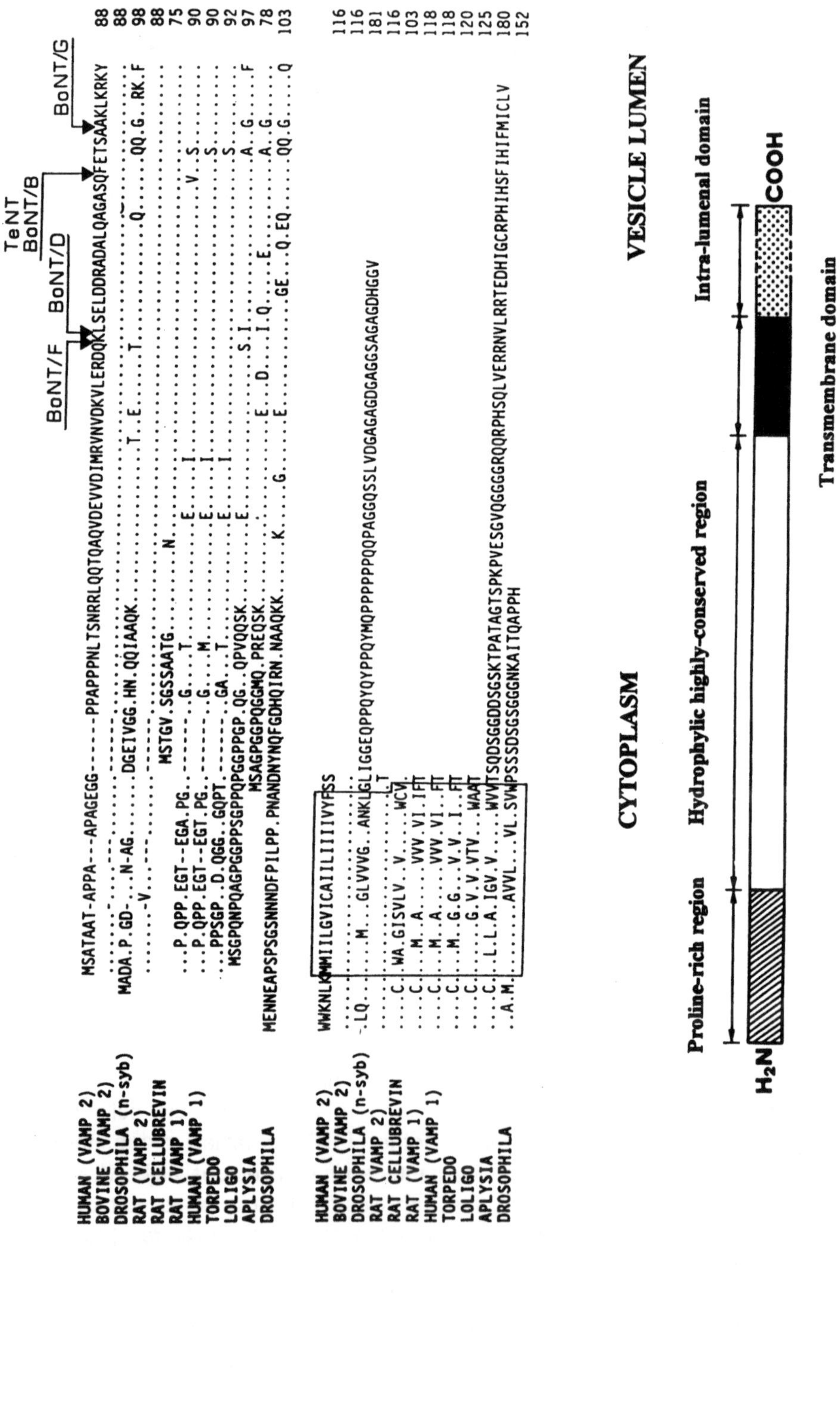
BoNT/F
BoNT/D
TeNT
BoNT/B
BoNT/G
HUMAN (VAMP 2)
BOVINE (VAMP 2)
DROSOPHILA (n-syb)
RAT (VAMP 2)
RAT CELLUBREVIN
RAT (VAMP 1)
HUMAN (VAMP 1)
TORPEDO
LOLIGO
APLYSIA
DROSOPHILA
CYTOPLASM
VESICLE LUMEN
Proline-rich region
Hydrophylic highly-conserved region
Intra-lumenal domain
Transmembrane domain
H₂N
COOH

residue present in human and mouse VAMP-1 (Fig. 4). This single amino acid replacement at the cleavage site of TeNT and BoNT/B has been proposed to be associated with the resistance of rats and chickens, and the sensitivity of humans and mice to tetanus and to type B botulism (PAYLING-WRIGHT 1955; PATARNELLO et al. 1993). This suggestion is supported by the demonstration that the receptor for BoNT/B is present in the rat nervous tissue (NISHIKI et al. 1994), and hence lack of receptor binding cannot be the cause of the resistance of rats to BoNT/B. The resistance of a rat VAMP isoform to cleavage by BoNT/B has already been exploited as a tool in cell biology to identify VAMP isoforms (BRAUN et al. 1994; STEINHARDT et al. 1994).

BoNT/D, /F, and /G cleave both VAMP isoforms with similar rates when the protein is in the synaptic vesicles (SCHIAVO et al. 1993b,c, 1994b) and indeed there appear to be no difference in species sensitivity to these form of botulism (PAYLING-WRIGHT 1955). By contrast, VAMP fragments derived from VAMP isoforms in solution are cleaved at different rates (YAMASAKI et al. 1994b), and these results may indicate that VAMP can adopt different conformations which are recognized and cleaved with different efficiency by these neurotoxins. Such an explanation would also account for the lack of neuroexocytosis inhibition of peptides derived from the same central part of the molecule (HUNT et al., 1994; POULAIN et al., in preparation).

6 SNAP-25

SNAP-25 (synaptosomal-associated protein of 25 kDa) was identified as a neuronal-specific protein composed of 206 amino acid residues (OYLER et al. 1989). It is present in all nerve cells, where it is believed to be anchored to the cytosolic surface of the nerve plasmalemma via palmitoylation of four cysteines, located in the center of the molecule (HESS et al. 1992). SNAP-25 is required for axonal growth during development and possibly nerve terminal plasticity in the mature nervous system (OSEN-SAND et al. 1993). The protein is highly conserved from *Drosophila* to mammals (BARK 1993; RISINGER and LARHAMMAR 1993; RISINGER et al. 1993; BARK and WILSON 1994a). There are two human isoforms differentially expressed during development: α is constitutively produced from embryogenesis, while isoform β begins to be expressed at birth and predominates in the adult. SNAP-25 α and β are generated by alternative splicing between two distinct exons. Although the two isoforms differ only for nine amino acids, these residues are located in the palmitoylated region and consequently

Fig. 4. Schematic structure and comparison of the amino acid sequence of known VAMPs. *Above*, their site of cleavage by clostridial neurotoxins. Rat VAMP-1 is not significantly cleaved by tetanus and botulinum B neurotoxins. *n-syb*, neuronal VAMP isoform of *Drosophila*; *box*, transmembrane segments. *Below*, the topology of VAMPs with the four protein domains

this difference may be relevant for regulating their membrane targetting (Bark and Wilson 1994b). Both isoforms are cleaved with similar efficiency in solution by BoNT/A and /E (Schiavo et al., unpublished results). Isoforms of SNAP-25 recognized by antibodies versus the neuronal β isoform and by BoNT/A and /E are present in other tissues, where they are involved in exocytosis (Sadoul et al. 1995).

BoNT/A removes only nine residues from the carboxy terminal of SNAP-25 (Fig. 5), and yet this is sufficient to block neuroexocytosis, thus indicating that this region is of critical importance in the function of the molecule (Schiavo et al. 1993d; Binz et al. 1994). The BoNT/A blockade of neurotransmitter release, but not that induced by BoNT/E, which removes 26 residues from the carboxy terminus (Fig. 5), is partially relieved by calcium and the calcium ionophore A23187 (Ashton et al. 1993). Moreover, it is completely reversed by α-latratoxin, which is believed to cause a massive calcium influx in the nerve terminal (Wanke et al. 1986). These results may be rationalized in terms of the model depicted in Fig. 6. It is proposed that SNAP-25 exists in two conformations: with or without bound calcium. SNAP-25 has a low calcium affinity and the calcium-bound form is the one active in neuroexocytosis. The BoNT/A-cleaved SNAP-25 is still able to convert to the calcium-binding active form but has a lower affinity for calcium and hence requires higher calcium concentrations to be present in a sufficient proportion. The BoNT/E-cleaved SNAP-25 has an even lower calcium affinity and its conversion to the calcium-bound form is negligible.

7 Syntaxin

Syntaxin is anchored to the nerve plasmalemma via a carboxy terminal segment and exposes most of its molecular mass to the cytosol. It is present at the active zones of the presynaptic membrane, where neurotransmitter release takes place. It interacts with both the synaptic vesicle protein synaptotagmin and the N-type calcium channel which is involved in the regulation of neurotransmitter release (Bennett et al. 1992; Inoue et al. 1992). Two isoforms have been identified in nerve cells, and several other isoforms are present in other tissues (Bennett et al. 1992, 1993). Recently, three syntaxin homologues have been identified in yeast, each of which plays a role at different levels of the secretory pathway, thus suggesting that these proteins might define the specificity of each vesicular traffic step (Bennett and Scheller 1994; Rothman and Warren 1994). As depicted in Fig. 7, BoNT/C cleaves both neuronol isoforms of syntaxin (syntaxin 1A and 1B) within the carboxy-terminal region, just before the membrane anchor (Blasi et al. 1993b; Schiavo et al. 1995). Thus, the BoNT/C proteolytic cleavage of syntaxin leads to the loss of its cytoplasmic domain with consequent impairment of function.

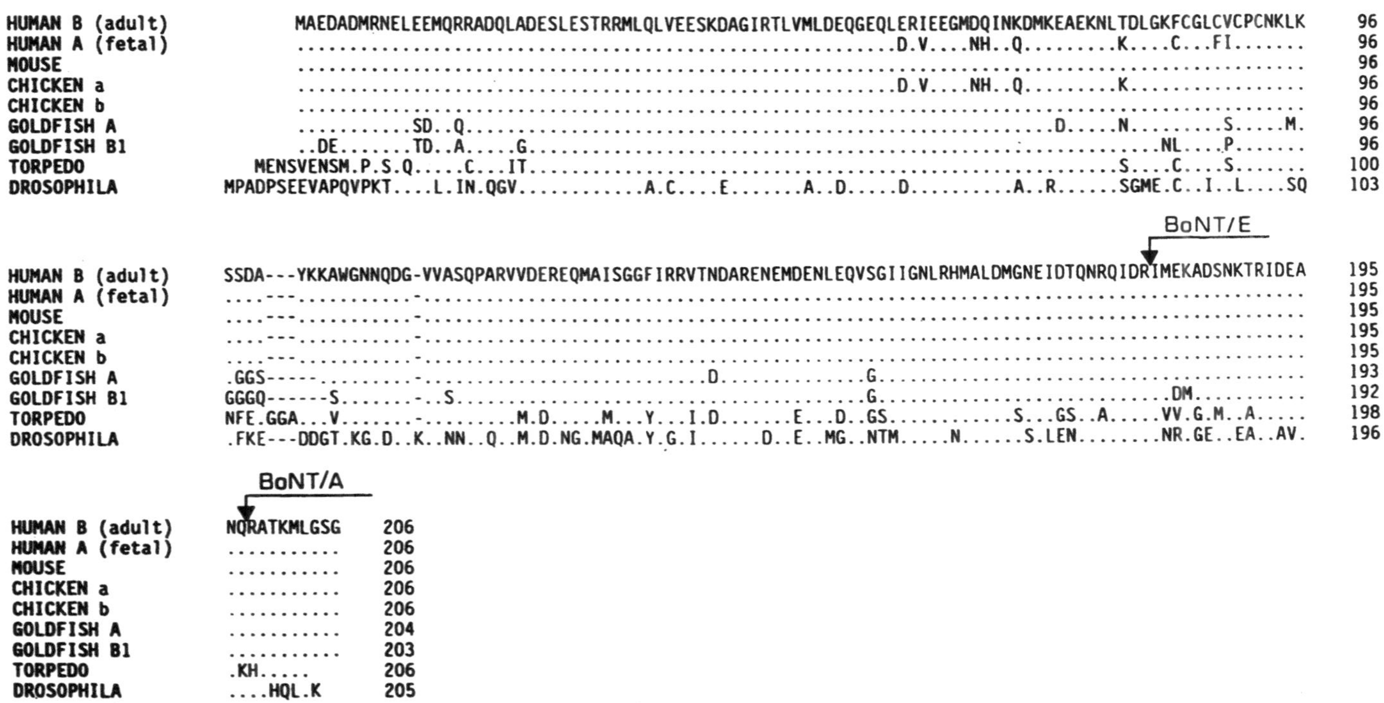

Fig. 5. Comparison of the primary structure of SNAP-25 from different origins and sites of cleavage with botulinum neurotoxins A and E

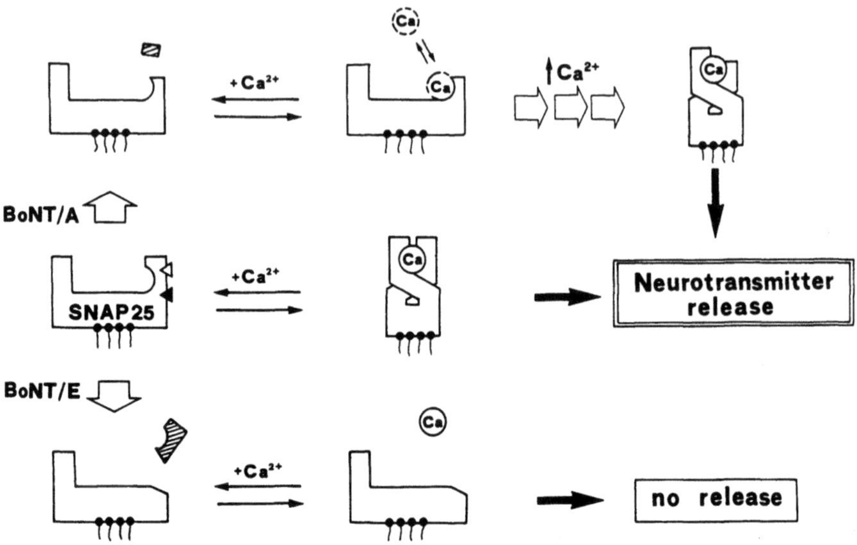

Fig. 6. SNAP-25 may be a low-affinity calcium binding protein. The partial restoration of the neuroexocytotic function by elevation of cytosolic calcium in neuromuscular junction poisoned with BoNT/A, but not BoNT/E, is explained by this model. SNAP-25 may exist in two conformations, with or without bound calcium, and only the calcium-bound form is active in neuroexocytosis. After BoNT/A removal of nine carboxy terminal residues (*open triangle*), SNAP-25 requires higher calcium concentrations to convert into the active form. The BoNT/E-cleaved SNAP-25 (*filled triangle*) has a even lower calcium affinity and is unable to convert into the active form

8 Neuroexocytosis Apparatus and Clostridial Neurotoxins

Recently SOLLNER et al. (1993a, b) have shown that VAMP, SNAP-25, and syntaxin together with a group of cytosolic proteins form a 20S multiprotein complex and have proposed that this multiprotein complex mediates vesicle docking and fusion at the synaptic terminal. The fact that cleavage of these three proteins leads to a persistent and sustained blockade of neuroexocytosis is a strong evidence in favor of the involvement of the 20S complex in neuroexocytosis. The cleavage of VAMP and of syntaxin by TeNT, BoNT/B, /D, /F, /G, and /C leads to the release in the cytosol of a large part of the cytosolic target molecule, and this is likely to impair the assembly of the neuroexocytosis apparatus. By contrast, BoNT/A removes only nine residues from the SNAP-25 carboxy terminus, a region essential for SNAP-25 folding and/or its participation in the complex.

Evidence that TeNT and BoNTs block neuroexocytosis via the specific cleavage of VAMP, SNAP-25, and syntaxin are rapidly accumulating, and can be summarized as it follows: (a) The intracellular activity of the toxins is inhibited by specific inhibitors of zinc endopeptidases such as phosphoramidon and captopril (SCHIAVO et al. 1992a; DE PAIVA et al. 1993b; DAYANITHI et al. 1994; Poulain et al., unpublished results). (b) VAMP or SNAP-25 or syntaxin are cleaved in

```
Rat   Syn 1A    MKDRTQELRTAKDSDDDDD---VTVTVDRDRFM-----DEFFEQVEEIRGFIDKIAENVEEVKRKHSAILASPNPDEKTKEELEELMSDIKKTANKVRSKLKSIEQS    99
Bovine    1A    ....................---..........-----.................................................................    99
Aplysia   1     MT...LAA.KA.-Q.....N-DD.A....SSG..-----E......D...EM.....S..D...K......SA.QT.D.M..........E...N.....A...V...N   101
Rat   Syn 1B    .........S......EEE---.-.H....H..-----...........C.E.LS.D..Q..KQ......A.........Q...D.TA..............A....    98
Rat   Syn 2     .R..LPD.TACRK...G.N--A.II..EK.H..-----.A..H......SS.AR..QH..D..KN..I..SA...EG.I.....D.NKE......RI.G...A....   100
EPIMORPHIN      .R..LPD.TACRTN..G.T--A.-.I.EK.H..-----.G..H......SS.AR..QH..D..KN..I..SA...EG.I.....D.DKE......RI.G........    99
Rat   Syn 3     ....LEQ.KAKQLTQ...T-DE.EIAI.NTA..-----....SEI..T.LN....S.H...A.KLY.I..SA.I.EP...DD..Q.TTE...R..N..N....S.KH    99
Rat   Syn 4     .R...H..RQGDNIS..E.EVR.ALV.HSGAARLSSPD....QK.QT..QTMA.LESK.R.LEKQQVT...T.L.E.SM.QG.QN.REE..QLGRE..AQ..A..PQ   106
Rat   Syn 5     MSCR-....FLS.CK.LQSRQ---NGIQTNKPALHATRQCS..TLMARR.GKDLSNTFAKL.KL-----T...------.R.SLFDDKAVE.EELTYIIKODIN.LNKQ   101

Rat   Syn 1A    IEQEEGLNRSSADLRIRKTQH------STLSRKFVEVMSEYNATQSDYRERCKGRIQRQLEIT-GRTTTSEELEDMLESGNPAIFASGIIMDSSISKQALSE----   194
Bovine    1A    .....................------...........................................................................----   194
Aplysia   1     .....HT.K............------A..........ND...E.I...........K...A..-.K...N.......I.........TQ....ETQQANET.AD----   196
Rat   Syn 1B    .....................------..........T.......K..D...D..........-.....N............KL...TDD.K...QMT....N.----   193
Rat   Syn 2     CD.D.NG..T.V.....R...------.V......D..T...EA.ILF...S............-.....D....E.....K.S..I.D..S..Q.TR...N.----   195
EPIMORPHIN      CD.D.NG..T.V.....R...------.V......D..T...EA.ILF...S............-.....DD...E.....K.S..I.D..S..Q.TR...N.----   194
Rat   Syn 3     ..ED.V--..........S..------.V.........TK..EA.V.F...S...........-.KK..D....E..........T....-..Q........----   193
Rat   Syn 4     K.EADE-.YN.VNT.MK....------GV..QQ...LINKC.SM..E...RNVE..R...K..NAGMVSD....Q..D..QSEV.V.N.LK.TQVTR...N.----   202
Rat   Syn 5     .A.LQDFV.AKGSQSG.HL.THSNTIVVS.QS.LASMSNDFKSLVEVRT.NL.QQRN.REQFSRAPVSALPLAPNN.GG.-.IVLGGESRASRDVAIDMMDPRTSQ   200

                                                                                             BoNT/C
                                                                                            ↓
Rat   Syn 1A    -----------IETRHSEIIKLENSIRELHDMFMDMAMLVESQGEMIDRIEYNVEHAVDYVERAVSDTKKAVKYQSKARRKKIMIIICCVILGIIIASTIGGIFG   288
Bovine    1A    -----------.........................................................S..........................V....V...V....F.....   288
Aplysia   1     -----------..A..ND.M...R...D...........................................Q....I.T.KM.......................LV.LA..I..LVGI...TL.   290
Rat   Syn 1B    -----------.....N.....T.........V.......................................S..............................VL.VVL..S...TL.L   288
Rat   Syn 2     -----------..S..KD.M...T......E........F..T....VNN..R..VNS.....H.KEE....I..........WI.AAVV.AVIAVL.LI..LSV.K   290
Rat   Sin 2'    -----------..S..KD.M...T......E........F..T....VNN..R..VNS.....H.KEE....I.........VMFVL..V.T.LV.LGIILATALS   289
Rat   Sin 2"    -----------..S..KD.M...T......E........F..T....VNN..R..VNS.....H.KEE....I........GVLCALGRQC   274
EPIMORPHIN      -----------..S..KD.M...T......E........F..T....VNN..R..VNS.....H.KEE....I..........WI.AAVA.AVIAVL.LI..LSV.K   289
Rat   Syn 3     -----------..G..KD.VR..S..K.........I.....N....L.N..L..M.T..H..K.RDE.R..M...GQ..K.L.I..VIV.V.LG.L.LI..LSV.LK   289
Rat   Syn 4     -----------.SA.....QQ..RT.....EI.TFL.TE..M.....R...K.ILSSA.....GQEHV.I.LEN.K...K..V..A..VSVTVL.L.VI..ITITVG   298
Rat   Syn 5     QLQLIDEQDSY.QS.ADTMQNI.ST.V..GSI.QQL.HM.KE.E.T.Q..DE..LG.QLD..R.H.EIL.YFQSVTSN.WLMVK.FLILIVFF..FVVFLA   301
```

Fig. 7. Comparison of the amino acid sequence of known syntaxins from different origins. Syntaxin 1A and 1B are expressed mainly in the central nervous system. Isoform 2 is the predominant isoform at the neuromuscular junction, and epimorphin is the mouse homologue. Syntaxin 2 is present in three different forms (*2, 2', 2"*) that differ only in the transmembrane region. Syntaxin 3 and 4 are plasmalemma proteins of nonneuronal cells, while isoform 5

synaptosomes intoxicated with TeNT or BoNTs (LINK et al. 1992; BLASI et al. 1993a,b; SCHIAVO et al. 1993c, 1994b; SCHIAVO and MONTECUCCO 1994) and in primary cultures of rat brain cortex neuronal cells (OSEN-SAND et al., in preparation) or spinal cord (WILLIAMSON et al., in preparation). (c) Peptides spanning the cleavage site of VAMP inhibit TeNT and BoNT/B in *Aplysia* neurons, chromaffin cells and neurohypophysial terminals (SCHIAVO et al. 1992b; HÖHNE-ZELL et al. 1993; DAYANITHI et al. 1994). (d) An antibody highly specific for VAMP prevents the blockade of neurotransmitter release in *Aplysia* neurons induced by TeNT and BoNT/B, but not that caused by BoNT/A (POULAIN et al. 1993). (e) An antibody raised versus the zinc-binding motif of TeNT blocks its activity in permeabilized chromaffin cells and accelerates the recovery of exocytotic function (BARTELS et al. 1994).

These results do not exclude that these toxins may act on other protein substrates inside neurons, but clearly indicate that VAMP, SNAP-25, and syntaxin are major targets of the intraneuronal activity of these neurotoxins in vivo, and that their proteolytic cleavage is linked to the blockade of neurotransmitter release.

9 Concluding Remarks and Future Perspectives

The neurotoxins produced by clostridia responsible for tetanus and botulism form a new group of zinc endopeptidases endowed with peculiar properties (MONTECUCCO and SCHIAVO 1993, 1994): (a) They are produced as inactive precusors which are activated by specific proteolysis followed by reduction of a single disulfide bond. (b) Their amino acid sequence around the zinc binding motif does not fit in any of the various groups of zinc endopeptidases. (c) They act in the cell cytosol and are very specific in terms of both protein target and peptide bond cleaved. (d) They cleave the folded target molecule but do not hydrolyze peptides spanning the cleavage site and hence they are conformational proteinases rather than sequential proteinases.

A future challenge is to define the structural basis of the specific recognition and cleavage of their substrates. This problem is related to that of finding specific inhibitors of their zinc endopeptidase activity that are membrane permeable. Such inhibitors would be potential therapeutic agents in the treatment of tetanus and botulism. Another important line of development is that aimed at determining the three-dimensional structure of these neurotoxins. Highly diffracting crystals of BoNT/A and BoNT/B are already under analysis (STEVENS et al. 1991). Another important development will be the determination of their receptors, which is related to the possibility of using the heavy chain of these toxins as carriers of biologics inside nerve cells.

Acknowledgments. We thank P. Washbourne for critical reading the manuscript. The authors are supported by CNR-Biotecnologie, Telethon-Italia grant 473 and MURST.

References

Allured VA, Collier RJ, Carroll SF, McKay DB (1986) Structure of exotoxin A from Pseudomonas aeruginosa at 3.0 Angstrom resolution. Proc Natl Acad Sci USA 83: 1320–1324

Anderson MD, Fairweather N, Charles IG, Emsley P,Isaacs NW, MacDermott G (1993) Crystallographic characterization of tetanus toxin fragment C. J Mol Biol 230: 673–674

Archer BT, Ozcelik T, Jahn R, Francke U, Südhof TC (1990) Structures and chromosomal localizations of two human genes encoding synaptobrevins 1 and 2. J Biol Chem 265: 17267–17273

Ashton AC, de Paiva AM, Poulain B, Tauc L, Dolly JO (1993) Factors underlying the characteristics inhibition of the neuronal release of transmitters by tetanus and various botulinum toxin. In: DasGupta BR (ed) Botulinum and tetanus neurotoxins. neurotransmission and biomedical aspects. Plenum, New York, pp 191–213

Bark C (1993) Structure of the chicken gene for SNAP-25 reveals duplicated exons encoding distinct isoforms of the protein. J Mol Biol 233: 67–76

Bark C (1994a) Regulated vesicular fusion in neurons: snapping together the details. Proc Natl Acad Sci USA 91: 4621–4624

Bark C, Wilson MC (1994b) Human cDNA clones encoding two different isoforms of the nerve terminal protein SNAP-25. Gene 139: 291–292

Bartels F, Bergel H, Bigalke H, Frevert J, Halpern J, Middlebrook JL (1994) Specific antibodies against the Zn-binding domain of clostridial neurotoxins restore exocytosis in chromaffin cells treated with tetanus or botulinum A neurotoxin. J Biol Chem 269: 8122–8127

Baumann U, Wu S, Flaherty KM, McKay DB (1993) Three-dimensional structure of the alkaline protease of pseudomonas aeruginosa: a two domain protein with a calcium binding parallel beta roll motif. EMBO J 12: 3357–3364

Baumert M, Maycox PR, Navone F, De Camilli P, Jahn R (1989) Synaptobrevin: an integral membrane protein of 18 000 daltons present in small synaptic vesicles of rat brain. EMBO J 8: 379–384

Bennett MK, Scheller RH (1994) A molecular description of synaptic vesicle membrane trafficking. Annu Rev Biochem 63: 63–100

Bennett MK, Calakos N, Scheller RH (1992) Syntaxin: a synaptic protein implicated in docking of synaptic vesicles at presynaptic active zones. Science 257: 255–259

Bennett MK, Garcia-Arras JE, Elferink LA, Peterson K, Fleming AM, Hazuka CD, Scheller RH (1993) The syntaxin family of vesicular transport receptors. Cell 74: 863–873

Binz T, Blasi J, Yamasaki S, Baumeister A, Link E, Südhof TC, Jahn R, Niemann H (1994) Proteolysis of SNAP-25 by types E and A botulinal neurotoxins. J Biol Chem 269: 1617–1620

Blasi J, Chapman ER, Link E, Binz T, Yamasaki S, DeCamilli P, Südhof T, Niemann H, Jahn R (1993a) Botulinum neurotoxin A selectively cleaves the synaptic protein SNAP-25. Nature 365: 160–163

Blasi J, Chapman ER, Yamasaki S, Binz T, Niemann H, Jahn R (1993b) Botulinum neurotoxin C blocks neurotransmitter release by means of cleaving HPC-1/syntaxin. EMBO J 12: 4821–4828

Bode W, Gomis-Ruth FX, Huber R, Zwilling R, Stocker W (1992) Structure of astacin and implication of astacins and zinc-ligation of collagenases. Nature 358: 164–166

Bode W, Gomis-Ruth FX, Stocker W (1993) Astacins, serralysin, snake venoms and matrix metalloproteinases exhibit identical zinc-binding environments (HEXXHXXGXXH and Met-turn) and topologies and should be grouped into a common family, the "metzincins." FEBS Lett 331: 134–140

Bode W, Reinemer P, Huber R, Kleine T, Schnierer S, Tschesche H (1994) The X-ray crystal structure of the catylytic domain of human neutrophil collagenase inhibited by a substrate analogue reveals the essentials for catalysis and specificity. EMBO J 13: 1263–1269

Braun JEA, Fritz BA, Wong SME, Lowe AW (1994) Identification of a vesicle-associated membrane protein (VAMP-like) membrane protein in zymogen granules of the rat exocryne pancreas. J Biol Chem 269: 5328–5335

Chin AC, Burgess RW, Wong BR, Schwarz TL, Scheller RH (1994) Differential expression of transcripts from syb, a Drosophila melanogaster gene encoding VAMP (synaptobrevin that is abundant in non-neuronal cells. Gene 131: 175–181

Choe S, Bennett MJ, Fujii G, Curmi PMG, Kantardjieff KA, Collier RJ, Eisenberg D (1992) The crystal structure of diphtheria toxin. Nature 367: 216–222

Corley Cain C, Trimble WS, Leinhard GE (1992) Members of the VAMP family of synaptic vesicles proteins are components of glucose transporter-containing vesicles from rat adipocytes. J Biol Chem 267: 11681–11684

DasGupta BR (1989) The structure of botulinum neurotoxin. In: Simpson LL (ed) Botulinum neurotoxins and tetanus toxin. Academic, New York, pp 53–67
Dayanithi G, Stecher B, Höhne-Zell B, Yamasaki S, Binz T, Weller U, Niemann H, Gratzl M (1994) Exploring the functional domain and the target of the tetanus toxin light chain in neurohypophysial terminals. Neuroscience 58: 423–431
de Paiva A, Poulain B, Lawrence GW, Shone CC, Tauc L, Dolly JO (1993a) A role for the interchain disulfide or its participating thiols in the internalization of botulinum neurotoxin A revealed by a toxin derivative that binds to ecto-acceptors and inhibits transmitter release intracellularly. J Biol Chem 268: 20838–20844
de Paiva A, Ashton AC, Foran P, Schiavo G, Montecucco C, Dolly JO (1993b) Botulinum A like type B and tetanus toxin fulfils criteria for being a zinc-dependent protease. J Neurochem 61: 2338–2341
DiAntonio A, Burgess RW, Chin AC, Deitcher DL, Scheller RH, Schwarz TL (1993) Identification and characterization of Drosophila genes for synaptic vesicle proteins. J Neurosci 13: 4924–4935
Elferink LA, Trimble WS, Scheller RH (1989) Two vesicle-associated membrane protein genes are differently expressed in the rat central nervous system. J Biol Chem 264: 11061–11064
Gomis-Ruth FX, Kress LF, Bode W (1993) First structure of a snake venom metalloproteinase: a prototype for matrix metalloproteinases/collagenases. EMBO J 12: 4151–4157
Hess TS, Slater TM, Wilson MC, Skene JHP (1992) The 25 kDa synaptosomal-associated protien SNAP-25 is the major methionine-rich polypeptide in rapid axonal transport and a major substrate for palmitoylation in adult CNS. J Neurosci 12: 4634–4641
Höhne-Zell B, Stecher B, Gratzl M (1993) Functional characterization of the catalytic site of the tetanus toxin light chain using permeabilized adrenal chromaffin cells. FEBS Lett 336: 175–180
Hunt JM, Bommert K, Charlton MP, Kistner A, Habermann E, Augustine GJ, Betz H (1994) A post-docking role for synaptobrevin in synaptic vesicle fusion. Neuron 12: 1269–1279
Inoue A, Obata K, Akagawa K (1992) Cloning and sequence analysis of cDNA for a neuronal cell membrane antigen, HPC-1. J Biol Chem 267: 10613–10619
Jiang W, Bond JS (1992) Families of metalloendopeptidases and their relationships. FEBS Lett 312: 110–114
Jongeneel CV, Bouvier J, Bairoch A (1989) A unique signature identifies a family of zinc-dependent metallopeptidases. FEBS Lett 242: 211–214
Krieglstein KG, Henschen A, Weller U, Habermann E (1991) Limited proteolysis of tetanus toxin. Eur J Biochem 202: 41–51
Link E, Edelmann L, Chou JH, Binz T, Yamasaki S, Eisel U, Baumert M, Südhof TC, Niemann H, Jahn R (1992) Tetanus toxin action: inhibition of neurotransmitter release linked to synaptobrevin proteolysis. Biochem Biophys Res Commun 189: 1017–1023
Lovejoy B, Cleasby A, Hassell AM, Longley K, Luther MA, Weigl D, McGeehan G, McElroy AB, Drewry D, Lambert MH, Jordan SR (1994) Structure of the catalytic domain of fibroblast collagenase complexed with an inhibitor. Science 263: 375–377
Matthews BW (1988) Structural basis of the action of thermolysin and related zinc peptidases. Acc Chem Res 21: 333–340
Matthews BW, Jansonius JN, Colman PM (1972) Three-dimensional structure of thermolysin. Nature New Biol 238: 37–41
McMahon HT, Ushkaryov YA, Edelmann L, Link E, Binz T, Niemann H, Jahn R, Südhof TC (1993) Cellubrevin is a ubiquitous tetanus-toxin substrate homologous to a putative synaptic vesicle fusion protein. Nature 364: 346–349
Miles EW (1977) Modification of histidyl residues in proteins by diethylcarbonate. Methods Enzymol 47: 431–442
Montecucco C, Schiavo G (1993) Tetanus and botulism neurotoxins: a new group of zinc proteases. Trends Biochem Sci 18: 324–327
Montecucco C, Schiavo G (1994) The molecular mechanism of action of tetanus and botulism neurotoxins. Mol Microbiol 13: 1–8
Montecucco C, Papini E, Schiavo G (1994) Bacterial protein toxins penetrate cells via a four-step mechanism. FEBS Lett 346: 92–98
Niemann H (1991) Molecular biology of clostridial neurotoxins. In: Alouf JE, Freer JH (eds) A sourcebook of bacterial protein toxins. Academic, London, pp 303–348
Nishiki T, Kamata Y, Nemoto Y, Omori A, Ito T, Takahashi M, Kozaki S (1994) Identification of protein receptor for clostridium botulinum type B neurotoxin in rat brain synaptosomes. J Biol Chem 269: 10498–10503

Osen-Sand A, Catsicas M, Staple JK, Jones KA, Ayala G, Knowles J, Grenningloh G, Catsicas S (1993) Inhibition of axonal growth by SNAP-25 antisense oligonucleotides in vitro and in vivo. Nature 364: 445–448

Oyler GA, Higgins GA, Hart RA, Battenberg E, Billingsley M, Bloom FE, Wilson MC (1989) The identification of a novel synaptosomal-associated protein, SNAP-25, differently expressed by neuronal subpopulations. J Cell Biol 109: 3039–3052

Patarnello T, Bargelloni L, Rossetto O, Schiavo G, Montecucco C (1993) Neurotransmission and secretion. Nature 364: 581–582

Pauptit RA, Karlsson R, Picot D, Jenkins JA, Niklaus-Reimer AS, Jansonius JN (1988) Crystal structure of neutral protease from Bacillus cereus refined at 3.0 A resolution and comparison with the homologous but more thermostable enzyme termolysin. J Mol Biol 199: 525–537

Payling-Wright G (1955) The neurotoxins of Clostridium botulinum and Clostridium tetani. Pharmacol Rev 7: 413–465

Poulain B, Rossetto O, Deloye F, Schiavo G, Tauc L, Montecucco C (1993) Antibodies against rat brain vesicle-associated membrane protein (sinaptobrevin) prevent inhibition of acetylcholine release by tetanus toxin of botulinum neurotoxin type B. J Neurochem 61: 1175–1178

Ralston E, Beushausen S, Ploug T (1994) Expression of the synaptic vesicle proteins VAMPs/ synaptobrevins 1 and 2 in non-neuronal tissue. J Biol Chem 269: 15403–15406

Risinger C, Larhammar D (1993) Multiple loci for synapse protein SNAP-25 in the tetraploid goldfish. Proc Natl Acad Sci USA 90: 10598–10602

Risinger C, Blomqvist AG, Lundell I, Lambertsson A, Nässel D, Pierbone VA, Brodin L, Larhammar D (1993) Evolutionary conservation of synaptosome-associated protein 25 kDa (SNAP-25) shown by Drosophila and Torpedo cDNA clones. J Biol Chem 268: 24408–24414

Robinson JP, Schmid MF, Morgan DG, Chiu W (1988) Three-dimensional structural analysis of tetanus toxin by electron crystallography. J Mol Biol 200: 367–375

Rossetto O, Schiavo G, Polverino de Laureto P, Fabbiani S, Montecucco C (1992) Surface topography of histidine residues of tetanus toxin probed by immobilized-metal-ion affinity chromatography. Biochem J 285: 9–12

Rossetto O, Gorza L, Schiavo G, Schiavo N, Neale EA, Scheller RH, Montecucco C (1994) VAMP/ synaptobrevin neuronal isoforms are ubiquitous and play a role additional to vesicle docking. (submitted)

Rothman JE, Warren G (1994) Implications of the SNARE hypothesis for intracellular membrane topology and dynamics. Curr Biol 4: 220–233

Sadoul K, Lang J, Montecucco C, Weller U, Catsicas S, Wollheim C, Halban P (1994) SNAP-25 is expressed in islets of Langerhans and is involved in insulin release. J Cell Biol (in press)

Schiavo G, Montecucco C (1994) Tetanus and botulism neurotoxins: isolation and assay. Methods Enzymol (in press)

Schiavo G, Papini E, Genna G, Montecucco C (1990) An intact interchain disulfide bond is required for the neurotoxicity of tetanus toxin. Infect Immun 58: 4136–4141

Schiavo G, Poulain B, Rossetto O, Benfenati F, Tauc L, Montecucco C (1992a) Tetanus toxin is a zinc protein and its inhibition of neurotransmitter release and protease activity depend on zinc. EMBO J 11: 3577–3583

Schiavo G, Benfenati F, Poulain B, Rossetto O, Polverino de Laureto P, DasGupta BR, Montecucco C (1992b) Tetanus and botulinum-B neurotoxins of block neurotransmitter release by a proteolytic cleavage of synaptobrevin. Nature 359: 832–835

Schiavo G, Rossetto O, Santucci A, DasGupta BR, Montecucco C (1992c) Botulinum neurotoxins are zinc proteins. J Biol Chem 267: 23479–23483

Schiavo G, Poulain B, Benfenati F, DasGupta BR, Montecucco C (1993a) Novel targets and catalytic activities of bacterial protein toxins. Trends Microbiol 1: 170–174

Schiavo G, Shone CC, Rossetto O, Alexandre FCG, Montecucco C (1993b) Botulinum neurotoxin serotype F is a zinc endopeptidase specific for VAMP/synaptobrevin. J Biol Chem 268: 11516–11519

Schiavo G, Rossetto O, Catsicas S, Polverino de Laureto P, DasGupta BR, Benfenati F, Montecucco C (1993c) Identification of the nerve-terminal targets of botulinum neurotoxins serotypes A, D and E. J Biol Chem 268: 23784–23787

Schiavo G, Santucci A, DasGupta BR, Metha PP, Jontes J, Benfenati F, Wilson MC, Montecucco C (1993d) Botulinum neurotoxins serotypes A and E cleave SNAP-25 at distinct COOH-terminal peptide bonds. FEBS Lett 335: 99–103

Schiavo G, Rossetto O, Benfenati F, Poulain B, Montecucco C (1994a) Tetanus and botulinum neurotoxins are zinc proteases specific for components of the neuroexocytosis apparatus. Ann NY Acad Sci 710: 65–75

Schiavo G, Malizio C, Trimble WS, Polverino de Laureto P, Milan G, Sugiyama H, Johnson EA, Montecucco C (1994b) Botulinum G neurotoxin cleaves VAMP/synaptobrevin at a single Ala/Ala peptide bond. J Biol Chem 269: 20213–20216

Schiavo G, Shone CC, Bennett MK, Scheller RH, Montecucco C (1994c) Botulinum neurotoxin type C cleaves a single Lys-Ala bond within the carboxyl-terminal region of syntaxin. J Biol Chem 270 (in press)

Shone CC, Quinn CP, Wait R, Hallis B, Fooks SG, Hambleton P (1993) Proteolytic cleavage of synthetic fragments of vesicle-associated membrane protein, isoform-2 by botulinum type B neurotoxin. Eur J Biochem 217: 965–971

Söllner T, Whiteheart SW, Brunner M, Erdjument-Bromage H, Geromanos S, Tempst P, Rothman JE (1993a) SNAP receptors implicated in vesicle targeting and fusion. Nature 362: 318–324

Söllner T, Bennett M, Whiteheart SW, Scheller RH, Rothman JE (1993b) A protein assembly-disassembly pathway in vitro that may correspond to sequential steps of synaptic vesicle docking, activation, and fusion. Cell 75: 409–418

Steinhardt RA, Bi G, Alderton JM (1994) Cell membrane resealing by a vesicular mechanism similar to neurotransmitter release. Science 263: 390–393

Stevens RC, Evenson ML, Tepp W, DasGupta BR (1991) Crystallization and preliminary X-ray analysis of botulinum neurotoxin type A. J Mol Biol 222: 877–880

Südhof TC, Baumert M, Perin MS, Jahn R (1989) A synaptic vesicle membrane protein is conserved from mammals to Drosophila. Neuron 2: 1475–1481

Thayer MM, Flaherty KM, McKay DB (1991) Three-dimensional structure of the elastase of Pseudomonas aeruginosa at 1.5 A resolution. J Biol Chem 266: 2864–2871

Trimble WS, Cowan DM, Scheller RH (1988) VAMP–1: a synaptic vesicle-associated integral membrane protein. Prot Natl Acad Sci USA 85: 4538–4542

Trimble WS, Linial M, Scheller RH (1991) Cellular and molecular biology of the presynaptic nerve terminal. Annu Rev Neurosci 14: 93–122

Vallee BL, Auld DS (1990) Zinc coordination, function and structure of zinc enzymes and other proteins. Biochemistry 29: 5647–5659

Wanke E, Ferroni A, Gattanini P, Meldolesi J (1986) Alpha-latratoxin of the black widow spider venom opens small, non-closing cation channel. Biochem Biophys Res Commun 134: 320–325

Wellhoner HH (1992) Tetanus and botulinum neurotoxins. In: Herken H, Hucho F (eds) Handbook of experimental pharmacology, vol 102. Springer, Berlin Heidelberg New York, pp 357–417

Wright JF, Pernollet M, Reboul A, Aude C, Colomb M (1992) Identification and partial characterization of a low affinity metal-binding site in the light chain of tetanus toxin. J Biol Chem 267: 9053–9058

Yamasaki S, Hu Y, Binz T, Kalkuhl A, Kurazono H, Tamura T, Jahn R, Kandel E, Niemann H (1994a) Synaptobrevin / VAMP of Aplysia californica: structure and proteolysis by tetanus and botulinal neurotoxins type D and F. Proc Natl Acad Sci USA 91: 4688–4692

Yamasaki S, Baumeister A, Binz T, Blasi J, Link E, Cornille F, Roques B, Fykse EM, Südhof TC, Jahn R, Niemann H (1994b) Cleavage of members of the synaptobrevin / VAMP family by types D and F botulinal neurotoxins and tetanus toxin. J Biol Chem 269: 12764–12772

Yamasaki S, Binz T, Hayashi T, Szabo E, Yamasaki N, Eklund M, Jahn R, Niemann H (1994c) Botulinum neurotoxin type G proteolyses the Ala^{81}-Ala^{82} bond of rat synaptobrevin 2. Biochem Biophys Res Commun 200: 829–835

Subject Index

Current Topics in Microbiology and Immunology

Volumes published since 1989 (and still available)

Vol. 151: **Jann, Klaus; Jann, Barbara (Ed.):** Bacterial Adhesins. 1990. 23 figs. XII, 192 pp. ISBN 3-540-51052-4

Vol. 152: **Bosma, Melvin J.; Phillips, Robert A.; Schuler, Walter (Ed.):** The Scid Mouse. Characterization and Potential Uses. EMBO Workshop held at the Basel Institute for Immunology, Basel, Switzerland, February 20–22, 1989. 1989. 72 figs. XII, 263 pp. ISBN 3-540-51512-7

Vol. 153: **Lambris, John D. (Ed.):** The Third Component of Complement. Chemistry and Biology. 1989. 38 figs. X, 251 pp. ISBN 3-540-51513-5

Vol. 154: **McDougall, James K. (Ed.):** Cytomegaloviruses. 1990. 58 figs. IX, 286 pp. ISBN 3-540-51514-3

Vol. 155: **Kaufmann, Stefan H. E. (Ed.):** T-Cell Paradigms in Parasitic and Bacterial Infections. 1990. 24 figs. IX, 162 pp. ISBN 3-540-51515-1

Vol. 156: **Dyrberg, Thomas (Ed.):** The Role of Viruses and the Immune System in Diabetes Mellitus. 1990. 15 figs. XI, 142 pp. ISBN 3-540-51918-1

Vol. 157: **Swanstrom, Roland; Vogt, Peter K. (Ed.):** Retroviruses. Strategies of Replication. 1990. 40 figs. XII, 260 pp. ISBN 3-540-51895-9

Vol. 158: **Muzyczka, Nicholas (Ed.):** Viral Expression Vectors. 1992. 20 figs. IX, 176 pp. ISBN 3-540-52431-2

Vol. 159: **Gray, David; Sprent, Jonathan (Ed.):** Immunological Memory. 1990. 38 figs. XII, 156 pp. ISBN 3-540-51921-1

Vol. 160: **Oldstone, Michael B. A.; Koprowski, Hilary (Eds.):** Retrovirus Infections of the Nervous System. 1990. 16 figs. XII, 176 pp. ISBN 3-540-51939-4

Vol. 161: **Racaniello, Vincent R. (Ed.):** Picornaviruses. 1990. 12 figs. X, 194 pp. ISBN 3-540-52429-0

Vol. 162: **Roy, Polly; Gorman, Barry M. (Eds.):** Bluetongue Viruses. 1990. 37 figs. X, 200 pp. ISBN 3-540-51922-X

Vol. 163: **Turner, Peter C.; Moyer, Richard W. (Eds.):** Poxviruses. 1990. 23 figs. X, 210 pp. ISBN 3-540-52430-4

Vol. 164: **Bækkeskov, Steinnun; Hansen, Bruno (Eds.):** Human Diabetes. 1990. 9 figs. X, 198 pp. ISBN 3-540-52652-8

Vol. 165: **Bothwell, Mark (Ed.):** Neuronal Growth Factors. 1991. 14 figs. IX, 173 pp. ISBN 3-540-52654-4

Vol. 166: **Potter, Michael; Melchers, Fritz (Eds.):** Mechanisms in B-Cell Neoplasia. 1990. 143 figs. XIX, 380 pp. ISBN 3-540-52886-5

Vol. 167: **Kaufmann, Stefan H. E. (Ed.):** Heat Shock Proteins and Immune Response. 1991. 18 figs. IX, 214 pp. ISBN 3-540-52857-1

Vol. 168: **Mason, William S.; Seeger, Christoph (Eds.):** Hepadnaviruses. Molecular Biology and Pathogenesis. 1991. 21 figs. X, 206 pp. ISBN 3-540-53060-6

Vol. 169: **Kolakofsky, Daniel (Ed.):** Bunyaviridae. 1991. 34 figs. X, 256 pp. ISBN 3-540-53061-4

Vol. 170: **Compans, Richard W. (Ed.):** Protein Traffic in Eukaryotic Cells. Selected Reviews. 1991. 14 figs. X, 186 pp. ISBN 3-540-53631-0

Vol. 171: **Kung, Hsing-Jien; Vogt, Peter K. (Eds.):** Retroviral Insertion and Oncogene Activation. 1991. 18 figs. X, 179 pp. ISBN 3-540-53857-7

Vol. 172: **Chesebro, Bruce W. (Ed.):** Transmissible Spongiform Encephalopathies. 1991. 48 figs. X, 288 pp. ISBN 3-540-53883-6

Vol. 173: **Pfeffer, Klaus; Heeg, Klaus; Wagner, Hermann; Riethmüller, Gert (Eds.):** Function and Specificity of γ/δ TCells. 1991. 41 figs. XII, 296 pp. ISBN 3-540-53781-3

Vol. 174: **Fleischer, Bernhard; Sjögren, Hans Olov (Eds.):** Superantigens. 1991. 13 figs. IX, 137 pp. ISBN 3-540-54205-1

Vol. 175: **Aktories, Klaus (Ed.):** ADP-Ribosylating Toxins. 1992. 23 figs. IX, 148 pp. ISBN 3-540-54598-0

Vol. 176: **Holland, John J. (Ed.):** Genetic Diversity of RNA Viruses. 1992. 34 figs. IX, 226 pp. ISBN 3-540-54652-9

Vol. 177: **Müller-Sieburg, Christa; Torok-Storb, Beverly; Visser, Jan; Storb, Rainer (Eds.):** Hematopoietic Stem Cells. 1992. 18 figs. XIII, 143 pp. ISBN 3-540-54531-X

Vol. 178: **Parker, Charles J. (Ed.):** Membrane Defenses Against Attack by Complement and Perforins. 1992. 26 figs. VIII, 188 pp. ISBN 3-540-54653-7

Vol. 179: **Rouse, Barry T. (Ed.):** Herpes Simplex Virus. 1992. 9 figs. X, 180 pp. ISBN 3-540-55066-6

Vol. 180: **Sansonetti, P. J. (Ed.):** Pathogenesis of Shigellosis. 1992. 15 figs. X, 143 pp. ISBN 3-540-55058-5

Vol. 181: **Russell, Stephen W.; Gordon, Siamon (Eds.):** Macrophage Biology and Activation. 1992. 42 figs. IX, 299 pp. ISBN 3-540-55293-6

Vol. 182: **Potter, Michael; Melchers, Fritz (Eds.):** Mechanisms in B-Cell Neoplasia. 1992. 188 figs. XX, 499 pp. ISBN 3-540-55658-3

Vol. 183: **Dimmock, Nigel J.:** Neutralization of Animal Viruses. 1993. 10 figs. VII, 149 pp. ISBN 3-540-56030-0

Vol. 184: **Dunon, Dominique; Mackay, Charles R.; Imhof, Beat A. (Eds.):** Adhesion in Leukocyte Homing and Differentiation. 1993. 37 figs. IX, 260 pp. ISBN 3-540-56756-9

Vol. 185: **Ramig, Robert F. (Ed.):** Rotaviruses. 1994. 37 figs. X, 380 pp. ISBN 3-540-56761-5

Vol. 186: **zur Hausen, Harald (Ed.):** Human Pathogenic Papillomaviruses. 1994. 37 figs. XIII, 274 pp. ISBN 3-540-57193-0

Vol. 187: **Rupprecht, Charles E.; Dietzschold, Bernhard; Koprowski, Hilary (Eds.):** Lyssaviruses. 1994. 50 figs. IX, 352 pp. ISBN 3-540-57194-9

Vol. 188: **Letvin, Norman L.; Desrosiers, Ronald C. (Eds.):** Simian Immunodeficiency Virus. 1994. 37 figs. X, 240 pp. ISBN 3-540-57274-0

Vol. 189: **Oldstone, Michael B. A. (Ed.):** Cytotoxic T-Lymphocytes in Human Viral and Malaria Infections. 1994. 37 figs. IX, 210 pp. ISBN 3-540-57259-7

Vol. 190: **Koprowski, Hilary; Lipkin, W. Ian (Eds.):** Borna Disease. 1995. 33 figs. IX, 134 pp. ISBN 3-540-57388-7

Vol. 191: **ter Meulen, Volker; Billeter, Martin A. (Eds.):** Measles Virus. 1995. 23 figs. IX, 196 pp. ISBN 3-540-57389-5

Vol. 192: **Dangl, Jeffrey L. (Ed.):** Bacterial Pathogenesis of Plants and Animals. 1994. 41 figs. IX, 343 pp. ISBN 3-540-57391-7

Vol. 193: **Chen, Irvin S. Y.; Koprowski, Hilary; Srinivasan, Alagarsamy; Vogt, Peter K. (Eds.):** Transacting Functions of Human Retroviruses. 1995. 49 figs. Approx. IX, 240 pp. ISBN 3-540-57901-X

Vol. 194: **Potter, Michael; Melchers, Fritz (Eds.):** Mechanisms in B-cell Neoplasia. 1995. Approx. 152 figs. XXV, 458 pp. ISBN 3-540-58447-1

MIX
Papier aus verantwortungsvollen Quellen
Paper from responsible sources
FSC® C105338
FSC
www.fsc.org

If you have any concerns about our products,
you can contact us on
ProductSafety@springernature.com

In case Publisher is established outside the EU,
the EU authorized representative is:
Springer Nature Customer Service Center GmbH
Europaplatz 3, 69115 Heidelberg, Germany

Printed by Libri Plureos GmbH
in Hamburg, Germany